AF344500

Ringing the Changes in Europe

de Gruyter Studies in Organization

International Management, Organization and Policy Analysis

An international and interdisciplinary book series from de Gruyter presenting comprehensive research on aspects of international management, organization studies and comparative public policy.
It covers cross-cultural and cross-national studies of topics such as:
- management; organizations; public policy, and/or their inter-relation
- industry and regulatory policies
- business-government relations
- international organizations
- comparative institutional frameworks.

While each book in the series ideally has a comparative empirical focus, specific national studies of a general theoretical, substantive or regional interest which relate to the development of cross-cultural and comparative theory are also encouraged.

The series is designed to stimulate and encourage the exchange of ideas across linguistic, national and cultural tradition of analysis, between academic researchers, practitioners and policy makers, and between disciplinary specialists.
The volumes present theoretical work, empirical studies, translations and 'state-of-the art' surveys. The *international* aspects of the series are uppermost: there is a strong commitment to work which crosses and opens boundaries.

Editor:

Prof. Stewart R. Clegg, Faculty of Business and Technology, University of Western Sydney, Macarthur, Campbelltown, Australia

Advisory Board:

Prof. Nancy J. Adler, McGill University, Dept. of Management, Montreal, Quebec, Canada
Prof. Richard Hall, State University of New York at Albany, Dept. of Sociology, Albany, New York, USA
Prof. Gary Hamilton, University of Washington, Seattle, Washington, USA
Prof. Geert Hofstede, University of Limburg, Maastricht, The Netherlands
Prof. Pradip N. Khandwalla, Indian Institute of Management, Vastrapur, Ahmedabad, India
Prof. Surenda Munshi, Sociology Group, Indian Institute of Management, Calcutta, India
Prof. Gordon Redding, University of Hong Kong, Dept. of Management Studies, Hong Kong

Adrienne Héritier · Christoph Knill
Susanne Mingers

in collaboration with Rhodes Barrett

Ringing the Changes in Europe

Regulatory Competition
and the Transformation of the State.
Britain, France, Germany

Walter de Gruyter · Berlin · New York 1996

This is the completely revised and updated publication of a book which was published before in German: Adrienne Héritier, Susanne Mingers, Christoph Knill, Martina Becka: Die Veränderung von Staatlichkeit in Europa. Ein regulativer Wettbewerb: Deutschland, Großbritannien, Frankreich. Opladen: Leske + Budrich 1994. *Adrienne Héritier*, Prof. Dr., Dept. of Political and Social Science, European University Institute, San Domenico di Fiesole, Italy. *Christoph Knill*, Dr., Dept. of Political and Social Science, European University Institute, San Domenico di Fiesole, Italy. *Susanne Mingers*, Dr., Faculty of Sociology, University of Bielefeld, Bielefeld, Germany. With 18 tables and 2 figures.

∞ Printed on acid-free paper which falls within the guidelines of the ANSI to ensure permanence and durability.

Library of Congress Cataloging-in-Publication Data

Héritier, Adrienne
  Ringing the changes in Europe : regulatory competition and the redefinition of the state : Britain, France, Germany / Adrienne Héritier, Christoph Knill, Susanne Mingers in collaboration with Rhodes Barrett.
  XIV, 364 p. 15,5 × 23,0 cm. − (De Gruyter studies in organization ; 74)
    Includes bibliographical references.
  ISBN (invalid) 31101476530 (alk. paper)
    1. Air quality management − Government policy − European Union countries − Case studies. 2. Air − Pollution − Law and legislation − European Union coutries − Case studies. 3. Industrial policy − European Union countries − Case studies. 4. Competition − Government policy − European Union countries − Case studies.   I. Knill, Christoph. II. Mingers, Susanne. III. Title. IV. Series.
HC240.9.A4W56   1996
363.73′9256′094−dc20   96-26447
                                      CIP

Die Deutsche Bibliothek − Cataloging-in-Publication Data

Héritier, Adrienne:
Ringing the changes in Europe : regulatory competition and the transformation of the State: Britain, France, Germany ; [with 18 tables] / Adrienne Héritier ; Christoph Knill ; Susanne Mingers. In collab. with Rhodes Barrett. − Berlin ; New York : de Gruyter, 1996
  (De Gruyter studies in organization ; 74 : International management, organization and policy analysis)
  ISBN 3-11-014765-3
NE: Knill, Christoph:; Mingers, Susanne:; GT

Foreword

This book was written in the context of a research project carried out at the Faculty of Sociology/Research Area Political Science at the University of Bielefeld in the period from March 1992 to April 1994 with the support of the German Science Foundation. Complementary field research and theoretical supplementation were undertaken during a one-year research period at the Max Planck Institute for Social Research in Cologne from October 1994 to August 1995.

The data underlying the report were gathered primarily by means of focused interviews with experts. In the course of extremely useful and stimulating conversations, over 170 people provided us with information on the issues and processes that determine clean-air policy at the national and supranational levels. As our work advanced, the complexity of the ways in which interaction between the European Union and its member states in the regulatory policy field modified state activities became more and more evident. Our survey of the Federal Republic of Germany, the United Kingdom and France revealed a variety of patterns: whereas Germany has set the pace in substantive measures, Britain has developed regulatory ambitions in shaping procedural rules. France in its turn has usually appeared in the supranational negotiating arena as a friendly observer and coalitionist. However, in this interactional process, it is not only the member states but also the EU Commission that has reoriented its substantive, strategic, and institutional interests and exercised a decisive influence on decisions.

Unfortunately we must spare the reader insight into the multifarious, impressive, enriching, and sometimes very amusing experiences we had in the course of our conversations in the countries under review. Indeed, the interview data recorded offers sufficient material to warm the heart of every scholar with a penchant for cultural anthropology. After more than fifty visits to a wide variety of institutions in Britain, for example, we would like once and for all to scotch the idea that the British are avid tea drinkers. Whereas in France we were offered not a single cup of coffee, in Britain we were practically inundated with the stimulating beverage.

Another assumption that needed revising concerned the temperament of Belgian taxi-drivers. A Hollywood car-chase is a country outing compared to what awaited us on our way to the EU Commission in Brussels. The intelligence that we had to keep an appointment and were pressed for time inspired the indubitably zealous cabby to exceed the speed limit by a hundred per cent, to take every curve on two wheels, to vault resolutely across traffic islands, and ignore the traffic lights.

In Germany we initially thought we would be safe from surprises and new cultural experiences. Far from it. Unaware of the festive mores at carnival time in the Rhineland, we were soon to gain new insight. Quite misjudging the dimensions of merry-making, we had unfortunately made various appointments in Düsseldorf on 'Old Wives' Day'. The first problem was to find our way through the administrative jungle of paper streamers, confetti, and false-nosed 'Karnevalisten' to our respondents, all of them besmirched with lipstick and with the ends of their ties snipped off. Once we had succeeded in finding them, our nonetheless serious conversations were repeatedly interrupted by some polonaise set off by another member of staff. It is no wonder that, given the background rumpus, transcription and evaluation of the recorded interviews presented certain difficulties.

How much more tranquil and relaxing our visits in England were! One member of staff of a utility insisted on rounding off the our visit with a tour of the local castle. We were treated not only to numerous historical works of art and other treasures but also to an exhaustive refutation of Einstein's theory of relativity.

After this brief excursion into 'cultural anthropology', we would like to express our gratitude to everyone who has contributed to this project. First of all we naturally thank our respondents in Germany, Britain, France, and Brussels, who so patiently allowed us to bombard them with questions, and who willingly supplied us with information and fruitful inspiration. We thank Helga Hollmann, Annette Spiering and Volker Verrel for the scrupulous checking and unwearying correction of the texts. Our thanks also go to our families, our partners and friends for their wide-ranging substantive, moral and other support.

The German Science Foundation has naturally been 'invaluable' in making the project financially possible, and — with the award of the Leibniz Prize — permitting us to continue our research into the changing nature of the state in Europe. A sojourn at the Max Planck Institute to which we were invited by Renate Mayntz and Fritz W. Scharpf provided a stimulating context for this purpose, proving extremely useful for the revision of the German version and theoretical discussion. We thank them heartily. We are grateful to Helmut Willke and his colleagues for innumerable constructive and critical suggestions, which were debated in joint project sessions — interspersed with heated games of basketball. Our thanks also go to the members of the 'new' research project, Dieter Kerwer, Dirk Lehmkuhl, Maria-

Elena Manfredini, Florence Rudolph and Michael Teutsch, who have accompanied the revision of the German version — on request and otherwise
— with unflagging and enthusiastic criticism.

Not least of all, we thank the University of Bielefeld, especially Otto
Lüke and Angelika Jehring, who have supported our work with such efficiency. The Faculty of Sociology, the Rectorate, the North Rhine-Westphalian Science Ministry and the German Science Foundation have made
possible our present stay at the European University Institute in Florence,
which provides particularly favourable conditions for the further pursuit of
research into the issue of changes in the nature of the state.

Florence in September 1995   Adrienne Héritier,
                                                                 Christoph Knill,
                                                                 Susanne Mingers

Table of Contents

Abbreviations

ABl.	Amtsblatt
ADEME	Agence de l'Environnement et de la Maîtrise de l'Energie
AFNOR	Association Française de Normalisation
AG	Amtsgericht
AGU	Arbeitsgemeinschaft Umweltfragen
AI	Alkali Inspectorate
AIRPARIF	Association Interdépartementale pour la Gestion du Réseau de Mesure de la Pollution Atmosphérique et d'Alerte en Région d'Ile-de-France
AMA	Association of Metropolitan Authorities
APPA	Association pour la Prévention de la Pollution Atmosphérique
AQA	Agence de la Qualité de l'Air
AREMA	Association pour la Mise en Oeuvre du Réseau d'Etude, de Mesure et d'Alerte pour la Prévention de la Pollution Atmosphérique
ASPA	Association pour la Surveillance et l'Etude de la Pollution Atmosphérique
B.A.U.M.	Bundesdeutscher Arbeitskreis für umweltbewußtes Management
BAnz.	Bundesanzeiger
BATNEEC	Best Available Techniques Not Entailing Excessive Cost
BBauG	Bundesbaugesetz
BBU	Bundesverband Bürgerinitiativen Umweltschutz
BCC	British Coal Corporation
BDI	Bundesverband der Deutschen Industrie
BGB	Bürgerliches Gesetzbuch
BGBl.	Bundesgesetzblatt
BImSchG	Bundesimmissionsschutzgesetz
BImSchVO	Bundesimmissionsschutzverordnung
BMU	Bundesministerium für Umwelt, Naturschutz und Reaktorsicherheit
BMWi	Bundesminister für Wirtschaft

BPEO	Best Practicable Environmental Option
bpm	Best Practicable Means
BS	British Standard
BSI	British Standard Institution
BUND	Bund für Umwelt und Naturschutz Deutschlands
C.U.S.	Communauté de Strasbourg
CAA	Clean Air Act
CBI	Confederation of British Industry
CDU	Christlich Demokratische Union
CEGB	Central Electricity Generating Board
CITEPA	Centre Interprofessionnel Technique d'Etudes de la Pollution Atmosphérique
CNPF	Confédération Nationale du Patronat Français
CO_2	Carbon dioxide
COREPER	Comité des Représentants Permanents
CPA	Control of Pollution Act
CFSP	Common Foreign and Security Policy
CSU	Christlich Soziale Union
DEn	Department of Energy
DGB	Deutscher Gewerkschaftsbund
DIHT	Deutscher Industrie- und Handelstag
DIN	Deutsches Institut für Normung
DNR	Deutscher Naturschutzring
DoE	Department of the Environment
DRIRE	Direction Régionale de l'Industrie, de la Recherche et de l'Environnement
DTI	Department of Trade and Industry
EA	Environmental Assessment
ECJ	European Court of Justice
EDF	Electricité de France
EEB	European Environmental Bureau
EIA	Environmental Impact Assessment
EIS	Environmental Impact Statement
EP	European Parliament
EPA	Environmental Protection Act
EU	European Union
EURES	Institut für regionale Studien in Europa e.V.

FDP	Freie Demokratische Partei
FoE	Friends of the Earth
G.E.	Génération Ecologie
HMIP	Her Majesty's Inspectorate of Pollution
HSE	Health and Safety Executive
HSWA	Health and Safety at Work Act
ICC	International Chamber of Commerce
IEEP	Institute for European Environmental Policy
IEHO	Institution of Environmental Health Officers
IFEN	Institut Français de l'Environnement
INSEE	Institut National de la Statistique et des Etudes Economiques
IPA	Interparlamentarische Arbeitsgemeinschaft
IPC	Integrated Pollution Control
ISO	International Standard Organization
KU	Koordinationsstelle Umwelt
LAI	Landesanstalt für Immissionsschutz
LGMB	Local Government Management Board
LRTAP	Convention on Long Range Transboundary Air Pollution
MAFF	Ministry of Agriculture, Fishery and Food
MURL	Ministerium für Umwelt, Raumordnung und Landwirtschaft des Landes Nordrhein-Westfalen
NAGUS	Normenausschuß Grundlagen des Umweltschutzes
NCB	National Coal Board
NEPA	National Environmental Policy Act
NO_x	Nitrogen oxide
NRA	National Rivers Authority
NSCA	National Society for Clean Air
OECD	Organization for Economic Cooperation and Development
PC	Parti Communiste
PHA	Public Health Act
PS	Parti Socialiste

R.P.R.   Rassemblement pour la République
RAIN   Reversing Acidification in Norway
RCEP   Royal Commission on Environmental Pollution

SEA   Single European Act
SGCI   Secrétariat Général du Comité Interministériel pour les Questions de Coopération Economique Européenne
SMP   Solvent Management Plan
SO_2 Sulphur dioxide
SPD   Sozialdemokratische Partei Deutschlands
SPPPI   Secrétariat Permanent pour la Prévention des Pollutions Industrielles
SRU   Sachverständigenrat für Umweltfragen
SWAP   Surface Water Acidific Programme

TA   Technische Anleitung
TGA   Trägergemeinschaft Akkreditierung
TÜV   Technischer Überwachungsverein

UBA   Umweltbundesamt
UDF   Union pour la Démocratie Française
UIU   Unabhängiges Institut für Umweltfragen
UNECE   United Nations Economic Commission for Europe

VCI   Verein Chemische Industrie
VDI   Verein Deutscher Ingenieure

WZB   Wissenschaftszentrum Berlin

Introduction

The nature of state activities is undergoing rapid transformation in Europe. The European Union exerts influence on state practices in member countries, which in their turn shape European policy. In putting its stamp on European regulation, Germany, for example, has indirectly provoked modifications to state practices in other member states. 'Directives on air pollution have been subject to strong German influence' (interview with *CNPF*, June 1993). 'The influence of European environmental legislation for Britain cannot be overestimated' (interview with DoE, January 1993). The Federal Republic has been subjected to such influence in its turn, having to incorporate novel regulatory elements deriving from other traditions in administrative procedures and problem solving, and to modify German understanding of the state accordingly. To quote one prediction, '...the comprehensive and consistent use of the Eco-Audit Regulation will revolutionize the German industrial landscape and introduce quite new elements into the Federal German legal system' (interview with EIA association, October 1993). In short, the joint supranational policy-making framework and the development towards an integrated market have provoked a mutual, ongoing process of transformation between member states and European institutions. The shape of things to come in institutional and policy innovation is becoming progressively apparent.

How and why is this change taking place in the European Union? What are the underlying mechanisms and motive forces and how do they affect European policy? How does European legislation influence policy content, institutions, and relations between the state and society in member countries? We examine these questions with reference to a specific policy area and an illustrative selection of member states. We focus on one regulatory area, namely clean air policy, analysing the genesis of clean-air legislation in the interaction between supranational and national strategies in Britain, France, and Germany.

Our central finding is that European clean-air policy is the product of regulative contest between leading member states. Countries displace one another as front-runner, leaving the rest of the field to adjust as need be. Whether a country succeeds in assuming policy leadership depends on a number of factors. It is important whether a country already disposes of appropriate domestic regulatory instruments and how compatible proposed regulation is with overriding Commission objectives and strategies. Given the substantial powers of the Commission to initiate legislation and define policy, its goals and strategies are determined by dual, potentially contradictory interests, the extension of regulatory activity and regulative restraint. Individual member states have many reasons for attempting to impose their regulatory regimes, cultures, and practices at the European level; primarily the desire to preserve national problem-solving traditions and institutions — hence minimizing the cost of legal adjustment to European legislation — and to create or maintain national competitive and locational advantages.

In the eighties, for example, the Federal Republic managed — with the support of the Commission — to impose its regulative-law, technology-based approach to clean-air policy (stationary sources) at the EU level. Britain, with its different, much more flexible regulatory tradition, came under considerable pressure to modify its administrative practices and the wonted close and informal cooperation between industry and regulators. For some years now, however, the tide has turned. Germany is at present confronted by substantial regulatory innovations that are congruent with British practice and are accordingly being fostered and promoted by the United Kingdom. These arrangements involve opening up administrative and industrial decision-making processes to a large degree of public scrutiny. For the German regulatory tradition, greater 'transparency' in administrative licensing processes and the passing on of company-related environmental data to affected interest groups demands considerable administrative readjustment. In other words, there is no 'structural winner' in the regulatory contest.

In France the need to adjust to European policy has been mitigated by two circumstances. Since energy supplies are based largely on nuclear technology, France is only marginally affected by rules on reducing power station emissions. Moreover, it has long disposed of a 'requisite variety' of instruments, a broad range of multifarious environmental regulatory tools facilitating instrumental compliance with European requirements. France has accordingly taken little initiative to influence European policy and has needed little effort to conform with European rules.

We are thus examining a two-way process. European rules transformed into new policy strategies come up against established national problem-solving philosophies, control structures and institutional traditions, and against rooted interorganizational relations. These correspond more or less strongly with supranational strategies, so that state arrangements in member countries have to be adapted and altered to varying degrees. Redefining the state in Europe must, however, be seen not as a one-way process initiated by Brussels alone. There is double interlinkage: vertical interaction between

European institutions and member states, and horizontal interaction between the member states. National goals and strategies confront one another, and regulatory competition develops among leading member states. The contest is decided by *ad hoc* coalitions among member states, but also by the strategic interests of the Commission. The at times decidedly heterogeneous nature of EU regulatory policy is attributable to the meeting of many different traditions of the state in the European negotiating process and the shifting ascendancy of different national regulatory conceptions.

The first task in examining its changing nature is to define the significant dimensions of the 'working state'.[1] Fundamental state structures and organization are the first aspect to be considered; centralization or decentralization in political and administrative power, the nature and qualities of administrative practice, and the allocation of financial resources among the different political and administrative levels. A second dimension is the set of control instruments and legal rules available within a given country, and — third — the predominant problem-solving philosophy or ideology justifying state intervention in the given policy area, aspects which are often closely interwoven with existing institutions (van Waarden 1992; see also Döhler 1990, 27ff.). Next, insight into the qualities of the state may be provided by investigating the interface between the state and society: state mediation of interests, the process through which associations are brought into policy-making and implementation, which involves the form and quality of direct interaction with clienteles and individuals. Also included is the aggregation of societal interests via political parties and their effect on state action. In keeping with our proposition that processes of change arise in interaction between the state at the national level and 'state' elements at the European level, we also investigate these five dimensions within the European framework.

These levels of investigation are linked on a policy network analysis basis, bringing together the differing theoretical explanatory perspectives of resource-exchange theory and negotiation theory, of new political institutionalism, and of policy analysis. The aim is to explain why and how Germany and Britain came in turn to dominate clean-air regulation, and what effects this has had on policy and internal state structures in member countries of the European Union. The subject-matter of our network analysis thus includes institutional structures and informal interactions, exchange and negotiating strategies and problem-solving cultures. The changing nature of the state is comprehended as a mutual process of influencing structures, strategies, and ideas. Member states' institutional structures and rules and their problem-solving cultures commend specific strategies for action and definitions of issues, long-term goals, and consequent courses of action. They interact with the strategies of European institutions, which are likewise shaped by their own structures and traditions. Community strategies in their turn affect existing structures in member states and the Community itself, and bring about changes in these structures. In brief, structures and strategies (Chandler 1962), like problem-solving traditions, are interlocked in a process of mutual transformation.

Our exploration of how and why the quality and form of the state in the EU and within its member states have changed over the past decade and a half is structured as follows. In a first section we explain our theoretical frame of reference, our analytical approach, and methodological procedure. In a second section we outline the nature of the 'working' state to provide a background for comparison. In this context we elaborate the major differences between the three countries under review. Having established the principal characteristics of the three, we attempt to explain why and how they have sought to influence European policy. Since we assume that the transformation of the state in the EU is a mutual process of influence among member states and between member states and the EU, we also sketch the most important aspects of the supranational interactor. The hypotheses of our inquiry, formulated in the theoretical section with respect to this interaction between member states and the European level, are explored in the third section with the aid of qualitative data. For this purpose we consider the twelve most important clean-air policy (stationary sources) directives and regulations that have been adopted in the period, which are at present being implemented, or which are still being negotiated. At the same time we discuss the impact that Community policy in this field has had on state practices in individual member states.

Notes

[1] We are focusing here on the output — as opposed to the input — aspects of state activities. That is to say we are interested in changes in policy instruments and their application, and not in the general democratic legitimization patterns of these policies.

1 Theoretical Frame of Reference and Analytical Approach

1.1 Policy Network Analysis as Frame of Reference

If one takes a policy network analysis approach[1] to examining change in the nature of the 'state' in the European Union, state action taken in a specific policy area is to be interpreted as interaction among relatively autonomous actors in the public and private sectors. The concept of 'the clear division between the state and society and ... of the state as the highest social control centre' (Mayntz 1993, 41) is abandoned. This altered view of the state appears appropriate, given the increasing complexity and differentiation of society and the emergence of large organizations. If it is to attain its goals, the state must rely on their voluntary cooperation. The actors participating in network activities, mostly organizations but also individuals, pursue differing but interdependent interests in a given policy or issue area (Mayntz 1993, 12). Actors thus dispose of resources important for others, and they accordingly interact on a relatively autonomous basis, even though they may be embedded in a formal, hierarchical organization (Scharpf 1993, 67ff.). Asymmetrical exchange or power relations can exist between actors (Mayntz 1993, 47) as a result of differences in the level of resources at their disposal. Membership of the network depends on whether the action of an individual actor has significant consequences, is 'consequential' (Laumann/Knoke 1987) in relation to the policy to be formulated together.

Network analysis brings together various explanatory approaches to account for how actors perform in a policy network: rational-choice, institutionalist theory, and policy analysis. One variant of the rational-choice approach explains actors' actions on the basis of a rational, utility-oriented exchange of resources (Aldrich/Pfeffer 1976); in this interpretation, the influence and the centrality of an actor in the policy network depend on the extent and type of available resources that are important for others. Network interactions are determined by negotiations that seek to achieve a common result (Mayntz 1993, 15) or 'joint production'. The resources that are exchanged among network actors without market pricing in a direct or indirect negotiating process are tangible and intangible (Tichy/Fombrun 1979, 927), and include political support ('generalized political exchange' — Marin

1990). Each actor seeks to influence the outcome of the bargaining process in accordance with his own interests. At the same time, however, the parties involved are concerned to bring negotiations to a common conclusion. The more stable and continuous these exchange relations are, the more readily the interests of the negotiating partner are recognized as legitimate (Mayntz 1993, 49).

Importantly, negotiation among actors is limited and enabled by institutional realities, such as organizational structures, legal rules, etc. 'The basic characteristic of an institutional argument is that prior institutional choices limit available future options' (Krasner 1988, 71). Organizations and the longer-term exchange and conflict relations existing among them develop a certain selectivity towards options for action; they favour some, hamper others, and exclude still others from the outset (Scharpf/Brockmann 1983, 14). In this way, institutional 'path-dependent avenues for action' are created, along which choices can be made and exchanges effected. Hence, institutional structures do not determine actors' action but keep it within certain bounds while nevertheless opening up opportunities that can be taken. Whether and how this occurs depends in its turn on cost-benefit calculations, strategic considerations, and the value orientations of the actors concerned. For example, a federal constitution and a second chamber in which the subnational entities are represented at the central level provide subnational actors with opportunities for action and for exchanges unavailable in a unitary system. Their use is subject to political calculation and prevailing value orientations. Where institutional and statutory rules are changed as a consequence of European policy, the points of reference for exchange calculations and action strategies shift for all actors involved. Well-established 'interorganizational behavioural strategies' (Scharpf/Brockmann 1983, 15) modify, and the nature of the state changes.

An institutional perspective, moreover, implies that the activities of network actors are influenced by a superordinate problem-solving philosophy reflecting the general value consensus on the appropriate manner for dealing with problems. The stability of a network is hence also established by common problem perceptions (Döhler 1990, 31) or 'belief systems' (Converse 1964). These are ideological models and convictions that change only slowly as a rule, especially when they affect 'core beliefs' and do not simply represent instrumental action orientations (Sabatier 1993). The latter, merely pragmatic orientations can very well be contradictory under the selfsame umbrella of superordinate value orientations, and are more open to change.

Finally, the issue perspectives of policy analysis have their contribution to make to policy network analysis, i.e., policy area and issue-specific studies are carried out, and network action and actor constellations are examined for each of the various phases in the policy cycle, i.e., problem definition, policy formation, policy implementation, and evaluation.

1.1.1   Policy Network Analysis in the European Context

The EU, with its sectoralization, the functional differentiation and fragmentation of policies, as well as the dominance of corporate actors in a horizontal web of interorganizational relationships at the negotiating level (Mayntz 1993; Kenis/Schneider 1991), appears the almost ideal area of application for policy network analysis (Schumann 1993, 39). However, European policy networks have a number of special attributes: less stability than national policy networks; higher actor fluctuation; a lower degree of institutionalization; a markedly heterogeneous range of actors; and the associated centrality of the Commission as well as high policy segmentation. Lower stability and higher actor fluctuation are attributable first to the relative recency of the networks (in environmental policy they are only about twenty years old), and second to the fact that they are policy-making networks. Their function is to develop new measures in European policy, and by their very nature they are short-lived, reforming in terms of the issues on which decisions have to be made. The persistence of networks derives essentially from the well-established interaction patterns of the implementation phase. However, European policy is implemented largely by the individual member states.

In EU policy networks we typically find corporate actors of a bureaucratic or associational type. Much more rarely do we encounter democratically elected bodies. In contrast to national networks, the intergovernmental actor is also to be found at the European level, representing the interests of individual member states, and which develops no corporate identity of its own (Coleman 1972, Schneider/Werle 1992). Important European actors such as the Council of Ministers are composed of fifteen representatives of different national networks. This also applies with respect to non-state actors at the European level. European associations are forums but not corporate actors (Kohler-Koch 1992). The numerous commissions and working groups for the preparation of European legislation also frequently see themselves as intergovernmental institutions largely representing the interests of member states. At times, however, discussions are dominated by *'copinage technocratique'*.

The relative abundance of intergovernmental actors is directly attributable to the 'enhanced complexity' of European networks, for actors are not willing to waive their specific, member-state oriented identity at the supranational level. The heterogeneity and multiple layering of European networks go hand in hand with the relative preponderance or centrality of one actor, the Commission, constituting a further structural feature of European networks. The first two aspects and the last are in direct logical correlation, since the disunity and diversity of European networks lends salience to the Commission.

In diversity of problem perception, problem-solving philosophy, strategy, and interests, European networks outbid the relatively homogeneous national policy networks, where, although differing views about problems and problem solving and divergent interests exist, the differences are not as great nor as intense. In a national network, problem processing is facilitated by a time-honoured sense of identity. As a rule, approaches to problem-solving are 'firmly rooted in nationally specific legal, political and administrative institutions which have been outcomes of long historical processes and which have shown great persistence over time .... Culture is ... embedded in legal and administrative institutions and the latter in turn buttress these cultural values, making them so enduring' (van Waarden 1992, 23). These values prove all the more resistant to attempts at European approximation and levelling, especially since there is no specifically European problem perception nor a European public. A practically non-existent normative-ideological identification with Europe impedes efforts to achieve collective output in negotiations in a twelvefold (now fifteenfold) policy network (Wallace 1983, 420). The goal of an integrated market is obviously insufficient to engender normative integration. In view of this multinational heterogeneity, the Euro-networks fulfil an essential informational function and in the absence of a European public are a step towards compensating this structural deficit (Streeck/Schmitter 1991). However, since it is always a matter of highly segmented 'low politics' (in contrast to 'high politics' with its close concern with sovereignty), there are narrow limits. For how much programmatic political mobilization and how much identity can be generated in a debate on reducing sulphur emissions from power stations or on the price of bananas? Many European policy measures are not only specialized, fragmented, and technically complex: they also lack an 'interesting arena' where issues can be debated. For, given the relatively limited powers of the European Parliament, its capacity to legitimize European policy and develop European ideas and arguments is modest (Majone 1993). Rivalry in problem perception and solving is thus structurally inherent to European policy networks, where arduous compromises have to be sought, or — depending on the negotiating situation, public attention, and expertise — measures are imposed on some member states by others.

A further structural aspect of European policy networks distinguishes them from national networks: the centrality of one actor, the European Commission. In view of the heterogeneity and multiplicity of actors participating in European policy networks, with divergent attitudes towards problems, differing regulatory styles and dissimilar interests, the position of the Commission must necessarily be strengthened. It is the only corporate actor (with the exception of the Court of Justice) to regard itself as having genuinely European interests. This preponderance is emphasized by the Commission's right to initiate legislation: 'The initiative ... is in the hands of the Commission which gives it tremendous influence over the final shape of policy within the Community' (Peters 1992, 89). In national networks this power is more broadly distributed. Considerable weight accrues to the Commission as an actor able, in preparing legislation, to call in national experts at its discretion to participate in drafting, and which is furthermore the only entity with a good grasp of all pending legislation (Schumann 1993, 13). It can also set the agenda. It decides not only on the contents of a proposal, but also on when it is to be submitted and how it is to relate to other proposals. Especially the more frequent majority voting in the Council of Ministers enhances Commission influence (Schumann 1993, 13).

A final structural aspect of policy networks at the European level is their segmentation and fragmentation in relation to one another. This is due partly to the often technical and highly complex nature of the matters to be dealt with. Also important is the institutional circumstance that, in contrast to national policy networks, no policy integration via party platforms or coalition agreements takes place that is anywhere near as well developed. Although a certain degree of substantive coherence is sought via the European Council, this body is strongly intergovernmental in nature, with a markedly member state orientation. In effect, coherence is established *ad hoc*, arising among the various policy sectors during negotiation of package deals between the various policy sectors. It is based on negative coordination plus bargained compensation. In package deals, assent to certain measures is achieved by offering benefits differently distributed in temporal and sectoral terms. Each actor involved 'has to accept the package or kill the bill along with his own project. There is no possibility of defection by one without triggering the immediate and automatic retaliation by the rest' (Tsebelis 1990, 110).

1.1.2   The Policy Area: New Regulatory Policy

Our hypotheses on policy-making processes under the conditions of European interest multiplicity and on transformation processes at different levels of the state relate to new regulatory policy.

A large part of European policy is regulatory in nature. It is concerned firstly with the shaping of market processes, i.e., with defining conditions for market access and market operation (old or classical regulatory policy); and second with curbing negative external impacts on the public or the workforce from productive activities and individual consumption (new regulatory policy).[2] New regulatory policy thus pursues the goal of changing the behaviour of producers and consumers. It uses a range of instruments to achieve this. It imposes certain types of behaviour or prohibits certain types of behaviour, threatening penalties in the event of non-compliance with these prescriptions and prohibitions. Alternatively, the attempt is also made to achieve changes in behaviour by offering incentives, formulating recommendations, and providing information (education, persuasion). Together with health and safety at work, consumer protection, and sections of social policy, environmental policy — the subject matter of our investigation — is a central field in new regulatory policy. What, then, are the typical features of decision-making processes in the context of European regulatory policy making?

1.1.3   Coordinating the Multiplicity of European Interests

Acting in interlocking networks provides new opportunities for action but also imposes restrictions. In the shaping of Community policy, the multiplicity of regulatory goals, interests, and instruments that exist in member states of the European Union generates a substantial need for coordination, and complex processes for reaching agreement. It is necessary to balance multifarious interests at all stages of European policy formation in defining issues, setting agendas, preparing decisions, and formulating policies. In this process, typical patterns of informal interaction among actors emerge in coordinating interests, developing both horizontally and vertically. The patterns used in the regulatory contest are the 'strategy of the first move', 'problem-solving orientation' and the related patterns of 'negative coordination, negotiation plus compensation'.

The first aspect to mention is the dynamics of 'regulatory contest' among high-regulating countries with different traditions in the field. In the first stage of policy making, problem definition and agenda setting, this competition gives rise to a coordinative dynamic that can be described as a 'strategy of the first move' and as 'unilateral adjustment'.[3] The strategy of the first move is used by a member state to gain a lead over other — high-regulating — member states. However, it succeeds only with the concurrence of the Commission, which is the 'gatekeeper' in this coordination pattern. For it decides whether the proposal advanced by the 'first mover' is taken up, thus arousing fears of 'unilateral adjustment' among other member states, and obliging them to react to the proposed measure. In other words, under the governance of a third party, the majority of actors adjust to the initiative of a single actor. The strategy of the 'first move' gives the initiator the opportunity to participate in determining which problems require action from the European perspective, and thus to influence the European decision-making agenda. Moreover, in defining a problem, ways of dealing with it in conformity with national practice can be proposed in the same 'package'.

If the Commission maintains its support, the advantage of making the first move is then carried over into the ensuing phase where the draft proposal is scrutinized by national experts, whose informal coordination patterns are dominated by a 'problem-solving orientation'. Under these circumstances, the initiator country may manage to entrench his 'framing' (Tversky/Kahnemann 1981) of the problem in the draft phase, so that the experts' common search for a solution sets out in this direction. The country putting forward the proposal thus takes the lead in policy formation.

In the next formal political decision-making phase, the related patterns of 'negative coordination, negotiation, and compensation' dominate (Scharpf/ Mohr 1994, 26f.). At this stage the original strategic advantage of making the first move is most difficult to maintain, since distribution questions are to the fore. The negative and positive concerns of member states are brought into prominence and govern subsequent negotiations. Where unanimity is required, a negotiated decision is to be obtained only if the overall benefits of the measure for the beneficiaries appear to be greater than the damage to the 'injured parties', and the costs accruing to them can be set off through compensation or package deals. As a rule this means that the original proposal has to suffer considerable modification.[4]

The informal coordination patterns we have mentioned as influencing the production of regulatory measures in Europe are favoured by specific institutional conditions. Thus the role of the Commission first as 'gatekeeper' then as 'process manager' in formulating problems and setting agendas is determined by its right to initiate legislation and the option of withdrawing bills. Or, to give another example: the possibility of making a decision in the Council of Ministers by a qualified majority facilitates processes of reaching agreement by negotiation among member states. In the following section we will be investigating the informal coordination patterns that typically come to bear in the various phases of policy formation, and will briefly describe how they fit in institutionally. In the empirical part of our study they will be discussed in the context of material policy in an illustrative account of the genesis of individual directives and regulations.

'Strategy of the First Move' and 'Unilateral Adjustment'

In pursuing a 'strategy of the first move', a member state will propose adoption into European policy by the Commission of a regulatory measure conforming with the interests of the proposer, and which has proven its worth in the national context.[5] When a member state advocates regulation of an issue at the European level, it is seeking to extend the scope of European policy in accordance with its own interests, and to shape European intervention to conform with its own regulatory principles. There are several reasons why member states — always high-regulating countries — choose to compete on the first move in regulatory policy-making to win the Commission for their proposals. First, an interest in avoiding the cost of institutional and legal adjustment; second, the endeavour to obtain favourable competitive conditions for domestic industry; third, an interest in expanding environmental technology markets for their own industry; and, finally, the hope of sustaining public authority negotiating positions vis-à-vis industry.

A particular institutional rule in environmental policy has proved decisive in encouraging regulatory competition in European policy-making. Among the fundamental environmental policy decisions of the seventies was the obligation for each member state to inform the Commission about planned legal and administrative measures (notification) and to put national decision-making on hold until the Commission decides within a very brief delay whether it intends to act in the field concerned. If the Commission so decides, it is required to submit a draft proposal within five months (Weinstock 1984, 310).[6] This results in a systematic coupling of new national and European political initiatives that tends to promote the diffusion of national regulatory measures, or provide incentives for governments to put forward regulatory proposals themselves and to influence European policy rather than having to adjust to it.[7] What decides whether a member state has its way with the strategy of the first move is how the Commission reacts. Because of its right to initiate legislation — the Council cannot act except on a Commission proposal — it acts as 'gatekeeper' in agenda setting, and as 'process manager' in the further course of events up to the Council of Ministers stage. Numerous different regulatory proposals are continually submitted to it, of which member states initially have no mutual knowledge.[8] From among the many policy proposals put to it, the Commission then selects those it wishes to promote. Vice versa, the high-regulating member states behave more or less like innovative policy entrepreneurs in the European regulation market, offering their wares to the Commission as monopoly demander.

The Commission officials listen (in the committees as in informal preconsultations) to everybody, but are free to choose whose ideas and proposals they adopt. This behaviour opens up great chances of influence for certain individual experts who, because they present ideas which are in line with the Commission's interests, may thus act as 'partisans' (Eichener 1992, 54).

It is no act of friendship on the part of the Commission if it manifests openness towards such policy proposals. After all, it has a relatively small staff and little expertise at its disposal, so that it must rely on the knowledge and policy experience of member states (Eichener 1992, 50). It is, moreover, fundamentally interested in expanding its regulatory activities and hence its field of action, because its activity in policy areas that directly commit budget funds are limited due to the paucity of financial resources (Majone 1994). Whether a member states' regulatory initiative is given a hearing or not depends essentially on whether the proposal fits in with the superordinate interventionist philosophy of the European institutions; for example if it is compatible with the subsidiarity principle.

If in this regulative contest the 'first mover' is able to win the support of the Commission for its proposal, it gains the opportunity to add an item to the European agenda for which it has a relatively precise concept. If the proposal is compatible with the Commission views on the issue and how to tackle it, the initiator country has a good chance of seeing its national approach become the received view of the issue, the perception and evaluation pattern ('frame' — Tversky/Kahnemann 1981)[9] for the ensuing drafting phase. The strong interest in the policy proposal generally goes hand in hand with a higher degree of national expertise in the area to be regulated. Both aspects enhance the 'first-mover' influence. Influence gained in the problem-definition and agenda-setting stages can be regarded as a positional good (Hirsch). Once an actor has formulated a proposal and — in cooperation with the Commission — has occupied the problem-solving terrain in a specific manner, and has moreover collaborated in drawing up the agenda (profiting from the inactivity of those concerned) the others are necessarily in less favourable starting positions. Initially, where it is a matter of developing this specific proposal, other member states can only react to the initiator's project; they adjust unilaterally to the definition of the issue and the proposed solution. In other words, whoever takes the initiative and makes the proposal calls the tune in this phase of defining the problem and setting the agenda.

In the regulation of environmental protection, there is such a pattern of taking turns in making a strategic first move. A member state takes the initiative in defining a problem and in formulating a proposed course of action. In so doing, it points European policy in a certain direction and obliges other member states to respond. But the pace-setter has no tenure, for the Commission is concerned not to lend its ear to one party in permanence. This is favoured by the strength of competition between high-regulating countries. Each keeps a watchful eye on the others. There is no structural 'first mover', so that over a longer period of time and range of measures (directives and regulations) the benefits are spread around — at least among the ambitious regulators. The negative distributional effects (i.e., adjustment costs), which accrue only to reactive member states, are also correspondingly distributed. In the overall perspective, a diffuse cost-benefit reciprocity arises,[10] 'emphasizing that actors expect to benefit in the long run and over many issues, rather than every time on every issue' (Caporaso 1992, 602).

It is likely that institutional changes both intensify and check efforts by member states to make the first move in defining issues and framing European solutions. With the introduction and expansion of the qualified majority principle in the Council of Ministers, the risk of having to submit to a 'foreign' regulatory style has increased. The unanimity principle gives greater scope for negotiation, because a measure can be vetoed. Under the qualified majority regime, member states are correspondingly more interested to make the first move in the regulatory contest and formulate policy proposals.

It could be objected that pursuing a first-move strategy need not necessarily give the initiator an edge over others. On the contrary, it may induce the immediate development of an opposing coalition determined to counteract the initiative. However, this seems relatively unlikely under the current specific institutional conditions for defining issues and setting the agenda in regulatory policy-making. Institutional arrangements are characterized by the exclusion of the public from problem definition and agenda setting. In this phase of policy formation, as we have mentioned, the Commission in its capacity of 'monopoly demander' has great latitude in selecting regulatory proposals from among those submitted to it on the multifarious European 'policy market' (Peters 1992) without obligation to report just what has been submitted. It does not regard itself as obliged to assume responsibility at an early date and in a central arena for the distributional and regulatory implications of individual regulative measures. The seclusion of problem definition and agenda setting is a consequence of central institutional aspects of the European political structures. The Commission as executive organ is not bound by the legislative programme of a government responsible in its turn to a parliamentary majority.

Precisely this institutional isolation is, however, succumbing to the European Parliament's striving for emancipation. The EP demands to be informed at a very early date about measures planned by the Commission, so that the pros and cons of the measures can be debated publicly and thoroughly in the parliamentary arena.[11] Early politicization would diminish the chances of a 'first mover' getting its way and entrenching its own regulatory approach at the drafting stage, thus reducing the need for other member states to adjust. This would also mean that negotiation and compensation demands enter the decision-making process at an earlier stage, adding to its overall unwieldiness.[12]

From the point of view of low-regulating countries, the 'strategy of the first move' is less attractive. Institutional adjustment costs to be expected are not so high, because there are often no national rules. Frequently there are no relevant national implementation authorities whose behaviour would have to be adapted.

From the economic perspective the low-regulating countries have no interest in a stringent regulatory regime, because their mild national rules — where they exist — afford them competitive advantage in the European market place. For such countries it is therefore more useful to sit on the fence and observe the direction things take, even to manifest resistance and deny the need for regulation, only to reap rewards — in the form of special allowances or concessions in other policy areas — for jumping on the bandwagon. Sometimes willingness to agree is facilitated by counting on subsequent non-implementation.

In considering the 'strategy of the first move' from the perspectives of both high-regulating and low-regulating countries, the following preference ranking presents itself for a high-regulating country: best would be a solution in keeping with its own national solution; next best would be to impose a portion of one's own ideas at the cost of making concessions; least favourable would be the failure to bring about any solution at all. For low-regulating countries, by contrast, the most favourable outcome would be no solution; a mixed solution would rank second on the preference scale, and the solution conforming to the ideas of high-regulating member states would be the most expensive.

Problem-Solving

In the drafting phase of European legislation, a coordination pattern tends to manifest itself that is referred to in the literature (Mayntz 1994; Scharpf 1991; Scharpf/Mohr 1994) as 'problem-solving orientation'.

... the starting point and trigger for problem-solving processes is a situation that demands change — instead of offering only an opportunity for maximizing a defined benefit. Also characteristic is that, in problem-solving processes, problem analysis and definition of the goal regarded as the 'solution' are important phases in the process — whereas in rational decision-making models, goals (as types of desirable 'benefit') are given. Thirdly and finally, problem-solving processes are characterized by at least initial uncertainty about the path leading to the desired goals, so that the accent, in contrast to the rational decision making model, is not on the cost-benefit comparison of given alternative decisions but on the finding of possible ways to solve the problem, which often includes repeated starts and trial-and-error behaviour (Mayntz 1994, 22).

Distribution issues remain in the background (Scharpf/Mohr 1994, 17). During the joint search for new tools to solve a problem, which has initially been broadly defined in collaboration between the Commission and a high-regulating 'first mover' country, there are no 'diplomatic behavioural patterns' and no 'hidden power games' (IEP 1989, 107) that are predominantly governed by questions of distribution. A certain 'denationalization' (Bach 1992, 92) of regulatory policy formation becomes apparent. Technical, scientific, and legal experts with a greater interest in pragmatic solutions set the tone. The more complex and the more strongly technically the question to be regulated, the easier it is to seal it off from politicization in this phase, and the more strongly the whole matter becomes a discourse under the auspices of the Commission among country experts with regulative experience. What has been said of European industrial safety (Eichener 1995, 1992) can also be said of European environmental policy:

The debates tend to move quickly to a level of technical details (about what is technologically possible and at which costs) so that technical expertise is a crucial condition for effective participation ... The interest in the matter is an important corresponding variable, because the higher the interest is, the more resources will be invested in the committee work. Members report that delegates from low-level countries frequently prefer to listen to discussions to get early information on regulatory acts than to actively contribute (Eichener 1992, 52).

Especially when committees are longer lived, common learning processes are set in motion that lead to cognitive rapprochement among national experts and to the development of 'epistemic communities' (Haas), which feel bound together by professional knowledge and the employment of a common technical language. This permits greater ease of understanding and agreement across national interest perspectives.

'Problem-solving' takes place under the direction of the Commission in the expert committees responsible for preparing legislation (IEP 1989, 126) and is furthered by certain institutional conditions. Their permanent status (standing committees) facilitates rapprochement in problem perception. Since it is a matter of a consultative — not a decision-making — process,[13] problem-solving proposals can be collected and compared in the course of consultation (Eichener 1992, 52). 'With my experts I can get that [a draft proposal] done that quickly' (interview with DG VIII, June 1994). However, institutional changes are also becoming apparent in consultations as well, which may make peaceable problem-solving in the draft phase more difficult in future. If the European Parliament obtains the right to be informed from the outset about the work being done in committees on draft legislation in order to subject Commission activities to parliamentary control (Lodge 1994; The European, 29 Dec. 1994 — 4 Jan. 1995, 1), distributional aspects, which can lead to politicization, will be brought into play much earlier than has hitherto been the case. However, politicization can also direct attention to new solutions not considered by the experts.

Negative Coordination, Negotiation and Compensation

Distributional perspectives, weighing up the costs and benefits of a decision, are foremost in the process of adopting of a Commission draft proposal and in Council decision-making. The conflicts of interest that emerge in this phase are initially stressed with the aid of the coordination pattern 'negative coordination', and then — if all goes well — reconciled by negotiation and compensation deals. In the case of negative coordination, the underlying interaction mode is determined by an egoistic interest orientation on the part of the parties involved (Scharpf 1992, 53). The Commission's draft legislation is considered only from the point of view of the particular party's own interests. If more costs are expected than benefits, rejection is indicated. If it is only a question of competing interests (rather than antagonistic interests), where satisfying the interests of one actor implies that satisfaction of the interests of the others would be diminished (though not excluded), compromise can in principle be reached (Benz 1992, 160). Compensation can be offered to offset potential losses for actors involved, where the matter being negotiated allows such a deal; i.e., when the decision is not a simple yes/no one (Scharpf 1992, 68). Where in view of the matter at issue, monetary compensation to obviate a veto appears unacceptable for normative reasons (Scharpf 1992, 70), package deals can be concluded, in which the benefit perspectives of the differentially affected parties are offset. As a rule, the more different policy areas package deals involve, the higher up the ladder they have to be negotiated (up to the European Council). Where compensation is agreed in negotiations within a regulatory field (or even within a given directive), this can be done in a specialist Council or in COREPER. Here national 'specialist diplomats' initially make their policy statements, which — if this is not yet known — permits the national delegations to identify their main opponents. Consensus is then often reached on conflicting positions not in multilateral but in successive, bilateral negotiations. The principal opponents in a conflict either get together for direct, bilateral talks, or a proponent member state with a pronounced interest in regulation attempts to entice onto its side the actor with the position closest to its own, then tackling the next closest country and so on. By this step-by-step procedure, in certain circumstances backed by a policy of targeted isolation, pressure on rejectionist parties can be raised.

If, despite differential interests, a negotiated solution is found through compensation or package deals, the question arises as to the fairness of the agreement reached. Because benefits covering several different issue areas are not open to comparison (incommensurability), an overall balance on costs and benefits for all the parties concerned is difficult. It is therefore imperative that a 'space of mutual trust' be established among actors if negotiated solutions are to be found. For if one does not really know whether 'fair' compensation can be expected for concessions made, one must at least be able to have faith that in future negotiations concessions can be expected from other actors, too, who are willing to make concessions with respect to one's own interests.

Various institutional conditions are apt to boost trust in negotiating fairness valid beyond decisional matters and over time, so that negative coordination/negotiation/compensation are not governed alone by calculations of interest and power on the part of the proponents and opponents of a given regulatory measure. The relatively small number of parties involved facilitates mutual monitoring of advantages gained and disadvantages suffered over longer periods and in relation to a range of different measures, thus increasing the likelihood of entente on the basis of reciprocity, albeit diffuse, with regard to benefits. Negotiations take place under the 'shadow of the future', for one is aware that exchange relationships are permanent, and conscious that any ruthless attempt to carry off a special advantage will be filed away in the collective institutional memory.

The European Union has no institutional centre above the hurly-burly of competing interests vested with the hierarchical capability to impose fairness and justice or to ensure diffuse reciprocity. The Commission, the body most likely to have a grasp of cost and benefit distribution over longer periods and wider areas of policy, is itself no homogeneous actor. Divergent national and sectoral loyalties are quick to disaffirm the Commission's formal independence from national interests.[14] Nevertheless, there are informal mechanisms — grounded in their turn in competition between member-state interests — that in practice ensure that costs and benefits are balanced out. The institutional memory of the Council of Ministers functions astonishingly well when it comes to keeping tabs on advance concessions made by a member state (Scharpf 1992, 77). Jealous scrutiny of other countries' advantages, the reflex of a competitive attitude, acts as a functional equivalent to a centrally upheld precept of fairness and justice obeying the rules of diffuse reciprocity. COREPER, for example, the committee of permanent representatives, appears to have developed a peer-consciousness that is oriented on such ideas of fairness and keeps alive the memory of advance concessions.[15]

Formal institutional conditions such as the unanimity rule, but also the risk of vetoing minorities where decisions are made by a qualified majority, force actors at least partly to accept other points of view and to make corresponding concessions. Diffuse reciprocity is becoming more and more important, for qualified majority decisions no longer always require specific reciprocity as is the case with unanimity (Schmidt 1995, 4).

The payoff to each individual party need not be improved in all instances of cooperation. There is an institutionalised preference for more cooperation, and the belief that all participants will profit in the long term, since unilateral action would lead to welfare losses in view of existing political and economic interdependencies. With diffuse reciprocity, consequently, distributional issues receive less attention, facilitating negotiations (Schmidt 1995, 4).

Furthermore, the possibility of making qualified majority decisions has the effect of a 'rod in the window'. In the 'shadow of the hierarchy' (Scharpf 1991), agreement can be more rapidly reached in negotiations. 'After the introduction of the qualified majority rule the EC Councils of Ministers have thus not abandoned the practice of unanimous decisions, but negotiations have become appreciably shorter' (Scharpf 1992, 25).[16]

In a fixed institutional framework, every member state is more willing to make concessions in the realization that 'economical' use has to be made of resistance during negotiations. The 'economy of the veto' tells every member state that it cannot 'always be against everything' (Peters 1992). A country must dose opposition to single measures across policy fields and over the medium term. It is advisable to give strategic support to top priority interests. Member states consequently think very carefully when preparing negotiations in the Council about where they are not prepared to make concessions under any circumstances, and about what can be sacrificed or bartered away in negotiations in the absence of top priority.

Apart from the institutional factors favouring the maintenance of diffuse reciprocity in the distribution of benefits, there are various possibilities in this phase for mobilizing institutional resources to steer negotiations in a desired direction. The presidency in the Council is one such possibility. If a member state holds the presidency, it can give specific issues priority over others in drawing up the Council agenda. Control of the agenda also gives the incumbent member state influence on shaping potential package deals.

Cooperation with the European Parliament is another, increasingly important institutional resource. By coordinating with their national members in various party groupings, individual countries can use specific Parliamentary objections or demands in the negotiating phase to lend weight to their standpoint in the Council of Ministers. However, this possibility is also open to countries in a potential counter-coalition, so that it can lead to the neutralization of positions.

Negotiations in this phase involve more than finding a 'fair' balance between material interests in a general institutional setting. Because the negotiating parties are national governments, divergent national institutional constraints come into play. The decision-making rules at subnational levels have to be taken into consideration in the negotiations — as a problem of interlinked arenas (Benz 1992, 160). A member state as an actor in the process cannot remain 'unresponsively inconsistent' (Zintl 1992, 129) because it has to justify itself 'internally'. 'One does not commit oneself somehow or other but in a process that not only takes place in public but is also normatively highly charged. This gives greater impetus to self-commitment' (Zintl 1992, 129). This institutional obligation can also lend smaller countries greater weight in decision-making ('paradox of weakness' — Grande 1994). Vice versa, the pressure and commitment to decisions produced in European negotiations can give the government more scope for action at home vis-à-vis the national parliament.

The Long-Term Perspective on Coordination Patterns

If one considers informal coordination patterns in their institutional context over the entire problem-definition period up to the adoption of a measure, three typical courses of events can be distinguished. Which course is taken depends first on whether the Commission finds it can share the proffered view of the issue and is willing to take it on, and — once this hurdle has been taken — on the characteristics of the policy to be elaborated, its complexity, its technicality or ease of comprehensibility, and its aptness for politicization. Third and last, the degree of politicization is determined — leaving aside the question of comprehensibility — by whether an easily appreciable redistribution of costs and benefits can be expected from the proposed measure.

Taking the first hurdle, i.e., bringing a problem to the attention of the Commission, having it recognized as needful of regulation at the European level, and getting it on the agenda is already a significant and far-reaching move in the formation of European policy. This remains true even if the initiator fails to carry its concrete conception of the measure through to adoption of the draft without numerous substantive modifications.

Always on condition that a first mover succeeds in persuading the Commission that a specific issue has generated an urgent need for European action and ought to take the form proposed by the initiator, a first course of action can permit the 'advantage of the first move' to be sustained. The chances of bringing the substantive concept more or less intact through to the decision-making stage are better where the issue to be regulated is a relatively narrowly defined matter accessible only to technical or scientific specialists and where costs and benefits are difficult to assess during the problem-solving and negotiation phases.

Always assuming that the problem to be dealt with is a complex, inaccessible one, a second typical course of events confronts the first mover immediately in the problem-solving phase with counter-proposals advanced by experts from member states with different regulatory practices, who would like to take quite different paths to the same goal. Either a new problem-solving 'frame' is developed together, the problem re-assessed, and solutions proposed, or diverging proposals are placed on offer side by side as an additive range of solutions. Even where experts submit two complex regulatory proposals, the material and immaterial follow-up costs and benefits are usually not clearly apparent owing to the inaccessibility of the subject matter. However, if a counter-proposal is tabled, at least the institutional adjustment costs are addressed at an early stage.

A third typical course of events involves easily comprehensible issues with substantial distributional dimensions that rapidly become evident. Under these conditions, the chances that a first mover will succeed in carrying its own regulatory proposal through to an advanced stage in policy development are less favourable. The typical mechanisms for negative coordination and compensation take early effect.

The Consequences of Regulatory Competition: Policy Attributes

If it is true that new regulatory measures in Europe are frequently initiated on the basis of regulative competition between member states operating under the given institutional conditions in the informal self-coordination modes described, the subject-matter of European regulatory policy will display certain typical attributes. Hypothetically, it can be assumed first that regulative competition is tending to make European regulation more and more dense and detailed, because it is in the interests of certain member states to introduce their own ideas into European policy.

On the other hand, the Commission — as we have shown — is also keen to expand regulatory policy. The assumption that regulatory competition leads to increasingly stringent regulation is, however, countered by pressure from the subsidiarity principle. Under the impression of this precept and the declining acceptance of more and more detailing regulation by EU institutions, the Commission is setting new accents in control instruments, giving member states more room for manoeuvre.

Secondly, the absence of a structural first mover means that no one problem-solving approach predominates. Different countries at different times draw on the wide variety of national regulation to influence how European issues are defined and put on the Commission agenda. At the concrete and instrumental levels, European regulatory policy — under the umbrella of vague superordinate control philosophies like 'subsidiarity' — is a patchwork of regulatory traditions and approaches taken from national regulatory experience to be stitched together into European legislation

(Richardson 1995). Not only within a single policy area but sometimes within one and the same directive, divergent approaches can be on parallel offer if it has proved impossible to negotiate a consensus; compromises on formula are agreed.

Thirdly, it can also be assumed that where a higher degree of politicization exists, difficulty in reaching agreement in the adoption phase can land the process in the classical 'joint decision trap' (Scharpf 1985), where decision-making dynamics leads to deadlock permitting only framework legislation. This gives member states a great deal of scope in implementation. One example that could be mentioned is the renouncement of emission standards for certain pollutants requiring precise compliance in favour of vague legal concepts such as 'economically feasible' or 'best practicable means'.

However, there is no denying that, from the perspective of the 'gatekeeper' with an eye to its own interest in expanding regulatory activities, even framework legislation is a first step towards progressively augmenting European influence in a policy sector. The practice of 'mother' and 'daughter' directives serves this end. In this first step — where member states manifest more resistance to new regulatory activity — the 'mother' lays down only a general framework, whereas the 'daughters' that follow in the ensuing years define obligations in much greater detail. A 'policy of self-commitment in small steps' is thus pursued, and a 'momentum of acceptance in sequences of decisions' is generated (Eichener 1995, 38), where the necessity to accept each successive stage is justified on the grounds of the commitment already made in preceding stages. The typical patterns in coordinating the multiplicity of European interests and their consequences for the general characteristics of European policy are discussed in chapter 2 in the context of the empirical analysis of the development of directives and regulations.

1.1.4 Hypotheses on Interaction Between National and Supranational Actors

The following issues are addressed in examining the processes of coordinating the multiplicity of European interests and their effects. In relation to member states: (1) Why do individual member states attempt to raise their own regulatory solutions to the European level? (2) What negotiating strategies do they employ in this endeavour at the supranational level? (3) Under what conditions are member states successful in promoting their interests in the European regulatory contest? We will advance a number of hypotheses on these questions, which are to be theoretically justified and empirically underpinned.

(1) Member states are categorized as high-regulating or low-regulating countries, each obeying a quite different rationale for action. Countries are regarded as being high-regulating that traditionally have at their disposal a highly differentiated legal regulation of environmental pollution in all media accompanied by just as highly differentiated state implementation arrangements. Regulatory philosophy and the regulative instruments may differ greatly from one high-regulating country to another; they may, for example, be more strongly concerned with environmental quality or with individual sources (i.e. emissions). By contrast, low-regulating countries are those that have either adopted no strongly differentiated environmental legislation, or have an extremely underdeveloped implementation infrastructure. Our analysis focuses on the 'big three' member states, the Federal Republic of Germany, the United Kingdom, and France — all high-regulating countries — but which differ markedly in their regulatory philosophy, their regulative concerns, and in the instruments they prefer to use. The low-regulating countries are only implicitly taken into account.

As far as negotiating interests are concerned, it is our assumption that high-regulating countries have a pronounced interest in bringing their own national arrangements and supranational regulation into line 'qualitatively' (content) and 'quantitatively' (level), and in expanding joint legislation. Member states with regulative ambitions face opposition in regulative competition primarily from member states that also dispose of a well-developed — but divergent — national regulatory tradition, and which for this very reason resist new regulative moves. These are the real rivals in regulation. The regulative contest has various causes, deriving a) from the particularities of a member state and the handling of these particularities in political and administrative processes; b) from the political and administrative rationale and dynamics of the interactional process at the supranational level; and c) from national economic considerations. The distinctive characteristics of a member state and the internal rationale of political and administrative processes give rise to the following causal factors:

- Member states interested in regulation see themselves strongly affected by environmental problems and subject to corresponding internal political pressure owing to own emissions and long-range pollutant transportation.
- The tradition of high-quality environmental protection is regarded in a country as a social good and an intrinsically desirable political goal.
- There are political institutional conditions in a member state enabling political attention to be drawn to environmental interests and facilitating access to the political agenda.
- Lower national legal and institutional adjustment costs accrue to both high-regulating and low-regulating countries if their own regulatory ap-

proach and the pertinent institutional measures are carried over to the European level.

- The environmental protection authorities in high-regulating member states are concerned to bring regulatory provisions at the European level into line with their own stringent requirements because they otherwise lose negotiating power and clout vis-à-vis their own industry.

The rationale and dynamics of the interaction processes at the supranational level give rise to a further cause:

- In order to be better equipped in the European regulatory contest, 'prudent politics' of an anticipatory nature is practised. Relevant national regulations are elaborated so that they can be thrown into the scale in the European regulatory process. This procedure is defended internally with the argument of being better able to influence legislation in Brussels, and of sparing national industry and administration the necessity of frequent and changing adaptations.

Finally, there are economic grounds for high-regulating countries to raise their own arrangements to the European level:

- For competitive reasons, every member state has a general interest in not burdening domestic industry alone with additional costs.
- High-regulating countries are furthermore keen to impose the environmental standards that apply for domestic industrial production processes on European industry as a whole. Production costs can thus be approximated and locational disadvantages for national industry avoided.
- Capital goods with a low standard of environmental protection can to be kept out of the home market.
- For a member state with a developed environmental technology industry of its own, new markets will be created if it succeeds in embodying technical provisions in European legislation that can be met only with products of the country's own industry.

It is furthermore assumed that low-regulating, frequently late industrializing countries have little interest in approximating national arrangements and in expanding Comunity legislation. The reasons are once again to be sought in the distinctive features of these countries and their internal political and administrative rationale, as in economic considerations. In the political and administrative perspective of low regulating countries, the following aspects are significant:

- Because of their geographical distance from the highly industrialized 'core' of the EU, the low-regulating countries are less affected by the impacts of massive pollution like 'acid rain'.
- Protection of the natural environment as a societal objective is only weakly rooted in the public mind. Those in positions of political responsibility are therefore under less pressure to take action in the environmental field.
- The low-regulating countries are as loath as other member states to introduce new, unfamiliar legal rules, because doing so generates high transpositional and institutional adjustment costs.

Economic considerations play an important role:

- For these countries less exigent regulation constitutes a comparative locational advantage.
- Satisfying stringent environmental protection requirements by technical means places a heavy financial burden on industry, which hampers efforts to 'catch up' on industrialization.
- Since they dispose of no highly developed environmental technology industry, low-regulating countries have no incentive to support strict European measures with the intention of improving sales opportunities. On the contrary, environmental technology has to be imported.

(2) With regard to member states' negotiating strategies at the supranational level, we postulate that low-regulating countries are willing to accept more stringent regulations if 'in exchange' they obtain concessions in other policy areas and are able to conclude package deals.

Some low-regulating countries go along with policy formation at the European level, but fail to implement it effectively. This opting out of practical performance 'facilitates' their concurrence within the context of European policy formation.

In order to assert their regulatory ideas in European negotiating processes, member states seek to exchange substantive measures for procedural arrangements. This mixing of procedural or constitutional with substantive issues is typical of European network processes, because the process of distributing decision-making among actors within the relatively recent structures has still not been completed.

In developing the above hypotheses on negotiating strategies, the member states have be treated as unitary actors, the perspective of interactional dynamics shifts if subnational and sectoral actors within member states are taken into account. Taking the network approach effectively forces us to abandon the 'layer-cake' view of levels in the supranational network, and to direct our attention to the wide range of horizontal and diagonal relationships. This broadened perspective permits further assumptions on member-state negotiating strategies in the regulatory contest.

In the supranational network, new possibilities for interaction and coalition arise that extend to all levels, but also across all functional differentiations and national borders. They provide opportunities to attain policy goals out of reach in the national context alone. The interlinkage of national and supranational networks can be interpreted as 'nested games' (Tsebelis 1990). 'The game in the principal arena is nested inside a bigger game, where the rules of the game themselves are variable; in this game, the set of available options is considerably larger than in the original one. The actor is now able to choose from the new set a strategy that is even better than his best option in the initial set' (Tsebelis 1990, 8).

This sort of cooperation is to be found in the form of collaboration among experts, among 'regulatory zealots' across national borders, who are able to assert themselves against other parties involved in the preparatory committees who have neither their superior expertise nor any time to invest. This provides the prospect of transcending a merely national-interest position in the sense of negative coordination (Eichener 1992).

Furthermore, vertical coalitions in European policy formation and implementation are more and more frequent. Active contacts are established between subnational actors and those at the supranational centre, i.e., the Commission or the European Court of Justice, to win support in Brussels against national central government for political innovation and conscientious implementation.

In the event of pronounced conflict between various actors and interest groups in the national network, there is a trend towards shifting conflict resolution to the supranational level.

The formation of cross-border expert coalitions and the relatively passive role taken by the low-regulating countries in the regulatory contest mean that the European regulatory process tends to culminate in a 'policy of outbidding' (*'politique de la surenchère'*).

(3) The regulatory contest will be won by the high-regulating country that offers not only the requisite national preconditions but also a well-developed national regulatory regime and expert experience permitting a proposal to be submitted at the European level. The prize is the 'advantage of the first move' over other member states in an expanded regulatory context.

For its parts, existing national regulations must (if European legislation is to be influenced) find the support of industry and environmental groups in the national network of the country concerned.

The chances of success at the supranational level increase if professional coalition partners from other member states can be won in the committees and political coalitions formed at Council level, and if the regulatory proposals of a member state can be brought into line with the Commission's superordinate goals and strategies.

The following hypotheses are advanced on the Commission's negotiating interest (1), strategic considerations, negotiating advantages (2), and prospects of success (3).

(1) The Commission's negotiating interest is to expand Community legislation. It has the following political reasons:

- Under the Single European Act and the Treaty of Maastricht the Commission is obliged to ensure a high level of environmental protection in harmonizing national regulatory arrangements.
- It is also subject to corresponding pressure to take action, because it bears responsibility for compliance with international environmental protection treaties signed by the EU.
- As a corporate European actor, the Commission is concerned to further the supranational integration of policies.
- As such, it also has a vested institutional interest in expanding its activities. Since the high cost of agricultural and development policies prevent it from doing so in areas that directly bind financial resources, regulatory policy appears a promising field of action to cultivate (Majone 1989).

The interest of the Commission in expanding Community legislation is further explained by economic motives:

- The Commission is concerned to promote market integration.
- In view of growing environmental regulation in member states, it wishes to prevent barriers to trade arising from differential product standards and locational disadvantages accruing to high-regulating countries.

(2) As far as the European Commission's negotiating strategies are concerned, we postulate that the Commission is concerned to achieve environmental standards with the aid of regulatory strategies that are politically acceptable to the member states. Under the influence of the subsidiarity principle, the Commission has in recent years sought more and more to find strategies that safeguard autonomy while remaining compatible with Community aims (Scharpf 1993b). In the early eighties, by contrast, the Commission operated more by means of detailed regulation.

The Commission welcomes regulatory initiatives from individual countries, because it has inadequate personnel resources of its own, and is therefore grateful to make use of member states' practical experience and expertise. The precondition for doing so is, of course, that the legislation proposed by member countries is consistent with the superordinate objectives of the Commission.

(3) Satisfaction of the Commission's negotiating interests depends — at the national level — on the existence of similar national arrangements. They contribute to legitimizing European legislation because political support already exists at the national level. At the supranational level, the attainment of regulatory goals is easier if there are already international agreements that can exert pressure. An essential role is also played by the extent to which a coalition of strong negotiating member states as advocates or pace-setters support a regulatory proposal and — backed by important national industrial associations — endorse the measure in the Council of Ministers and the European Parliament.

1.1.5 Redefining the State: Analytical Dimensions and Hypotheses

What impact does the European policy of abating atmospheric pollutant emissions have on political and administrative practices, that is to say on the nature of the state, in the member countries of the Union? A number of typical national behavioural profiles in reaction to the Europeanization of policy are be observed:

- It is conceivable that common European policy formation and implementation take place in some countries without noteworthy adaptation of state arrangements. This is the case where — on the basis of existing 'slack resources' — 'old structures' are adapted to 'new rules' by, for example, activating existing but only latent instruments and strategies in response to European requirements.
- There is also the possibility of purely formal and not substantive adjustment. In this case, the transposition of European legislation into national law occurs without the relevant authorities being set up and the resources necessary to implement the laws being allocated.
- By contrast, there are cases where both formal and substantive changes in the state are to be observed at various levels, involving the reorganization of structures and the re-orientation of strategies and instruments.
- Finally, it is conceivable that the pressure for European-induced change triggers internal reforms more comprehensive than absolutely necessary to meet the requirements of European legislation. This anticipatory adjustment to European legislation can be used to push through internal policy transformations that would otherwise have been difficult to achieve; or national policy innovations are undertaken with the intention of taking the lead in European policy formation.

In the context of our inquiry, we are interested primarily in the final two possibilities. The question of actual changes in the nature of the state raises the issue of the choice of analytical dimensions in terms of which the transformation in member states is to be measured. An analysis of such dimensions is always selective and cannot cover all conceivable aspects of 'the state'. For our purposes, the central aspects to be analysed emerge from the interpretation of policy formation outlined above as a policy-network process. From this point of view, state action is seen as determined by processes of exchange and negotiation among state and societal actors within the context of institutional limits and possibilities. The actors involved share an interest in a given policy topic. The manner in which the policy issue is dealt with is, in a very general sense, governed by a fundamental problem-solving concept common to all actors. The general policy goals can, however, be attained by means of very different problem-solving instruments. Such a policy-network-related understanding of the state thus differs from a classical constitutional definition concentrating on formal, institutional aspects of state structures and processes. The state as we understand it here is a dynamic web of interactional processes among governmental and societal actors working together in giving shape to the substance of policy, in the sense of both formulation and implementation.

From this analytical perspective on the state, we discern the following five central dimensions: the general institutional setting, policy instruments, problem-solving philosophies, the collaboration between the state and associations and clienteles, and the role of political parties in aggregating interests.

- Exchange and negotiating relationships in policy networks are embedded in institutional structures that both make them possible and place constraints on them. We assume that these institutional factors change as the substance of policy is Europeanized. On the vertical axis political and administrative powers and financial resources may become centralized or decentralized in relation to the various system levels. On the horizontal axis, the result may be a concentration of powers or their multiplication or segmentation.
- Actors' activity in policy networks is determined essentially by the policy instruments used in the policy area. The use of such instruments is intended to induce certain behaviour in actors by influencing incentive structures. Some instruments prescribe or proscribe specific acts. At the same time they penalize any failure to conform with prescriptions and prohibitions. Alternatively, positive incentives, such as tax concessions, can be offered to encourage certain modes of behaviour, or reliance might be placed on information and education alone. Since Europe

abounds in such policy instruments, it can be assumed that European policy formation will confront every member state with the necessity of enlarging and modifying its own range of instruments.

- A further level of the state to attract attention where the genesis and implementation of policies is interpreted as a policy network analysis is the superordinate problem-solving 'philosophy'. It provides orientation for action or constitutes a 'belief system' (Sabatier 1993), guiding the activities of policy network actors. Guiding orientations reflect fundamental, deep-seated convictions (core beliefs — Sabatier 1993) on how a problem ought to be resolved. Given the cultural multiplicity of the continent, very different orientations confront one another in Europe, which in the event of European unification will require corresponding widening or modification of the individual national problem-solving horizon. For instance, it might prove necessary to abandon one's fundamental conviction that state intervention is needed only where scientific evidence on the impact of certain pollutant emissions is available, and to adopt a precautionary mode of action.

- Constitutive to an understanding of state activities within a policy network is the aspect that state and private, usually corporate actors interact in elaborating policy contents, both at the formation stage and in the implementation phase. The transformation of state arrangements is therefore also to be observed primarily at the interface between the state and society, i.e., in the interaction between governmental and societal actors.

  At this nexus, widely differing forms of interaction between the state and associations are to be seen, ranging from self-regulation by associations (with reservation of the right of state intervention), via the 'agency capturing' of public authorities by associations, the bipartite collaboration of associations under state auspices, to state functionalization of associations. In this context, Europeanization can bring about shifts in accent or drastic changes.

- A further aspect of the interface between the state and society is the function of political parties and the role they play in aggregating political interests. Policy formation at the European level and the circumstance that specific areas of decision-making have shifted to the European level confront national parties with new conditions for action.

1.2 Sets of Variables in the National/European Context and in International Comparison

The present study primarily addresses interaction between individual member states and the European Union. However, comparison between member states is always implicit in so far as they are in competition for influence at the European level. Our point of departure for comparing practices in the three countries under review is the question how and why they tackle industrial air pollution. We go on to investigate why individual countries succeed in transferring solutions to the European level, and what consequences the specific form of policy 'Communitization' has for state activities in individual member countries.

The first step being a comparison of countries, we begin by outlining the analytical considerations underlying this project. Essentially, comparative policy research faces two problems. It can compare aggregate data such as expenditure in a policy area, unemployment figures etc., for a large number of countries. At least for the OECD countries these data are relatively easy to come by and easy to compare. But statistical data alone provide no insight into the processes that generate such data. These processes can, however, be recorded and analysed in country case studies, though at the risk of becoming mired in a mass of detail, however interesting, and producing an idiosyncratic profile of a policy area not comparable with others.

Since the issue we are dealing with is strongly process-oriented, we have opted for qualitative network analysis. We thus run the risk of becoming so involved in national particularities that we lose sight of the comparative dimensions. In order to reduce this risk, we have chosen to examine countries — Britain, France, and Germany — at a similar level of economic and technological development but with very different political and administrative institutional structures. Whereas Germany has a federal structure and a two chamber system providing for representation of subnational entities, and thus a politically decentralized structure, Britain is a unitary state with a strong executive and a strong concentration of power in London; France is a unitary state with a strong executive, but the regionalization process initiated in the eighties is shifting political weight towards the *conseils régionaux* and *conseils généraux*, which creates particular conditions for handling political conflicts.

In order to separate constant from variable influences, a distinction is made between the two sets of independent variables: transnational factors (common influence) and country-specific underlying conditions (varying influence). It is our assumption that the specific national setting and the specific differences in the political and administrative systems explain the differing approaches to problem-solving in clean-air policy and the corresponding efforts to embody them in European legislation. The distinction drawn between common and divergent qualifying factors permits systematic comparison, at least up to a certain point. There is a limit in that different underlying conditions, such as geography or the institutional configuration of the political and administrative system, do not of themselves determine policy solutions but merely open up certain options and 'avenues for action'. Along these avenues there is a margin for action to be exploited by political calculation in network processes. However, knowledge of the common and specifically national background conditions, of the central aspects of national networks permits relative precision in predicting how a country will react to a European policy proposal.

Our dual concern with the generation of European regulatory decisions by member states and the impact of these policy decisions on the redefinition of 'the state' in member countries and the European Union is reflected in a twofold set of dependent variables: (1) the role a member state plays in European policy-making, and (2) the changes in administrative patterns and policy practices in member states due to European legislation. In the first part of the analysis, we explain why and which member states play an active, passive, or dilatory role in regulatory policy formation at the European level. The attitude of pace-setter, passive spectator, or brakesman is thus considered a dependent variable requiring explanation. Of course, many factors are responsible for a certain view of an issue and specific policy options developing in the countries concerned, and for these being introduced into European policy formation. To reduce the complexity of independent variables, a distinction is drawn for analytical purposes between, on the one hand, external, transnational underlying conditions and longer and shorter-term national conditions and, on the other, aspects of network structures and network processes in the countries under review.

The international influence factors shared by all the member states investigated, include:

- international treaties (UNECE, OECD);
- dramatic environmental events (ecological disasters such as Chernobyl, Seveso) or newly perceived problems (acid rain and forest dieback);
- new scientific findings;
- global economic cycle developments;
- technological innovations and innovations in problem-solving approaches and their model function (Japan, USA);
- global ecological technology competition.

At the same time, specifically national influence factors come to bear and influence network processes at the national level:

- the geographical situation;
- settlement patterns, population and industrial density;
- national economic cycles and structural crises;
- the public financial situation;
- structures and ownership in the energy sector;
- nuclear policy and energy policy in general;
- key national policy developments (privatization, centralization in Britain, or reunification in Germany, regionalization in France);
- general relations with the EU.

These underlying conditions, it is assumed, act on network processes in member states and are indirectly responsible for certain policy solutions developing in individual countries. Underlying conditions are subject to political interpretation and modification in national policy networks and in the dynamics of political processes within these networks. Only in terms of these processes can we directly explain why and which underlying conditions lead to what policy measures.

Since national strategies can succeed only in interplay with European institutional interests, European actors are also taken as independent variables in explaining Community decisions. European-level actors, too, are subject to the external influence factors mentioned (international treaties, environmental incidents, etc.). However, external factors specific to the EU also affect them and influence their behaviour. They include:

- changes in the Community's self-conception, such as acceptance of the subsidiarity principle;
- constitutional reforms (Single European Act, Maastricht Treaty);
- enlargement of the Community;
- the financial situation of the EU.

These factors affect European actors and their calculations, and contribute to explaining why certain decisions are made in the policy area under review.

Turning to our second perspective of inquiry, we examine how the state is undergoing redefinition in the member states[17] as a result of the Community assuming competence in regulatory policy. The dependent variables to be explained are the changes occurring in state arrangements in the five dimensions outlined above, namely in political and administrative structures, in policy instruments, in problem-solving philosophy, in collaboration between the state and associations and clienteles, and in the role of political parties in aggregating interests. Our independent variables are the positions and negotiating strategies of the countries under review in decision-making on all important directives in the clean air/stationary sources area, which we examine from the perspective of their impact on political and administrative practices.

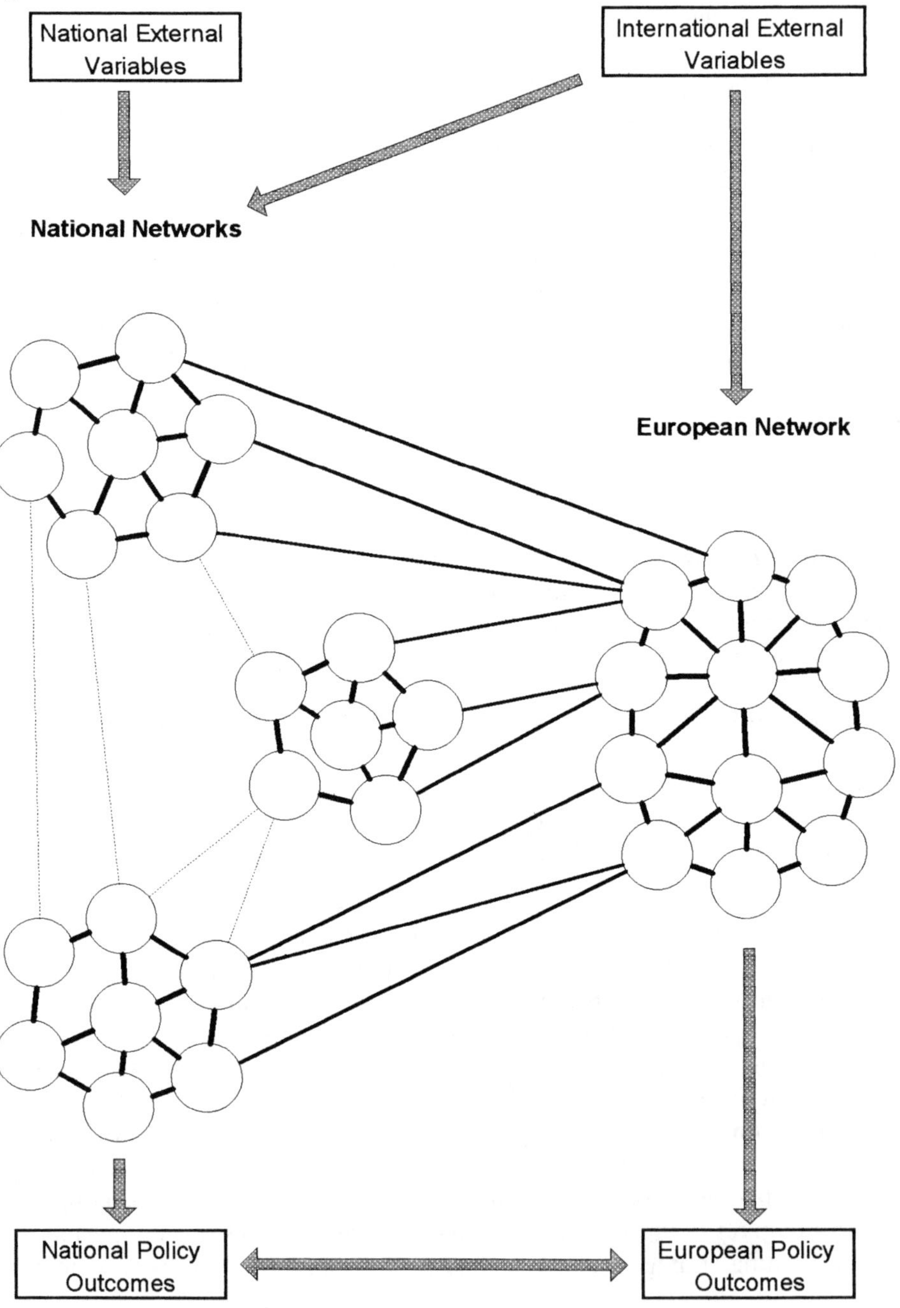

Figure 1: European Policy-Making: An Interactive Process Among National and European Actors

1.3   Procedure

The data needed within the context of a qualitative network analysis to discover the conditions for and forms taken by changes in the nature of state activities were collected by means of intensive, exploratory interviews, and document analysis. In order to cover all relevant network positions in the individual countries, interviews were conducted with 57 actors in Germany, 47 in Britain, 39 in France, and 25 in European institutions.

The focus of the exploratory interviews was on the institutions, processes, and subject-matter of clean-air policy in the countries concerned, the influence they exerted, and the strategies pursued to this end in decision-making processes at the supranational level, as well as the impact of European policy on policy practices in the three countries.

The interviews were conducted over a period of four years (from September 1991 to February 1995) with members of the following institutions (see appendix): national parliaments, competent government departments in central government, specialist agencies, implementing administrative authorities, specialist institutes, expert bodies, commissions of enquiry, competent state government departments, local government authorities and local associations, representatives of the most important political parties, industrial trade associations; with polluting industries and the environmental industry; with individual firms, semi-public bodies such as chambers of trade and industry, employers' associations, trade unions; with representatives of the scientific community and the general public, i.e., citizens' action groups. With 16 exceptions, all interviews were tape-recorded and subsequently subjected to qualitative content analysis in relation to the topic of investigation.

Notes

1   A distinction is generally made between sociological network analysis as a form of social structure analysis and policy networks as certain forms of state guidance. Sociological network analysis describes and measures relations between social entities within a social system (Pappi 1993, 85). The term policy network refers to negotiating relations between a plurality of state and private organizational actors that reach a collective decision in a common problem area (Mayntz 1993, 39ff.).

2   The literature distinguishes an additional type of social-regulatory policy covering the exertion of influence on behaviour in interpersonal relations without market or productive relations been affected. The regulation of abortion is one example.

3   Lindblom speaks of 'parametric adjustment' when ' ... in a decision situation, a decision maker X adjusts his decision to Y's decisions already made and to Y's expected decisions; but he does not seek, as a recognized condition of making his

own decision effective, to induce a response from Y; nor does he allow the choice of his decision to be influenced by any consideration of the consequences of his decision for Y' (Lindblom 1965, 37).

4 The coordination patterns of the strategy of the first move, of problem-solving, and of negative coordination/negotiation and compensation correspond to the three interaction modes 'competition orientation, cooperative orientation, and individual/ egoistic' distinguished by Scharpf with reference to the social-psychology theory of Kelley and Thibaut; see Scharpf 1993.

5 According to DG XI of the Commission, the vast majority of regulatory proposals are made on the initiative of member states (interview with DG XI, March 1993). The Commission is of course also induced to make use of its right to initiate legislation by its own action programmes, by memorandums from the Council of Ministers, by demands from the EP, and in meeting obligations arising under international treaties.

6 Article 102 of the EEC Treaty lays down that projects distorting the conditions of competition within the internal market are to be reported to the Commission.

7 A French interviewee described competition between high-regulating member states as follows: 'The Commission always looks for whatever is most stringent, whether it's in England or in Germany. ... It's a competition for the most stringent ... It's like being in a swimming pool and you have to be first to get to the other side. That what administrative competition is all about ...' (interview with CNPF, June 1993).

8 In some cases, bilateral coordination between two member states with similar regulatory or deregulatory goals is to be observed. A current example is the case of waste water policy, where Britain, in its endeavours to have deadlines extended by which the necessary infrastructure such as sewage treatment plants has to be constructed to meet EU standards, contacted like-minded countries like Greece.

9 The 'frame' adopted by a decision-maker in formulating a problem is determined by the norms, habits, and characteristics of the actor concerned (Tversky/Kahnemann 1981, 453).

10 Diffuse reciprocity, which includes low-regulating member states, is established for the latter through the compensation and package deals in the negotiation phase.

11 In the context of the mandatory confirmation of individual commissioners by the European Parliament, the newly elected Commission reported on the planned working programme of the Commission in February 1992. Early politicization can, however, also throw light on technical or general objective aspects that have hitherto played no role in the decision-making process (Fischer 1993; de Leon 1993).

12 'Regulatory zealots' in the environmental field who are unable to advance their projects in the national arena see this as an opportunity to assert themselves with the aid of coalition partners at the European level against their own government (Eichener 1995, 31; Kohler-Koch 1990, 224; see also the empirical section of the present study).

13 The directives and regulations examined in the context of the present study were prepared in the committees.

14 Although Commission officials are supposed to act without regard for national interests, cabinet circles and contacts with national members in other cabinets are especially apt to spark such national interests.

15 Our thanks to Dieter Freiburghaus for this information.

16 Of 233 internal market decisions made by the Council of Ministers over a five year period, 91 were adopted against the will of one or more member states. A formal

vote was taken only on these 91 decisions (Kevin Brown, Financial Times 13.9.1994).

[17] The nature of supranational arrangements naturally also changes as a consequence of policy becoming the concern of the Community. However, this is not the primary subject matter of the analysis.

2 Regulatory Principles at the National and Supranational Levels: Policy Patterns in Environmental Protection

Interaction between the European Union (EU) and its member states and the impact it has on the state and on policy making depend on a complex range of variables. The action taken by member states in the supranational arena correlates with factors that both determine specific perception of the clean-air issue and, as network variables, constitute the institutional and instrumental prerequisites for state action. We look at the basis for national clean-air policy in Germany, Britain, and France as a first step towards understanding the interactional processes in which national and supranational interests and strategies exert their effect and determine European policy. The supranational elements in their turn affect the nature of the state within individual European Union countries.

2.1 Federal Republic of Germany: Regulatory Law and Technological Capabilities

The ' "birth" of environmental policy' (Bechmann, 1984, 54) in Germany was marked by a number of pioneering, primarily conceptual and organizational measures that shaped environmental protection in the Federal Republic. In the 1969 Federal government statement of policy, environmental protection and nature conservation were given high priority. 1970 saw the concepts environmental protection and environmental policy added to the political repertoire (Keiter/Staupe 1991, 18). The Federal Minister of the Interior (department: *Bundesministerium des Inneren, BMI*) was called upon to submit a crash programme on environmental protection to the Federal Government (*BMI* 1970) and a Federal environment programme (*BMI* 1971). Important powers were also concentrated in a new Environment Directorate.[1] The establishment of the Environment Cabinet[2] in 1970 took account of the coordination made necessary by the principle of departmental autonomy. In 1971, the steering committee for the Environment Cabinet was transformed into a Standing Committee of Heads of Directorate on Environmental Affairs.[3] An amendment to the Basic Law gave the Federation concurrent legislative powers in the fields of waste, air and noise (see *BMI*, 1974, 4).[4] This precondition having been satisfied, the Federal Environment Office[5] was set up in Berlin in 1976, and 'as an answer to the reactor disaster in Chernobyl', the Federal Ministry for the Environment, Nature Conservation and Reactor Safety (*BMU*)[6] was established (interview Federal Environment Office, November 1992). Some states of the Federation had already set up environment ministries. However, these important stations in institutional change reveal nothing as yet of the political forces nor of the motives, obstacles and initial conditions shaping the history of German environmental policy. We address this question in the following section

2.1.1 Problem Perception as the Basis for State Action

The concept of environmental protection prevailing in the Federal Republic differs sharply from that to be found in Britain. In the United Kingdom, 'environmental protection' is understood in the sense of nature conservation, while in Germany the term is used primarily to mean technical protection of the environment. These contrasting interpretations have engendered diametrically opposite views of the issue and thus different approaches to problem-solving. The British are concerned with the quality of the environmental media, whereas the German approach is based primarily on pollutant emissions and on technological capabilities.

Strictly speaking, the Germans follow a cumulative principle. The Federal Pollution Control Act,[7] the aim of which is to 'protect human beings, animals, and plants, the soil, water, the atmosphere, as well as cultural and other physical assets from harmful environmental impacts [German: *Immissionen*], ... and to prevent harmful environmental impacts from occurring',[8] covers both emission-based and *Immission*-related measures. But the emphasis is clearly on emissions, since the Federal Government considers the ambient quality approach, 'which takes as its point of departure the load capacity of the environment ... , to be inadequate as a priority criterion, especially in view of the frequently still unknown repercussions for the environment of human activity' (*BMU* 1990,17).[9]

The German approach, with its primary emphasis on emissions and the best available abatement technology, obeys two characteristic policy principles underlying the Pollution Control Act: the precautionary or anticipatory principle and the polluter-pays principle. In contrast to the scientific and cost-benefit perspective taken in British clean-air policy, the maxim in Germany is precaution. The principle states that harmful environmental impacts are to be prevented, and that environmentally friendly variants are to be given preference already at the construction and production levels.[10] Whereas first steps are taken to reduce emissions in Britain as soon as scientific evidence indicates that they are the cause of environmental damage, and the measures envisaged are declared to be economically defensible measures are initiated in Germany despite scientific uncertainty over causes and effects.[11] However, in Germany, too, consistent application of the precautionary principle comes up against competing, especially economic interests. Many examples show that the precautionary principle can degenerate at times into meritorious theory with little practical relevance.

For plants requiring a permit, this precautionary obligation to keep to the best available technology is laid down 'as the hammer of the law' in the Federal Pollution Control Act.[12] According the Federal Ministry of the Environment, 'abandoning this principle would automatically weaken environmental protection. Our philosophy is to use the best available technology in every case' (interview with *BMU*, July 1993). This means that plants operating in less stressed or unpolluted areas also have to meet the best available technology requirement. In an approach taking account only of pollution concentrations and harmful impacts, pollutant emissions limits would be set not in accordance with the strict primacy of technical potential but with reference to prevailing ambient media quality. In plain terms, the cleaner the air, the more may be emitted on condition that the air quality standards are respected. However, this is not to say that a purely emission-related approach is the 'better' or environmentally more beneficial strategy, automatically ensuring clean air. Even where the best available techniques are deployed, air pollution can be considerable depending on the density and intensity of industrial development.

A further dictum of the Federal Pollution Control Act is that the polluter must pay. The polluter-pays principle provides that protective measures against harmful impacts 'are in principle to be taken vis-à-vis the polluter. For this reason, statutory requirements are primarily concerned with the emission source' (Hansmann 1992, 14). Taking impacts alone into account would contradict not only the precautionary principle but also 'the polluter-pays principle, because then everyone would simply build higher chimneys, and in the end no-one would know who was responsible for the higher pollution figures' (interview with City of Duisburg Environment Coordination Service,[13] August 1993).

The development and consolidation of this best available technology approach in Germany must be seen in the context of the following long, medium, and short-term influences.

Long-Term Factors in Problem Perception

Among the long-term independent variables that continue to shape clean-air policy in the Federal Republic are geography, the emission and pollution situation, settlement and industrial density structure, and energy policy.

The Geographical Situation: Emissions and Ambient Air Quality

Whilst Britain's insular position and prevailing winds facilitate the dispersion of polluted air and the influx of clean air, German geography and air quality incline one less to take a deep breath. The Federal Republic not only produces a notable amount of air pollutants but also involuntarily imports quantities from neighbouring countries, in pre-unification days especially from the former German Democratic Republic.

As table 1 shows, pollutant emissions in Germany are considerable. However, the emission of dust, sulphur dioxide (SO_2) and carbon dioxide (CO_2) have been reduced despite rising production.[14] Only nitrogen oxide emissions have grown by 27 per cent, because since 1970 the passenger car fleet almost doubled (*BMU* 1990, 139). The quantitative development of pollutant emissions shows 'that pollution by industry has fallen while transport pollution has increased dramatically. Transport is now polluter number one in the Federal Republic' (interview with Greenpeace, November 1992).[15]

The air in Germany is polluted not only as a result of domestic energy consumption and industrial processes but also by foreign emissions. As the following tables show, the Federal Republic is particularly affected with regard to SO_2 and NO_2 (emissions > 200.000 t/a) by transboundary emissions from Belgium (SO_2), France (SO_2, NO_2), Britain (SO_2, NO_2), the Netherlands (NO_2), Poland (SO_2), and the former Czechoslovakia (SO_2).

German emissions of the same pollutants are a problem especially for France (SO_2, NO_2), the Netherlands (SO_2), Austria (SO_2, NO_2), Poland (SO_2, NO_2), Sweden (NO_2), and the former Soviet Union (SO_2, NO_2). Immediately following unification, considerably increased quantities of sulphur dioxide were emitted in some areas, especially from East German plants, so that Italy (SO_2), the former Yugoslavia (SO_2), Romania (SO_2), and Sweden (SO_2) are now among the countries suffering under appreciable exports of air pollution from Germany.

Table 1:   Pollutant Emissions in the Federal Republic of Germany (Western states) —
1000 tonnes/year

Pollutant/year	1975	1980	1985	1990
Dust	810	690	580	450
– Energy consumption	527	467	420	330
= Power stations/distr.heating	160	130	90	25
= Industry	50	40	30	18
= Small consumer	17	15	9	6
= Households	80	50	40	25
= Transport	60	62	71	76
= Bulk goods shipments	160	170	180	180
– Industrial processes	280	220	160	130
Sulphur dioxide	3,350	3,200	2,400	940
– Energy consumption	3,225	3,095	2,285	850
= Power stations/distr.heating	1,750	1,900	1,500	320
= Industry	840	750	470	320
= Small consumers	200	140	100	50
= Households	300	200	130	85
= Transport	135	105	85	75
– Industrial processes	100	120	100	90
Carbon dioxide	738,000	805,000	738,000	725,000
– Energy consumption	716,000	783,000	722,000	708,000
= Power stations/distr.heating	235,000	274,000	249,000	255,000
= Industry	187,000	193,000	161,000	144,000
= Small consumers	67,000	62,000	55,000	47,000
= Households	112,000	117,000	115,000	94,000
= Transport	115,000	137,000	142,000	168,000
– Industrial processes	22,000	22,000	16,000	17,000
Nitrogen oxides	2,550	2,950	2,950	2,600
– Energy consumption	2,485	2,880	2,920	2,590
= Power stations/distr.heating	660	800	760	340
= Industry	390	350	270	240
= Small consumers	55	55	50	35
= Households	80	85	90	75
= Transport	1,300	1,590	1,750	1,900
– Industrial processes	40	45	25	16

Source: Federal Environment Office (1993), Berlin.

Table 2:   Sulphur Dioxide Balance for the Federal Republic[1] — 100 tonnes sulphur/year

Country	Imports			Exports		
	1985	1988	1991	1985	1988	1991[2]
Belgium	244	230	265	54	65	109
France	475	388	509	601	213	578
United Kingdom	562	468	467	132	34	175
Italy	85	83	76	180	81	329
Yugoslavia	31	21	26	162	67	383
Netherlands	143	149	117	289	141	201
Austria	19	13	19	278	138	503
Poland	279	140	381	412	260	2,524
Romania	12	9	13	91	43	234
Sweden	8	4	6	154	102	290
CSSR	410	256	868	300	158	1,233
USSR[3]	26	20	10	536	251	706

1) Only the most important countries are taken into account.
2) The sometimes striking rise in emissions is attributable essentially to pollution by plants in the eastern states of the Federation frequently obsolescent but still in operation.
3) For 1991 only Russia was taken into account.

Source: Compiled from EMEP/MSC-W Report 1/1992.

Settlement and Industrial Density Structure
Some sections of the public in Germany are particularly affected by the repercussions of air pollution, since, in comparison to other member states of the European Union, the Federal Republic has both a very dense settlement structure and (especially during the seventies and eighties) a highly concentrated industrial density structure. With a population in excess of sixty million (western states) and a total surface area of 250.000 km², the Federal Republic is very densely populated (244 inhabitants per km²), and, particularly in the state of North Rhine-Westphalia (NRW) (especially the Ruhr District) and in the Saarland, it has extensive industrial conurbations.

High population density accompanied by a high degree of industrialization brought air pollution issues to the fore at a relatively early date, and the gravity of problems demanded intervention. The state of North Rhine-Westphalia provides a telling example of this development. Public concern[16] and the consequent pressure on politicians was most pronounced in this state, so that long before the Federal Government took action, state legislation was being passed. The North Rhine-Westphalian Clean Air Act[17] was adopted as early as 1959. In 1962 the Pollution Control Act[18] came into force, on the model of which the 1974 Federal Pollution Control Act[19] was elaborated.

Table 3:   Nitrogen Oxides Balance for the Federal Republic[1] — 100 tonnes
           nitrogen/year

Country	Imports			Exports		
	1985	1988	1991	1985	1988	1991[2]
Belgium	92	107	131	56	50	57
France	313	321	449	412	296	343
United Kingdom	310	278	312	100	54	98
Italy	40	41	44	179	138	167
Yugoslavia	6	5	6	172	130	160
Netherlands	156	175	203	78	73	89
Austria	10	11	20	243	239	258
Poland	77	43	84	374	482	595
Romania	3	2	2	105	88	78
Sweden	12	9	14	169	211	193
CSSR	93	74	166	275	282	298
USSR[3]	10	8	3	670	562	368

1) Only the most important countries are taken into account.
2) The sometimes striking rise in emissions is attributable essentially to pollution by
   plants in the eastern states frequently obsolescent but still in operation.
3) For 1991 only Russia was taken into account.

Source: Compiled from EMEP/MSC-W Report 1/1992.

The Structure of the Energy Sector
The meshing of clean-air policy and energy policy in the Federal Republic is
a further element that has influenced perception of the issues and the way in
which it has been tackled. During the 1981 crisis on the deployment of
Pershing missiles in Germany, the crumbling government coalition between
Social Democrats and the liberal Free Democrats took the debate on forest
dieback[20] as an opportunity to distract and contain growing grassroots opposition to nuclear policy. In order to dispel political opposition to the mis-
sile deployment programme (and the construction of nuclear warheads) and
to the non-military use of nuclear power, the Federal government performed
a U-turn in the international arena. In 1982 the foreign minister of the day,
Hans-Dietrich Genscher (Free Democrats), took the Federal Republic into
the '30 per cent Club' whose aim it was to reduce SO_2 emissions by 30 per
cent by 1993 (base year 1980). Whereas 'Germany had always gone arm in
arm with Britain in clean-air policy' and had 'stonewalled from morning till
night' against a clean-air convention (interview with the European Environmental Bureau, March 1993), the Federal government now fundamentally
modified its stance, to the bitter disappointment of the British. Clean-air

policy was also pursued on the domestic scene, by a highly motley coalition: the Green party (*die Grünen*), rapidly growing in importance; the Free Democrats (FDP), in competition for votes with the Greens; and the regional Bavarian Christian Socialist Union (CSU), concerned about the health of Bavarian forests, 'which are the Bavarian farmers' savings bank' (interview with the European Environmental Bureau, March 1993) (Windhoff-Héritier 1993, 11f.; see also Boehmer-Christiansen/Skea 1991, 192; Weidner 1989).

In Germany, energy policy and clean-air policy are so closely intertwined because the supply and use of energy causes a high proportion of total emissions. If we take SO_2 emissions the share of the energy sector is still in excess of 95 per cent, and for CO_2 its contribution is around 98 per cent. While abating emissions of SO_2 and NO_x requires taking 'measures to influence the amount and structure of energy consumption,' investigations conducted by the Federal Commission of Enquiry 'Protection of the Earth's Atmosphere'[21] came to the conclusion that it would be possible to cut CO_2 emissions by 30 per cent (base year 1987) by 2005. This could be achieved through efficient energy use, the substitution of energy carriers, the use of renewable energy sources, and through waste management measures (*UBA* 1990, 129).[22] In November 1990, the Federal Government fixed the target for CO_2 emission reduction at between 25 and 30 per cent and adopted the measures mentioned above for meeting this target. However, it is a moot point whether the measures envisaged, which are now being put into effect, will suffice to bring about such a drop in CO_2 emissions. The actual reduction that has hitherto been achieved is in the order of 11 per cent (interview with *UBA*, October 1993). For a range of economic, social, and also political reasons, cutting back CO_2 emissions is difficult to achieve by expanding nuclear energy (14 per cent of current primary energy consumption) and simultaneously reducing energy output from hard coal and lignite (28 per cent of primary energy consumption). Mining, a sector on which a large number of jobs depend, is a strong social identification factor, especially in the Ruhr District. The environmental movement, too, has preferred coal-fired power stations 'for the tactical reason of not attacking the coal-fired stations too vigorously because the alternative is always nuclear power which is to be rejected before all else' (interview with Robin Wood, November 1992).

Medium and Short-Term Problem-Perception Factors

Perception of clean-air policy issues is also influenced by variables taking effect in the medium and short terms. These include the economic setting, environmentally relevant incidents, public environmental awareness, international environmental policy, progress in environmental technology, and changes in the market for environmental technology.

The Economic Situation

Over the past twenty years, the Federal Republic has experienced economic ups and downs. A glance at macroeconomic data (see table 4) gives an impression of the extent of cyclical fluctuations, which, as we will see, did not fail to have an effect on political commitment to environmental protection.

The '*Sturm und Drang*' period (Müller 1984, 126) in German environmental policy in the late sixties and early seventies[23] was brought to an initial end by the oil price shock at the beginning of 1974 (when the world market price for energy raw materials leapt by over 200 per cent over the previous year), and by the subsequent downturn in the economy. Resistance to state intervention grew in industrial circles, taking expression, for example, in the formation of relevant organizational entities in commercial and industrial federations and in the Federal Ministry of Economics (Posse 1986, 48).

In the further course of the seventies, the continuing economic crisis left still less room for environmental matters. The impotence of environmental interests in the face of government concern with savings and management and labour concern with jobs, income and profits became particularly evident at the so-called 'Gymnich Talks' (3 July 1975) when industry succeeded in pushing through a slowdown in environmental policy to diminish the alleged detrimental repercussions for the economy, such as a logjam of investment projects, endangered jobs, etc. (see Drexler 1980, 8; Bechmann 1984, 67).[24]

In the course of the eighties the economy prospered. Economic growth accelerated, unemployment and inflation fell, while the propensity to export rose. Economic success was accompanied by environmental policy success. In the mid-eighties environmental ministries were set up by both Federal and state governments. Important amendments were adopted to statutes, rules, and regulations, such as the Large Combustion Plant Regulation.[25] It was only at the end of the eighties that economic considerations once again took priority over protection of the environment in the course of the reunification of East and West Germany in 1989/90. To facilitate economic recovery in the eastern states of the Federation, the 'Investment Facilitation and Residential Development Land Act'[26] was passed, coming into force in May

1993, which, among other things, shortened regional planning procedure, restricted public participation in planning, and 'thinned out' the catalogue of plants requiring authorization (Töpfer, quoted in *Der Spiegel* 1/1993, 33). Many environmentalists strongly criticized these measures, considering them a retrograde step in environmental policy:

All waste disposal plants ... are now no longer subject to public-works planning procedure as they had been under the law relating to waste. Now they're all subsumed under the Pollution Control Act with the result that anyone who wants authorization for a waste incineration plant needs no planning justification, nor must he demonstrate that there is any need for the plant at all. If the plant meets the standards laid down in the technical guidelines on clean air, the operator of the plant is entitled to be granted a permit. ... This is an instance of deregulation that should be considered extremely critically (interview with the Bielefeld Environment Department,[27] August 1993; interview with Independent Institute for Environmental Affairs (*UIU*),[28] November 1992).[29]

Table 4:   Outline Data on the Economic Situation in the Federal Republic of Germany — Percentage Change from the Preceding Year

	1974	1977	1980	1983	1986	1989[1]	1992[1]
Economic growth[1]	0.24	2.92	1.27	1.50	2.25	3.29	2.90
Unemployment rate[2]	2.56	4.34	3.63	8.87	8.50	7.57	6.16
Inflation rate[2]	6.96	3.68	5.45	3.30	-0.13	2.78	4.46
Energy prices[3]	220.30	8.88	65.09	-12.09	-35.53	16.37	0.58
Exports of goods[4]	232.00	272.70	350.30	430.80	528.50	641.60	664.80
Balance of trade[4]	51.20	38.10	8.80	41.70	113.60	135.10	36.00
Budget balance[4]	-12.14	-29.14	-43.00	-42.80	-24.50	4.07	-36.47

1) Western states of the Federation
2) Percentage
3) Growth rate in the world market price for energy raw materials
4) In DM billion

Source: German Institute for Economic Research (*DIW*)

Not only reunification of the country but also the world-wide recession had a negative impact on environmental protection. Industry's willingness to agree to environmental protection measures had declined (interview with the Confederation of German Industries [*Bundesverband der Deutschen Industrie, BDI*], March 1993) and demands by sections of the industry, 'to take a break in environmental protection' (interview with *B.A.U.M.*, January 1993), became louder. Politicians, too, have back-pedalled and sought ways to save money. Since 1992, 'there have been instructions from the Office of the Federal Chancellor that all future measures like environmental protection that cost money are to be checked out and "approved" at the very outset by the Chancellor's Office' (interview with the Federal Environment Office, November 1992). The 1993 budget cut funding for environmental protection by 6 per cent, 'more than for any other category of spending.' The opposition considers that environmental policy has 'ground to a halt,' and Environment Minister Töpfer had been 'shunted onto a siding in Cabinet'[30] (Interview with SPD, March 1993).

As these examples show, environmental policy in Germany competes primarily with financial, industrial, and labour market interests. Spill-over from the area of economic policy (as well as energy policy) is accordingly most frequent and pronounced. This empirical circumstance can be explained theoretically by the so-called capacity thesis. The development of environmental policy is governed not only by the extent of environmental policy problems but also depends on the available capacities — socio-economic, political-institutional, and technical-economic (see von Prittwitz 1990, 107ff).

National Environment-Related Events
While environmental policy had been weakened in Germany by economic problems, a series of events (external shocks) had an impact on the environment that gave new momentum to this policy area. Among the most important of these phenomena to affect the Federal Republic has been forest dieback, which in the early eighties roused the public and the political class more thoroughly than in other European countries, giving new impetus particularly to clean-air policy and later to soil conservation. In major hearings on clean-air standards during the seventies, the causes for forest dieback were the subject of intense discussion. At first the government hesitated to introduce more stringent environmental quality standards, especially for sulphur dioxide. In an initial phase, environmental organizations and scientists whose 'theories posited that forest damage was primarily due to air pollution were systematically excluded from the hearings on clean-air policy formation' (interview with Weidner, November 1992). Power station operators were to the fore in collecting counterinformation to contradict such theories and promote other explanations such as the effects of dry periods, inappropriate farming methods, and infestation by engraver beetles (ibid.).

A wide variety of theories and practical methods of investigation have been pursued in the debate on forest dieback and its causes (see *SRU* 1983; *FBW* 1986). Although science has yet to produce incontrovertible evidence on what is killing the forests or to establish a causal connection between air pollution and onset of tree damage, forest dieback 'gave environmental policy a mighty boost,' contributing decisively to the taking of precautionary legislative measures like the amended Technical Guidelines for the Maintenance of Clean Air (*Technische Anleitung zur Reinhaltung der Luft [TA*

Luft])[31] and the adoption of the Large Combustion Plant Regulation[32] in 1983 (interview with *SRU*, March 1993).

Public Environmental Awareness and the Environmental Movement
The public debate in Germany on environmental protection began in about the mid-seventies. At that time the predominant topic was nuclear energy, the opponents of which engaged to some extent in militant confrontation with its advocates. The subject of environmental protection gained in political explosiveness. Numerous citizens' action groups formed, joining forces in the Federal Association of Citizens' Groups for Environmental Protection (*Bundesverband Bürgerinitiativen Umweltschutz* [*BBU*]) and the German Federation for the Environment and Nature Conservation (*Bund für Umwelt und Naturschutz Deutschland* [*BUND*]).

In the early eighties, with the advent of the anti-nuclear power movement and the Greens, a first environmental movement came into being that benefited from the progressive forest dieback. New environmental organizations such as Robin Wood were founded, already existing organizations shifted the emphasis of their activities to combating forest dieback and acid rain (interview with Greenpeace, November 1992). The threat to the forests, with their high symbolic value in German culture,[33] had a lasting effect in concentrating the public mind on the dangers of air pollution. In the eighties the growing urgency of the problem and increasing public awareness created the socio-political background against which more stringent environmental measures could be taken than in the previous decade (see Dierkes/Fietkau 1987). With help from the media, environmental protection became firmly entrenched in the discourse and profile of industry, unions, parties, and churches. Environmental organizations reaped the benefits of growing public attention to environment issues in the eighties as membership gathered pace, donations mounted, and their propensity to 'multiply' flourished. There are now all shapes and sizes of environmental organization, covering the whole spectrum from classical nature conservation associations, trusts, consumer organizations, scientific ecological institutes and action groups. Most environmental organizations are largely financed by member contributions. There is little non-project-linked public sector support. In 1992, funding of environmental organizations by the Federal Ministry of the Environment totalled DM 4.6 million, of which DM 2.28 million was tied to specific projects and the remaining DM 2.72 million was earmarked for environmental information campaigns (Hey/Brendle 1992, 77). While some environmental organizations 'have been demanding for years that associations be provided with the funds necessary to appoint full-time staff' (interview with *BUND*, March 1993), the majority of organizations reject state subsidization for fear they could become dependent on and be influenced by the bodies holding the purse strings (interview with Greenpeace, November 1992).

In so far as environmental organizations are primarily financed by their members, they are obliged to win adherents and to 'keep them on the ball' through politically spectacular demands and activities rather than compromise. Both this compulsion to radicalism and the traditionally hands-off relationship between environmental organizations and the political and administrative authorities determine the pressure-group function and lobbying deficit of German environmental associations. Like the largely regional French environment groups, but in complete contrast to the 'success-oriented', pragmatic, and integrationist environmental associations of the traditional type to be found in the United Kingdom, the 'value-oriented' German associations, with their stress on policy independence, exercise extremely little direct influence on political decision-making. However, in making environmental protection a public issue, the German environmental organizations, particularly through the Green party, have made a crucial contribution (see in greater detail Hey/Brendle 1992, 75ff., 98f. and 273ff.).[34]

International Environmental Policy
At first, international developments generated no changes in environmental policy behaviour in Germany. From the mid-seventies the Scandinavian countries took the lead in pressuring the Federal Republic and other industrial countries to cut back emissions emanating from industry (coal), especially outputs of sulphur dioxide and nitrogen oxides. This pressure was increased by a series of voluntary international agreements. Until 1982, however, international pressure failed to evoke the slightest positive response from the Federal Government. In international negotiations both the United Kingdom and the Federal Republic were among the countries urging less stringent rules on the reduction of transboundary emissions (see von Prittwitz 1984). Germany, for example, refused to accept the reduction of SO_2 emissions under the 1979 Geneva Convention on Long Range Transboundary Air Pollution (LRTAP) and, in cahoots with Britain, saw to it that the best available technology was to be required for emission abatement only where economically defensible.

The Federal Republic was induced to execute a U-turn in the international arena only by the cumulative effect of the growth of national environmental problems, exemplified by forest dieback; of increasing public pressure from the local population; of political party rivalry playing the numerous political arenas in Germany's federal structure; and of the proliferation of technical potential in environmental protection. The fresh stance taken by the Federal Government was first displayed to the surprise of everyone concerned at the June, 1982 Stockholm Conference on 'Acidification of the Environment', when the German delegate advocated the formulation of clear and verifiable goals for a programme of international emission abatement. The Federal Republic supported the Helsinki Protocol, approved in July 1985 by the Executive Body of the Geneva LRTAP Convention, which prescribed the reduction of national SO_2 emissions by at least 30 per cent between 1980 and 1993 at the latest. The German government went still further with the announcement of a 60 per cent reduction in SO_2. Germany also joined a number of other countries in pronouncing particular commitment in the question of the nitrogen oxide abatement. Members of the UN Economic Commission for Europe (UNECE) proved capable, under the so-called 'NO_x Protocol' signed on 1 November 1988, only of agreeing to freeze national NO_x emissions until 1994 at the 1987 level. In Sofia on 31 October 1988, the Federal Republic and eleven other European countries consequently signed a joint declaration on a 30 per cent reduction in NO_x emissions by 1998 at the latest (taking any base year between 1980 and 1986) (see Weidner 1989, 8ff.).

Technological Progress

The influence of technological progress on environmental protection in Germany is evident in the reduction rates of pollutant emission volumes. The Federal Republic has the largest number of flue gas desulpherization and denitrogenization installations in operation in Europe (Weidner 1989, 13). Although gross national product (at constant prices of 1980) rose by 49 per cent between 1970 and 1987, it proved possible to reduce sulphur dioxide emissions by 40 per cent, particulate matter emissions by 57 per cent over the same period (*BMU* 1990, 137), and carbon dioxide emissions by, to date, 11 per cent (interview with *UBA*, October 1993). According to a representative of the North Rhine-Westphalian State Institute for Pollution Control,[35] greater reductions are hardly feasible: 'There's not much left to sort out. If you look at pictures of the Ruhr District, we've dealt with the kilos, the grams, and even the milligrams. We're now into scales recalling the famous lump of sugar in Lake Constance' (interview with the State Institute for Pollution Control, February 1993).

From a financial point of view, German industry's — politically opportune — commitment to environment technology has had an appreciable impact. Between 1975 and 1990, DM 153 billion was spent on environmental protection by manufacturing industries. Spending over this period thus increased by about 250 per cent (110 per cent adjusted for price rises).[36] Since the beginning of the eighties, the ratio of environmental protection-related investments to total investments in manufacturing has more than doubled (per firm in 1980: DM 0.5 million; 1985: DM 1.3 million; 1989: DM 1.6 million) (*DIW* 1993, 201). The share of environmental protection spending in GNP in the Federal Republic is at a level (1991: 1.7 per cent) similar to that in a number of important competitor countries (Austria: 1.9 per cent; Netherlands: 1.5 per cent; United States: 1.4 per cent). Only in Japan, Britain and France is the ratio distinctly lower than in the Federal Republic of Germany[37] (ibid., 200).

Higher spending on environmental protection need not automatically mean greater or more efficient protection for the environment. Thanks to their product and fuel-related measures, the French, for example, have succeeded in abating crucial environmental burdens at relatively low cost. To achieve the same effect, the Germans, with their primarily technology and plant-related measures have had to cough up far more: 'Reducing pollution at source costs a lot of money and is a great deal more expensive than merely maintaining air quality' (interview with an environmental consultancy firm, September 1993).

In Germany, progress in environmental protection technology and perception of environmental problems are more closely intertwined than in Britain and France. Both Germany's emission-based approach to environmental issues and the categories of indeterminate reference (e.g., 'best available technology' and 'latest scientific findings') with which German environmental law operates encourage interaction between the two aspects. The politically imposed reliance on the principle of combating air pollution at source incites firms to develop and deploy the best possible environmental techniques. Moreover, the use of indeterminate categories of reference makes it possible to stay at the cutting edge of scientific and technical progress in implementing environmental regulations[38] (interview with Association of German Engineers [*Verein Deutscher Ingenieure, VDI*], January 1993). The (ideal-typical) 'synchronous' development of environment technology and environment law should be evident both in conversion from down-stream cleansing methods to integrated and environmentally favourable techniques and in legislative endeavours to protect the environment not only by regulating what comes out of the pipe but also intervening legally in the production process to regulate material inputs and production procedures. This 'synchronous' interplay between technology and law gets out of step as soon as external factors, such as economic troubles, bring one party (usually the regulator) to drag his feet.[39] Environmental protection in Germany with its 'end-of-pipe' techniques is still in its infancy.

As far as environmental protection measures are concerned, we still have a long way to go. So far we've done nothing except tacking elaborate filters onto conventional processes. We have to get away from this abatement method and move towards avoidance techniques (interview with steel producer, March 1993).

The development of new technologies that 'stop emissions before they start' seems wise not only from the point of view of preventive environmental protection. It also brings economic benefits.

Filters work only when you pump in a lot of energy. We have installations that need 11.5 megawatt hours to precipitate one tonne of dust. That's an awful lot of energy that has to be converted somewhere or other. We won't get much further with traditional abatement techniques. They take too much energy (interview with steel producer, March 1993; interview with *EURES*, August 1993).

Transformation of the Environmental Technology Market
Not only technological developments but also economic reasons favour continuing the pursuit of environmental protection. Relatively early on, Germany discovered new and lucrative sales opportunities to be won through a strong position on the environmental technology market, both at home and abroad. By the mid-eighties, gross output of environmental protection goods and services was already over DM 21 billion (*BMU* 1990, 60) and had created over 430,000 jobs (Sprenger 1989). Germany has also been successful in the export of environmental protection equipment. The share in total exports of environmental protection goods is now at about 25 per cent (*BMU* 1990, 60), and the export ratio of environmental technology is around 40 per cent (Sprenger 1989, 194f.). In Europe at least, German environmental technology can thus be said to dominate the market.

In the fields of mechanical engineering, pipe production, steel production, we're ahead of the field as far as emission-reduction technology is concerned. German filter technology is really the best available on the world market, if you disregard the Japanese (interview with steel producer, March 1993; interview with pollution protection equipment industry, July 1993).

Conclusion

Since about 1982, the Federal Republic has been the pace-setter in technical environmental protection for important aspects of clean-air policy at both the national and international levels — with, among other things, the Large Combustion Plant Regulation, the 1985 amendment to the Federal Pollution Control Act, and the 1983 and 1986 amendments to the Technical Guidelines On Air Quality.[40] However, progress in environmental policy since the early eighties cannot be attributed to the change in government in Bonn from a Social Democrat/Liberal (SPD/FDP) coalition (1969-1982) to an alliance between FDP and the conservative CDU/CSU. In political style, in the environmental policy strategies and instruments deployed by the Conservative-Liberal coalition, 'things were left essentially the way they had been under the preceding government' (Weidner 1989, 51). The larger-scale effects attained since 1982 are to be regarded not as achievements of the government of the time. They can be attributed to 'generally more favourable underlying conditions for environmental policy action' (ibid., 47).

These improved underlying conditions can be described in terms of the problem-perception factors mentioned above. Transformations in policy — both progressive and regressive with respect to environmental protection — correspond essentially with three factors: the increase in problematic environmental policy situations ('problem-pressure thesis,' Jänicke/Mönch 1988), the growth in the importance of environmental protection issues in the public mind ('postmaterialist change in values', Inglehart 1971, 1977), and the parallel increase or restriction in the political options for processing and overcoming environmental problems.[41]

As the long, medium, and short-term problem-perception factors indicate, pressure in Germany was generated by rising air pollution caused by the country's geographical position, by domestic and transboundary emissions, and by dense settlement and industrial concentration structures. It was to become most evident with the advent of forest dieback. The increasing public attention given environmental protection is attributable to no small degree to the strong impact the issue has had on the German population. It has manifested itself especially in the growth of environmental groups, in the establishment of the anti-nuclear power movement, and in the popularity of the Greens.[42] The opportunities for dealing with environmental problems are determined primarily by underlying economic conditions. A stable economic situation and financial incentives (new sales and export opportunities in the environmental technology market) foster environmental policy, while an unstable economic situation (due to the oil price shock, a downturn in the economy, the costs accruing from the reunification of Germany, recession) brings environmental policy to a standstill.

Considered as a whole — and especially in comparison with other European countries — economic, political, and technical conditions in Germany favour public ecological awareness, willingness in industrial circles to take action, and readiness and determination among politicians to make the necessary decisions. In its particular way, an environmental policy has emerged in Germany that is characterized by active measures to achieve 'technocratic, down-stream' (Hey/Brendle 1992, 52) emission abatement.

2.1.2 Institutional and Instrumental Preconditions for State Action

Regulatory action by individual governments in the clean-air field is not only determined by the degree of problem awareness but are also governed by a range of institutional and instrumental variables. Within a policy analysis framework, the changes in national clean-air policy are explained by reference to traditions and processes within the country concerned. Investigating these developments and their causes will help us to understand interactional processes between the EU and member states and their national and supranational consequences and to test hypotheses on the subject.

Political Initiative and State Action: Institutional Prerequisites

It is our assumption that the potential for political energy to develop, which in turn produces policy innovations, depends institutionally on the range of available political arenas, the democratic institutions, and the role of the courts. These variables are to be examined in the following sections.

The Multiplicity of Political Arenas
In contrast to the British and French unitary and centralist constitutions, the Federal Republic of Germany is characterized by federal structures. Each of the three levels, federal, state, and local, has its own political jurisdiction. Each has its own areas of competence, tasks, and finances subject to the constitutional order, the fiscal system, and federal legislation. The powers of the state in the Federal Republic are thus not concentrated at the centre but are shared out among the Federation (*Bund*) and the eleven, now (since reunification) sixteen states (*Bundesländer* or *Länder*).[43] The states provide a multiplicity of political arenas where a profusion of political initiative and activity can unfold. In their individual constitutions, the states can autonomously shape their constitutional order subject to the principles of the rule of law. The ruling principle is that the states have legislative, executive, and juridical competence unless the federal constitution, the Basic Law (*Grundgesetz*), provides otherwise.

The Federal Council or *Bundesrat* is an effective mechanism in federal dynamics, where the states exercise decisive influence on the formulation and adoption of legislation. Government and opposition thus often meet in battle not only in the Federal Parliamentary Assembly, the *Bundestag*, but also in the *Bundesrat*. If the *Bundesrat* is dominated by state governments formed by parties belonging to the opposition in the *Bundestag*, projects of the Federal Government and the *Bundestag* may by modified or even thrown out by the refusal of the Council to give its consent: 'The minority in an interlinked decision-making system is obviously able not only to block new initiatives by a majority government. In this constellation impulses for reform can just as easily be generated in response to government inertia' (Czada 1993, 73). Moreover, interests can be asserted in the *Bundesrat* that, especially in relation to spheres of competence and to finance, where irrespective of party allegiance, the states find themselves at variance with the Federation (see Lehmbruch 1976).

Within the structure of the states, administrative counties (*Kreise*) and municipalities (*Gemeinden*) are public corporations (*'Gebietskörperschaften'*), enjoying their own constitutionally protected legal status under the federal constitution. Counties and municipalities have the right within the framework of the law to regulate local community affairs in their own name, under their own responsibility, and through elected bodies (Article 28 (2) of the Basic Law). This guaranteed local self-government, which has a long tradition in the Federal Republic, is exercised by counties and municipalities in the field of environmental protection primarily through local bye-laws (such as those imposing waste-disposal or sewage charges); within the scope of its planning authority, for example town and country planning; or by exercising its organizational powers, for instance by setting up local environmental authorities.[44] (Keiter/Staupe 1991, 15; Salzwedel/ Preusker 1983, 28f.; Jaedicke et al. 1990, 34). Regional problems, such as heavy air pollution in industrial centres, are more likely to find a place on the political agenda and be dealt with closer to home than in centrally governed countries, where, as in Britain, local scope for action is largely determined by central government.[45] However, staffing and financial constraints limit local government commitment to environmental protection in Germany and Britain alike.

The federal structure of German government, containing elements both of cooperation and of competition, provides a multiplicity of political levels, which, although in great measure autonomous, are closely interlinked in a wide variety of ways. In the clean-air field, important vertical and horizontal forms of political linkage operate at the government level (Environment Ministers' Conference), the parliamentary level (Environment Cabinet), in the administration (State Committee for Pollution Control, Standing Federal/state head of directorate committees for environmental affairs), and at the party political level.[46]

Although the federal structure of the Federal Republic of Germany offers a greater variety of opportunities than unitary structures as in France and Britain for mobilizing political energy and government activity, a decentralized political system nevertheless runs the risk of responsibilities shifting between legislature and executive, and thus of decision-making deadlock. At the federal level, the objectives of environmental policy are set in statute form. However, it is the job of the states to fill these statutes with detailed content and implement them. The 'dependence of the political field on implementation' is extremely high, since regulatory instruments (prescriptions and prohibitions) are the preferred *modus operandi* (Hey/Brendle 1992, 56). Since implementation is generally the task of the states (*Vollzugsföderalismus* or executory federalism), they have a say in federal legislation (*Zustimmungsgesetze*, legislation requiring approval by the *Bundesrat*). In so far as 'harsh executive instruments in air and water pollution control and in nature conservation are lacking, there is a discrepancy between goals and instruments' (interview with Bielefeld Environmental Protection Directorate, August 1993).

Administrative federalism is accompanied by specific implementation features. There may be a contradiction between central goal setting and practical implementation at the regional level. On the other hand, there is the real possibility that states with particular ambitions, like North Rhine-Westphalia in environmental protection, may go beyond Federal legislation in the stringency of their implementation.

Local authorities are only 'supporting' actors in the clean-air field in most states of the Federation. Among the traditional areas of responsibility in environmental protection assigned to local authorities are physical planning, roads and transport, parks and open space planning, cleansing and sewage disposal. Other tasks have accrued in the course of the eighties and nineties: developing waste recycling, cleaning up old contaminated sites, rehabilitating old sewerage systems, remediating and preventing noise nuisances, intensifying nature and landscape conservation, monitoring the environmental compatibility of energy supplies, environmental consultancy, keeping environmental records and registers, and preparing environmental reports (see *BMU* 1990, 38; Keiter/Staupe 1991, 26; see also compilation by Jaedicke et al. 1990, 31).

As the expanded catalogue shows, the states have since the mid-eighties been delegating more and more supplementary responsibilities and powers to the local authorities to fulfil administrative, planning, operational, and advisory functions in the environmental field. The prevailing view during the seventies had been that the most important environmental matters should be in the hands of special authority at the state level (North Rhine-Westphalian model) instead of being assigned to agencies at the local level (Bavarian model) (Kessler 1984), since the local authorities, being dependent on trade tax revenues, were suspected of dragging their feet when it came to active environmental policy (Lahl 1988).

An important mode of environmental policy making in the Federal Republic is bi-, tri- or multipartisan decision-making. Sailing under the flag of delegated powers, this interaction is based on cooperation between private and public-sector actors, a typical and traditional element in German government. In environmental protection the bodies involved are institutions (generally private registered associations) that are responsible, to give some examples, for defining indeterminate categories of reference (Association of German Engineers); for plant inspection (Technical Control Association [*Technischer Überwachungsverein, TÜV*]); for the supervision of plant or standardization (German Institute for Standardization [*Deutsches Institut für Normung, DIN*]). Industry is the primary beneficiary of the influence that membership in such institutions confers. Thanks to adequate financial and human resources, industry, in contrast to the environmental associations, is able to secure its dominant presence in the relevant bodies and working committees.

In summing up, it can be said that, due to the multitude of political arenas and of forms of public-private cooperation, political, economic and private actors have a variety of institutional opportunities to articulate issues and to initiate government activity than in centralized countries such as Britain and France.

The Electoral System and Political Party Access to the Political System
Access to the political system and the possibility of asserting political demands are essentially determined by, among other factors, the electoral system. The 1956 Federal Electoral Act (*Bundeswahlgesetz*) lays down that the 518 members of the *Bundestag* (since reunification 662) are elected by a system of proportional representation slightly modified to allow an element of personal election ('semi-personalized proportional representation' (Ismayr 1992).[47] In contrast to the British relative majority system, the electoral arrangements in Germany foster a more diverse party landscape enlivened by competition, favouring the establishment and growth in influence of smaller or new parties — in spite of the five per cent hurdle — and enhancing the scope for effective political action by the opposition, which is obliged to consider a range of coalition options (Czada/Lehmbruch 1990, 61).

In the field of environmental protection, the advent of the 'Greens' in the late seventies put a decisive spur to environmental policy. The setting up of the party at the federal level, which occurred in the spring of 1979, was primarily motivated by the nuclear policy then pursued by the Federal government. The new party was composed of adherents of citizens' action groups and environmental movements, which were seeking at the parliamentary level to assert the 'post-materialist' objectives of the 'new social movements' — ecological, feminist, and anti-nuclear movements — and to provide a counterweight to established and professionalized politics. In October 1979 the Greens managed for the first time to take the five-per-cent hurdle in a state parliament. With 6.5 per cent of the second votes, the Greens took their place in the parliamentary assembly of the city-state of Bremen. They failed to win any seats in the general election to the *Bundestag* in 1980. However, they established themselves as the fourth parliamentary party in the *Bundestag* in 1983 with 5.6 per cent of the vote, which they raised to 8.3 per cent in 1987. Once the Greens were in Parliament, the Social Democrats, in particular, began to make advances to them, so that 'red-green' coalitions came into being at the state level in Hesse, Berlin, and Lower Saxony, and representatives of environmental organizations were given the environment portfolio in states governments.[48]

Table 5:   Results of the Bundestag Elections 1972-1990 — as percentage of valid second votes cast

	1972	1976	1980	1983	1987	1990
CDU/CSU	44.9	48.6	44.5	48.8	44.3	43.8
SPD	45.8	42.6	42.9	38.2	37.0	33.5
FDP	8.4	7.9	10.6	7.0	9.1	11.0
The Greens	-	-	1.5	5.6	8.3	3.9
NPD	0.6	0.3	0.2	0.2	0.6	-
DKP	0.3	0.3	0.2	0.2	-	-
PDS	-	-	-	-	-	2.4
Alliance 90/Greens	-	-	-	-	-	1.2
Republicans	-	-	-	-	-	2.1
Other parties	0.0	0.3	0.0	0.0	0.8	0.0

Source: Compiled from the Statistical Yearbooks for the Federal Republic of Germany

In 1990, however, the Greens suffered a setback in the elections to the *Bundestag*. The reasons for this reverse were complex and controversial. It was doubtless attributable mainly to the conflicts raging within the party, splitting it into a 'fundamentalist' and a '*realpolitik*' camp. Moreover, the established parties, out to woo young voters, have exhibited increasing commitment in the environmental field, thus lowering the profile of the Greens as the ecological party *par excellence*. The Greens 'fell victim to their policy of "non-colonization",' because, had they merged with the Greens in the former East Germany in time, they would have been able to take the five-per-cent hurdle and gain representation in the Bundestag (von Beyme 1991, 91). The Greens were to amalgamate with the East German party Alliance '90 (*Bündnis '90*) some three years later in 1993, winning access to the Bundestag in 1994 and achieving remarkable electoral success in North Rhine-Westphalia and Bremen in 1995.

Thus, especially before the greening of the other parties, the Greens made a major contribution towards raising public awareness, to confronting the large parties with environmental issues, and to formulating more stringent environmental legislation.

The Role of the Courts

The courts are a further important source of new state activity. In the Federal Republic, it is the task of the Federal Constitutional Court to ensure that legislation conforms as far as possible with the Basic Law. The jurisdiction of the Court is as to the constitutionality of laws. The Court may not initiate judicial review in its own right but only on submission of a complaint. Every citizen who feels that his basic rights have been infringed is entitled to make a constitutional complaint. The decisions of the Constitutional Court are binding on the constitutional organs of the Federation and the states and on all public authorities and courts. Although only a small proportion of constitutional complaints in the Federal Republic have resulted in statutes being declared void, lower court decisions being reversed or administrative acts set aside (some one to two per cent), and although the legislature occasionally fails in its duty to provide constitutionally acceptable arrangements, the effect of constitutional jurisdiction ought not to be underestimated, since the mere prospect of an action often induces the legislator to modify or drop a project[49] (von Beyme 1991, 375). In the clean-air field, constitutional jurisdiction plays a major role especially in the process of normative law-making. By bringing an action before the Federal Constitutional Court, German citizens can test the constitutionality of rules laid down by government. This power of judicial review, which is entrenched in the Basic Law, gives the Federal Constitutional Court direct influence on political decision making.

In addition to the Constitutional Court, the administrative courts represent a further instrument of intervention and influence that the citizen can use. Just as statutes passed by the *Bundestag* and the regulations issued by the government can be voided by the Federal Constitutional Court if they fail to conform with the constitutional order, the administrative courts have jurisdiction over the activities of administrative bodies with respect to their legality. In the context of licensing procedures, the administrative courts exercise significant influence to the extent that the binding effect of technical standards (*VDI* guidelines and *DIN* standards pertaining to air pollution, for example) is limited. The majority of courts do not recognize the technical rules as anticipated expert opinions, since they neither take account of the individual case nor possess the current relevance of expert opinions. The administrative courts are thus competent to decide what limit values are to apply[50] (interview with Commission on Clean Air in the *VDI* and the *DIN*, January 1993).

The utility for citizens' action groups of taking a matter to the administrative courts is not only to obtain review of procedural legality, for example, in the approval of large-scale projects. In many cases it is 'not decisive whether government action (authorization procedure) is legal or not. The delay and the expression of public resistance considerably reduces the chances of such projects being realized,' since for financial and planning reasons private investors are frequently guided 'in the first place by the attitude of the public to the project, secondly by the attitude of the local authorities, and only in third place by other locational factors' (Hey/Brendle 1992, 61).

Since, unlike France, the Federal Republic knows no general right of associations to take legal action, only persons directly affected by the matter in question, such as emissions emanating from a neighbouring industrial plant, may do so. There is opposition in government, administration and industry to introducing such a right for associations since further complications and delays are feared in approval procedures (interview with Duisburg Factory Inspectorate, March 1993; interview with steel producer, February 1993).

A further limitation to the jurisdiction of the administrative courts in the environmental protection field has recently arisen with the coming into force of the Investment Facilitation and Residential Development Land Act, which provides that cases are no longer to be decided by state administrative courts but by the Federal Administrative Court or the Federal Constitutional Court. In important issues citizens are now 'denied recourse to the first two levels of the court hierarchy' (interview with UIU, November 1992).

Despite these restrictions on the powers and jurisdiction of the courts, there is no doubt that, as third-party 'neutral actors', they are in a much stronger position than their British counterparts in their ability to change the underlying conditions for political action. In many cases it is the courts, especially the administrative courts, that oblige the politicians to take up environmental issues or that win the day for environmental interests against competing concerns. The responsive attitude of the courts to environmental protection may be attributed to the circumstance that, 'due to their institutional independence, the courts are less subject to the pressure of organized lobbies than are other institutions of the state' (Hartkopf/Bohne 1983, 136).

The Mode of State Action and the Style of Regulation

On the assumption that member states of the European Union seek to impose their style of regulation at the supranational level and that the nature of the state modifies as policy and politics shift to the Community level, it is essential to analyse the structures and instruments of policy formulation as well as the modes of implementation.

Legal Regulatory Structures
In the Federal Republic of Germany, where administrative action is typically guided by the law, law-making goes beyond legislation in the sense of statutes enacted by Parliament to include secondary legislation in the form of regulations (*Rechtsverordnungen*). In matters that are not fundamental, the legislature may delegate its law-making powers (in obedience to the so-called 'essentiality principle')[51] to the Federal government, to a Federal minister, or the governments of the states. This delegation, too, has to be sanctioned by an enabling Federal Act stating the content, objectives and scope actually delegated. The same holds for delegation of law-making powers from state parliaments to state administrative bodies. In environmental law, numerous regulations have been passed on this constitutional basis complementing and concretizing Federal environmental legislation. There are, for example, nineteen regulations to date dealing with implementation of the Federal Pollution Control Act. Further implementing provisions on environmental acts are to be found in administrative directives (*Verwaltungsvorschriften*), which are issued by public authorities only for internal use or to instruct subordinate bodies. These provisions have external effect at least indirectly by imposing uniform application of laws or setting standards in the exercise of administrative discretion. Although administrative regulations or directives are not directly binding on the courts, they have practical significance similar to that of regulations (Keiter/Staupe 1991, 15; Salzwedel/Preusker 1983, 31).

The Federal government's First Environment Programme in October 1971 described environmental policy in its first thesis as the totality of measures necessary and suitable to guarantee human beings an environment appropriate for a healthy and decent life, to protect the soil, air and water, plant and animal life from deleterious effects of human intervention, and to eliminate the damage caused by human intervention (*BMI* 1971).

To achieve these goals an environmental law regime has been created containing preventative and protective environmental provisions and providing for the elimination of environmental damage. German environmental law embraces provisions under public and private (civil) law. The public law provisions go beyond the individual case, seeking to realize protection at source. The provisions at private law aim to remedy subjectively suffered violation of an object of legal protection (see Schäfer 1977, 4f.).

The foundations of a free-standing law relating to pollution control were laid in 1959 with the amendment of the General Trade Regulations and the supplementation of the Civil Code. At that time the Federation lacked the legislative competence to introduce a comprehensive, uniform set of rules. Some states consequently adopted their own pollution control acts, among them North Rhine-Westphalia, a state particularly badly hit by air pollution problems, which introduced legislation in 1962. After three years of negotiation in the Interior Committee, the Bundestag passed the Federal Pollution Control Act in January 1974, which, after having obtained the concurrence of the *Bundesrat* in February, came into force in April of the same year.

In addition to the two general principles governing environmental policy, prevention and placing the cost burden on the polluter, there is a third important guiding principle characteristic of German policy-making, namely the so-called cooperation principle in the context of public hearings. This is a 'political procedural principle,' with the aid of which the Federal government 'seeks to obtain the broadest possible participation of all groups in society in the conception and implementation of environmental objectives and measures' (*BMU* 1990, 20). The principle is applied by submitting all important legislative projects and amendments in the preparatory phase to so-called hearings, and by setting up committees and working groups within the framework of which selected interest groups may participate in policy and decision making and submit opinions (see Keiter/Staupe 1991, 7f.).

In contradiction to the formal intention of the cooperation principle, the weight of influence accorded the various lobby groups, such as industrial and environmental associations, in the policy formation process is not well balanced.

Not all groupings are granted access to hearings[52] on request. Especially critical scientists or less well established environmental groups that take a radical stance are frequently excluded (interview with Weidner, November 1992) or given far less opportunity to exert influence than representatives of industrial associations (interview with Greenpeace, November 1992).[53] According to *BUND* it is 'very easy' at hearings for the big industrial associations, 'to get proposed amendments to bills recorded directly word for word in the minutes, whereas environmental organizations must, if at all, submit their proposals in writing, and then they're expected to formulate them with great precision. ... The fundamental aspects are put forward by the industrial associations and find a hearing' (interview with *BUND*, March 1993). The unpaid environmentalists suffer a further disadvantage vis-à-vis the industrial associations with their 'highly paid consultants' when the Ministry 'sets very short deadlines. When there are often at most two or three weeks to gets to grips with bills of eighty, a hundred or even more pages in length (ibid.). By contrast, the Federal Ministry of the Environment claims that 'all the relevant groups are invited, including environmental organizations'. Unfortunately, the Ministry states, 'the environmental groups often fail to appear although we'd be happy if they did, because we'd then have more support against the objections of industry'. The Ministry does not see that representatives of industry enjoy preferential treatment and stress that everyone, industrial associations and environmental organizations alike are given an equal hearing.' Moreover, 'in the vast majority of cases' bills 'are sent to all the parties concerned six weeks before the date of the hearing and no later' (interview with *BMU*, January 1994).[54]

If it is true that the environmental groups enjoy more limited access and influence than the industrial associations, this is attributable not only to the attitude taken by public authorities towards them.[55] In stark contrast to the British environmental associations, which have been integrated in the policy network for decades now, and unlike the French environmental groups, which participate at least at the local level within the framework of consultative bodies, German environmental associations typically take a defensive attitude towards government. The reasons are technical, ideological, historical, and political. While the industrial associations have staff with excellent legal and substantive expertise, 'among environmentalists there are few good environmental lawyers' and technically knowledgeable experts (ibid.).

Moreover, environmental organizations are dependent on public favour from both a personnel and financial point of view. If the environmental associations show willingness to compromise and thus to cooperate with the political system, they earn less applause than if they advance concrete demands for more sweeping environmental protection with little real hope of political success. Exerting concrete influence on policy-making seldom brings clearly visible and communicable success. 'Lobbying is not very spectacular and bears the stigma of having to play by certain rules of the game (ability to compromise, loyalty, etc.), which make it more difficult for members to identify with the organization' (Hey/Brendle 1992, 77). The environmental associations are consequently forced into assuming a pressure group function, thus finding themselves in a radical and strongly oppositional position vis-à-vis governmental actors. Environmental organizations tend to regard the environmental authorities 'as opponents rather than teammates in an "environmental policy community" ' (ibid.). Contact with politicians and public authorities is consciously avoided so as 'not to be subject to any coercion' and 'to avoid pursuing *realpolitik* so as to be able to represent radical positions' (interview with Robin Wood, November 1992). Both the authorities and the industrial associations correspondingly fail to accept the environmental organizations — with the notable exception of *BUND* — as partners in negotiation and cooperation, regarding them as unwilling to compromise and too radical in their demands, without adequate competence in the matter, and with no understanding of political processes (ibid., 78).

The cool relationship prevailing between environmental associations and government, or the 'tradition' of 'top-down' environmental politics is not a recent phenomenon. The 'representation of interests in the environmental protection field has no traditional home in the political arena. The administrative authorities remain relatively impervious to environmental interests' (Hey/Brendle 1992, 78). The endeavours of public authorities to keep public participation to a minimum,

> ... has to do with the German tradition of subservience to the state. ... Environmental protection has been developed technocratically without the participation of those who have been demanding it. ... Two things come together in this context: the predominant influence of conservative constitutional theories, which regard the state as something standing above the particularist interests of society, and the view of the private association inculcated in the established civil service mind as a disruptive particularist interest rather than a corrective instance. This is strongly anchored in the civil servant mentality (interview with *EURES*, August 1993).

In recent years, however, favourably influenced by a change in generation within public authorities, the picture has begun to evolve.

> Whereas it never used to occur to the authorities to include environmental organizations as clients or allies, but only to see them as a disruptive factor, there is now a new type of environmental official. They're younger people graduating from the technical universities or coming themselves from environmental organizations (interview with Weidner, November 1992).

If one is to understand the reasoning behind the diametrically opposite strategies pursued by German and British environmental associations, it is necessary to look at the differences between the political systems in the Federal Republic and the United Kingdom.

> The British strategy is the answer to a political system that permits no aspirations to power: the "first-past-the-post system". The two party system has no room whatsoever for ecological interests, whereas in Germany the political system, because it is so fragmented — municipality, region, state, federation — leaves far more room for action, and because the electoral system gives newcomers a chance (interview with *EURES,* August 1993).

From this point of view the adversative strategy pursued by the German ecological movement like the cooperative strategy followed by the British movement is necessary and rational.

Policy Instruments Under the Federal Pollution Control Act

The Federal Pollution Control Act contains rules dealing with the major sources of harmful effects on the environment: industry, private households, and light industry / trades as well as transport. The measures prescribed in these fields are plant-specific, product-specific, or area-specific. However, the arrangements laid down by the Act are not comprehensive but exclude certain areas of regulation (in industry, for example, the risks of nuclear energy) or are complemented by supplementary provisions under state law (e.g., smog regulations). With regard to clean air and to industry, the Pollution Control Act regulates the construction and operation of plants, the properties of plants, materials, products, fuels, and lubricants, and the monitoring of air pollution. The Act also contains various so-called 'common provisions', for example on the appointment of a works pollution control officer.

The Pollution Control Act distinguishes between plants subject to authorization and those which are not. Like the British Alkali Works Regulation Act and the French law on *installations classées*, the German Act lists plants that have to obtain individual authorization or which are subject to inspection before construction and commissioning. However, whereas British plant operators are subject only to an extremely limited extent to concrete requirements, such as statutory quality and emission standards, there is a series of regulations pursuant to the German Pollution Control Act, which set emission limits, prescribe the measurement of emissions and pollution concentrations, and impose safety inspection of plants. The regulations that have been issued to date are the 1989 Hazardous Incident Regulation, the 1983 Large Combustion Plant Regulation, and the 1990 Regulation on Incineration Plants for Wastes and Similar Combustible Substances. In addition to these statutory instruments, the first administrative directive pursuant to the BImSchG, the so-called Technical Guidelines for the Maintenance of Clean Air of 1964, further emission and pollution-concentration limit values, and various implementing regulations, for example on the granting of planning permission for industrial plants.

The Federal Pollution Control Act, which the Federal Ministry for the Environment holds to be 'probably the most modern pollution control act in Europe', is based essentially on four requirements that have to be met by plant operators: compliance with emission and pollution concentration standards; orientation on the best available technology even in unpolluted areas, the avoidance, recycling, or elimination of wastes, and the utilization or passing on of waste heat to third parties to reduce CO_2 emissions (interview with *BMU*, July 1993).

Plant operators are not only obliged to fulfil the conditions for licensing, they also have to observe subsequent directives issued by the authorities, if for example the pollution burden changes in the impact area of a plant, advances are made in technology, or more stringent standards come into force. However, the authorities may issue subsequent directives only if the effort required to satisfy them stands in reasonable proportion to the results to be attained (principle of proportionality of means and ends). This latter principle has applied only since the amendment of the Pollution Control Act. In the original version of the Act, the principle laid down was that of economic justifiability, which, like the British bpm (best practicable means) principle, took account of the financial burden on the individual plant operator, and unlike the principle of proportionality, does not weigh up means and ends regardless of the costs accruing to the individual plant operator (interview with Federal Ministry for the Environment, July 1993).

Besides the emission-oriented supervisory measures for industrial plants, the Pollution Control Act provides for further measures that are primarily concerned with harmful impacts on the environment. The Act empowers state governments to take preventative measures for areas where serious deleterious impacts on the environment have occurred or threaten to occur. These preventative measures include the designation of areas as polluted zones, smog zones, or clean-air zones;[56] the establishment of pollution and emission registers and the adoption of clean-air plans, which include appraisal of existing emissions and pollution and projects to reduce air pollution. The third amendment to the 1990 Pollution Control Act reorganized the law relating to clean-air planning. Instead of polluted zones there are now to be inspection zones.[57] In the inspection zones, the competent state authorities are required to monitor the type and extent of harmful air pollution and to take steps to reduce the levels of pollutants in the air. The authorities have to keep an emission register for inspection zones, recording the type, amount and geographical and temporal distribution, as well as the discharge conditions of air pollutants from certain plants and vehicles.[58] Under the amended Section 27 of the Act, the operator of a plant requiring authorization is furthermore obliged to submit a so-called emission declaration to the authorities supplying information on the emissions caused by the operator's plant (type, quantity, and distribution, as well as discharge conditions). If the permissible pollution levels in the inspection zone are exceeded or it is expected that they will be exceeded, the authority has to draw up a clean-air plan justifying and describing clean-up and preventative measures.

To keep track of and control emissions and pollution in the territory of the Federal Republic, measurements are taken by the Federal Environment Office, by state authorities (in North Rhine-Westphalia the State Institute for Pollution Control), by approved institutes (e.g., the German Institute for Standardization) and by industrial enterprises themselves. What is to be measured, as well as procedure, evaluation methods, and publication of the results are laid down in administrative directives issued by the Federal Minister of the Environment with the approval of the *Bundesrat*.

In contrast to Britain and France, but also to most other members of the European Union, the legal system in the Federal Republic of Germany, dominated by a formalist interpretation of the principle of the rule of law, is characterized by a 'classical', highly detailed, and comprehensive formulation of law (see Mayntz et. al. 1982). The clean-air field, too, has been especially subject to a conservative, administrative regime leaving little scope for exceptions to the rule and for cooperative arrangements between administrators and plant operators ('regulative interventionist philosophy'). However, the use of general terms such as 'best available technology' leaves some room for interpretation and 'bargaining' between regulator and operator.

Industry is still particularly critical of the lack of flexibility and incentive for self-regulatory action in environmental policy as regards the use of market and quasi-market strategies. 'There's practically no room for personal initiative. In many cases we could do a lot for environmental protection at very little cost regardless of administrative directives. But that brings us no credit whatsoever. So we leave it be. It's as simple as that' (interview with steel producer, February 1993).

However, since prices do not 'speak the ecological truth', and 'ecologically damaging behaviour is rewarded with low production costs,' the 'self-control effect' is zero, and as a result comprehensive, expensive administrative controls have to be deployed. 'But if a plant manager got wind of the fact that he could cut costs through ecologically correct behaviour, you wouldn't need any more monitoring' (interview with steel producer, March 1993).

Notwithstanding these doubtless convincing arguments in favour of the efficient employment of economic instruments for environmental protection, it is a moot point to what extent the 'self-regulating forces of industry really' suffice 'precisely in a period of economic recession.'

Though some might claim that all the rules and regulations are unnecessary in environmental protection, everybody in industry knows it's not true. That's the way it is. People see it's going to cost something, and at the moment we've got other problems than spending money on environmental protection. And if something is really to be achieved, it's simply not enough to rely on the self-regulating forces of industry, you have to have rules. Only when there's a law on the table does anything get done (interview with environmental consultancy firm, September 1993).

The Federal Ministry for the Environment believes that there is a widespread conviction among the public that concrete regulatory arrangements are the only certain guarantee for efficient environmental protection, which makes it more difficult to deploy economic instruments.

The reason why we have our problems with economic instruments is that they automatically imply that, where costs are particularly high, less is done, and where costs are lower, more is done over and above the best available technology. That's the principle of economic efficiency, which from the environmental point of view isn't necessarily a bad thing. But in the current climate of opinion in Germany, any deviation from the state of the art is automatically understood as easing off (interview with *BMU*, July 1993).

Industry even 'often complains about stringent administrative rules, but basically they're happy that the strict regulative regime provides *a high degree of legal predictability and clarity in outline conditions* [emphasis added]' (interview with *EURES*, August 1993).

Thus, since about the mid-eighties, a fundamental debate has arisen in the Federal Republic on the adequacy of existing and alternative control instruments. Especially during the seventies, environmental policy was, 'owing also to its predecessors (for example in factory and trade supervision) concerned initially with the impositions and prohibitions most appropriate for combating dangers' (*SRU* 1987, 67). The search for new, more suitable control instruments was provoked by the implementation gap apparent in many environmental fields (see details in Mayntz et al. 1978), the economic inefficiency of regulatory measures, and efforts to place more emphasis on prevention.[59]. The Advisory Council on Environmental Questions judges that these efforts have not been crowned with success because, 'in contrast to the scientific discourse, practical environmental policy continues to place comparatively little emphasis on more strongly economically oriented solutions' (*SRU* 1987, 24). A compensatory arrangement was found in 1986 in connection with the Technical Guidelines on Clean Air, which took economic aspects into greater account, but which has never been applied in practice, since 'the authorities shy away from such an arrangement, preferring hard and fast rules' (interview with steel producer, February 1992) and 'the complexity of the provisions make it impossible to apply' (interview with Duisburg Factory Inspectorate, March 1993; interview with Advisory Council for Environmental Questions, March 1993).

Implementation: Structure and Mode
The states of the Federal Republic of Germany bear large responsibility for the implementation of laws. In implementing environmental legislation, the states follow different organizational models. In North Rhine-Westphalia and in most other states, the factory inspectorates (*Gewerbeaufsichtsämter*) are the authorities responsible. In Bavaria and, since 1986 in Baden-Württemberg, licensing and supervision of industrial plants is generally the task of county (*Landratsamt*) or county borough (*kreisfreie Stadt*) (Jaedicke et al. 1990, 34f.).

In North Rhine-Westphalia the regional administrative authorities (*Regierungspräsidien*)[60] and the factory inspectorates are the supervisory and licensing authorities with competence in environmental protection matters.

The regional administrative authorities in the five administrative districts (*Regierungsbezirke*)[61] Düsseldorf, Cologne, Münster, Detmold and Arnsberg have supervisory functions vis-à-vis the factory inspectorates as well as licensing authority for large industrial plants. Technical supervision in the field of pollution control is the task of the Ministry of Labour, Health and Social Affairs[62] in the state capital Düsseldorf. For small and medium-sized plants, the factory inspectorates have their own independent approval procedure and collaborate in preparing regional administrative authority approval proceedings.[63]

Depending on a range of factors such as the size of plant, production methods, etc., the Federal Pollution Control Act provides for two different approval procedures: a formal procedure and a simplified procedure. Under the current provisions, the formal authorization procedure is initiated on written application by the plant operator, who is to submit all the necessary information and documents. The project is then made public by releasing the application and documents not constituting trade secrets for inspection for a period of one month. Objections can be filed until two weeks after expiry of the inspection period, which are then discussed with the applicant and the objectors after expiry of the objection deadline. If the conditions for authorization have been met, the *Regierungspräsidium* or the factory inspectorate issues an authorization. The formal approval procedure is then concluded. In the simplified procedure, the applicant's plans are not made public. The application documents are not made available for inspection, there is no deadline for objections, nor is there any discussion. (greater detail in Hansmann 1992, 24ff.).

Within the framework of licensing procedure pursuant to the Federal Pollution Control Act, local authorities dispose of three different forms of participatory rights. The first of these, deriving from communities' right of self-government, is the possibility of issuing an opinion within the context of authorization procedure. A second right, also deriving from the self-government guarantee, is to refuse consent. Because local authorities are vested with sovereign planning powers, they are entitled to be consulted in the course of the decision-making process on the permissibility of certain projects.[64] The local authority can refuse to give its consent if the project contravenes valid regulations or is contrary to matters of public interest that affect local planning powers, and has local impact.[65] The third participatory right vested in local authorities is that of entering an objection. Without having to assert infringement of its own rights, local authorities as legal persons (public corporations) may take action and enter objections of any sort (greater detail in Ernst 1989, 13ff.).

As regards the practical aspects of implementation, the duration of licensing processes has been castigated. For years industry has been complaining that authorization '*takes far too long* [emphasis added] and is too costly because the authorities expect perfect documentation with an extremely copious dossier' (interview with chemical company, February 1993). Especially representatives of industry who 'have to do with production methods, would wholeheartedly welcome a speeding-up of licensing procedure. It's no use if we've developed something and have to wait five years until we can start production only because a comma is out of place somewhere' (interview with *VCI*, March 1993). Industry's demand for approval procedures to be accelerated, which became more vehement in the late eighties, was reflected in the 1990 9th Regulation Implementing the Federal Pollution Control Act, which sets forth procedural rules for the licensing process. In 1993 the Regulation on Plants Requiring Authorization was amended. The construction and operation of plants listed in the appendix require approval only if they are operated at the same place for a longer period than twelve months (formerly six months). Moreover, permits can now be granted under the simplified procedure for pilot plants for three instead of two years. Modification of a pilot plant for purposes other than the approved development or testing is now no longer subject to formal but to simplified procedure. However, environmental organizations believe that 'a small number of problematic proceedings determine how industry represents the matter externally, while the vast majority of cases are *relatively unproblematic*' [emphasis added] (Führ 1991, 25). An empirical study showed that on average licensing took 6.9 months, and that 75 per cent of all cases were below this average (Steinberg 1990). On the basis of this study and evaluation of the North Rhine-Westphalian trade supervision statistics, the Darmstadt Ecology Institute's Licensing Procedure Coordination Service comes to the conclusion that in North Rhine-Westphalia (1989-1991) about 45 per cent of authorizations take less than six months, about 35 per cent between six and twelve months, and about 21 per cent over twelve months. Delays are attributed not to public participation, but to inefficient work flows in the administration, to inadequate manpower, technical, and financial resources, and to the lack of opportunity to involve potential objectors before planning actually begins (Küppers 1993, 61).

The shape and form of cooperation between the authorities and the emitter is a further determinant of the course and duration of approval proceedings. The detailed environmental protection legislation with its orientation on regulatory law (*Ordnungsrecht*) limits the scope for cooperative negotiated solutions and informal agreements between the licensing authority and plant operators. Within certain limits, however, 'informal forms of cooperation between administration and emitters is quite usual' in Germany (interview with Weidner, November 1992). Both the administrative authorities and industry have an interest in such cooperation, mostly agreed in the context of preliminary negotiations. Since factory inspectorates lack man-power and financial resources, and controls can therefore not be carried out in a comprehensive manner the inspectorates have to rely on collaboration with plant operators. Since they 'don't make unreasonable proposals', 'cooperation with industry is good. They have understanding for the special wishes of the authorities as long as they don't involve too much expense' (interview with Duisburg Environment Coordination Service, August 1993). Industry, for its part, is interested in retaining the goodwill of the authorities to improve the chances of obtaining concessions, for example when a plant operator is unable to reduce emissions by the prescribed deadline (interview with the Confederation of German Industry, March 1993). In some states, the technical control associations (*Technische Überwachungsvereine* [*TÜV*]) play a less formal, because not legally provided, intermediary role between the administrative authorities and plant operators:

Over the past twenty or thirty years this has become customary. And in this capacity [as licensing inspectors] we mediate between plant operators and authorities, ... if, for example, the authority discovers that plant emissions are too high, we think things through together to see how the requirements can be met (interview with the Federation of Technical Control Associations, February 1992).

Leaving aside 'regulatory freedom' in the relationship between administration and industry, both sides — albeit for different reasons — judge narrow legal requirements to be disadvantageous. The administrative side complains that the profusion of rules leads to the overloading, unintelligibility, even inapplicability or impracticability of regulation (see the example described earlier of the compensation options provided under the *TA Luft*) (interview with Federal Environment Office, November 1992). Industry, for its part, protests that laws are too restrictive and inflexible, frequently imposing measures that seem 'senseless from both environmental and economic points of view' (interview with Confederation of German Industry, March 1993; interview with chemical company, February 1993).[66]

Given the large number of rules that may be applied and the practice of bargaining between regulator and operator, administrative practice differs widely 'from state to state, from regional council to regional council. Plant operators often have to go from site to site' and sort out exactly 'what licensing practice is on this point, and what laws, administrative directives, and regulations apply in the particular instance' (interview with pollution protection equipment industry, July 1993).

Whilst cooperation between the licensing authorities and industry is quite wide-spread, environmental organizations are only marginally involved in decision-making processes. As late as the early eighties, environmental organizations were 'almost completely excluded' from negotiations between the administration and plant operators, 'because the environmental organizations were regarded as a disruptive factor' (interview with Weidner, November 1992). Relations between the licensing authorities and environmental groups, like those between environmentalists and industry are still far from close. Both administration and industry try to keep confrontation with environmental groups to a minimum to avoid disrupting the course of authorization proceedings and having to accept delays.

The limited opportunities enjoyed by environmental organizations to access and influence approval procedure is apparent in the restrictive operation of the right to inspect records in administrative practice. It was discovered that between 1980 and 1984 in North Rhine-Westphalia 72 per cent of all licensing proceedings pursuant to the Federal Pollution Control Act were not open to the public (Führ 1989, 35f.). Moreover, where in formal licensing procedure the opening of application documents to public inspection is announced, it was found on the basis of 67 interviews with objectors (confirming Haussmann-Grassel [1985, 50]) that inspecting the records was fraught with difficulty. Even within the licensing authority anyone wishing to inspect files is confronted by the problem of locating them (Führ 1989, 76f.). Administrative officials, especially in rural areas, often claim that there is no right of inspection. Where inspection is in principle allowed, problems arise in working through the files, since there is frequently no special room set aside for the purpose, and the documents have to be perused at the desk of a member of the authority staff in the midst of normal office activity. It is not always possible to copy documents. What is more, in most states access to records is not permitted outside normal office hours, so that people who have a full working day have to take leave in order to inspect licensing documents. Often there is a lack of information on expected emissions into the air from the plant concerned and on its safety specifications, although the competent authority is not allowed to withhold these data from publication in assertion of the protection of commercial confidentiality vis-à-vis the applicant (Haussmann-Grassel 1985, 50). Like this information, opinions issued by the public authorities involved or commissioned expertises often fail to reach the light of day or do so belatedly, i.e., after expiry of the inspection period or after conclusion of the discussion process. Administrative practice thus poses a number of obstacles to comprehensive, well-informed discussion in the real sense of the word (Führ 1989, 70; interview with the German Federation for the Environment and Nature Conservation, March 1993; interview with the Darmstadt Ecology Institute, August 1993).

As regards the role of local authorities in the implementation process in the seventies, they did indeed feel that the 'topic of environmental protection was an imposition' and that environmental protection measures were primarily an expense, both aspects have since gained new salience, not least of all because of 'the change in generation among authority staff towards more progressive and environmentally aware officials' (interview with Weidner, November 1992). Moreover, the more effective solution has proved to be to give greater clout to local authorities, and thus to decentralize environmental protection functions (see Mayntz 1987, 98). Thus in the eighties, a more comprehensive environmental protection concept more broadly anchored in society, 'makes apparent the complex influences that local government exercises on the environment. Where the environmental policy espoused by Federal and states governments fails to live up to the aspirations of committed environmentalists, much hope is pinned on local authorities' (Jaedicke et al. 1990, 36).

The enhanced importance of local authorities is reflected at the institutional level. As numerous surveys have shown (see Jaedicke et al. 1990, 57 for an overview), local authorities have carried out drastic organizational innovations in the environmental protection field. Almost all large cities and rural counties now have their own organizational units for environmental protection. According to a survey carried out by the University of Hanover in 1985 in cities with more than 100,000 inhabitants (Fiebig et al. 1986), the majority of municipalities surveyed (75 per cent) had set up directorates or sections for environmental affairs, or had appointed environment commissioners. A survey conducted two years later by Hofjann (1988) showed that powers were no longer being transferred only to individual divisions or persons, but that the predominant trend was to establish environmental offices or directorates. Similar developments and restructuring trends are apparent not only in larger cities but also in rural counties (Seele 1987).

However, the initiative for setting up new agencies or directorates at the local level generally comes, as Jaedicke, Kern and Wollmann conclude from their case studies, 'from the political sphere; the administration mostly persists in established structures. "Red-Green" majorities [Social Democrats plus the Greens], either in stable cooperation over a legislative period or in *ad hoc* alliance, are particularly apt to initiate environment offices or directorates' (Jaedicke et al. 1990, 61).[67]

In the field of protection against ambient pollution, the local authorities in North Rhine-Westphalia have two major functions. Under the state Pollution Control Act, local authorities have to investigate the pollution situation and record harmful environmental impacts. Local authorities participate in procedures under the Federal Pollution Control Act as representatives of the public interest, giving their opinion on pollution issues. Since 'pollution values are more likely to involve the local authorities than the factory inspectorates,' local authorities have a vital 'pollution monitoring function' to fulfil, a far from unproblematic task, however, because 'industries and the substances they discharge through their chimneys' are 'a black box for the local authorities'. Since the quantities of harmful substances produced by individual plants are subject to data protection, and 'the factory inspectorates would never disclose emission data', the administrative authorities required to deliver an opinion in municipal committees come with empty hands. Trade supervision officials are seldom willing to attend meetings of the committees and to report, because — and this is the other side of the coin as far as the political autonomy of factory supervision is concerned — 'they are not elected' (interview with Bielefeld Environment Directorate, August 1993).

Although the final decision on the authorization of a plant is made by the factory inspectorate or the regional council, local authority scope for influencing procedure is relatively great. Even if all conditions formulated by the local authority are not always taken into account, 'it can be said that, all in all, no factory inspectorate would license a technical plant in the face of massive resistance from the local authority' (interview with Bielefeld Environment Directorate, August 1993; interview with Duisburg Environmental Protection Coordination Service, August 1993).

Conclusion

Given the multiplicity of German political arenas where parties can compete for public favour, the relatively good prospects for newcomers to catch the spotlight, and the open attitude taken by the courts, it can be said that, compared with Britain and France, the opportunities for mobilization in the environmental field are extremely favourable in the Federal Republic. Whereas in the early seventies governmental bureaucracy was largely true to the classical pattern of interventionist administration seeking to avert danger by taking direct and short-term action, by the early nineties administrative structures had become established that sought increasingly to pursue strategies of systematic risk management ('prevention'). Such horizontally and vertically 'differentiated centralization' (von Prittwitz 1990, 199) in administrative structures, finding expression in the establishment of environment ministries and offices at the federal, state, and local levels, forestalls control deficiencies, thus obviating 'incongruity between decision-making structures and problem structures' and the phenomena attendant on 'tardy and perfunctory control' (von Prittwitz 1990, 199). Public authorities still tend to seal themselves off from the public, but as a new generation of administrative staff arrives this proclivity is progressively weakening. In view of the multiplication of cooperative negotiation approaches, the strict regulative intervention philosophy is being mitigated by more flexible elements.

In the debate on European clean-air policy, Britain emerges as Germany's chief opponent and antipole. Why this should be will become clear when we consider the specific conditions for action prevailing in the British policy network.

2.2 Britain: Chumminess and Secrecy

The British look back on a long tradition of clean-air policy. The Alkali Act was passed as long ago as 1863 to combat industrial emissions, the world's first essay in legislating against air pollution. The policy instruments provided under the Alkali Act and the implementation practice that developed on this basis have evolved over time, and until the late eighties of the present century they were a keystone in British regulation. They have been shaped by central elements in the British tradition of the state and administration, and by institutional singularities of the British governmental system. British clean-air policy has been conditioned by the traditional informality and consensus-seeking in administration; the procedural bias of the legal system and the almost total lack of codification; the absence of judicial review; the undeveloped administrative law control of the exercise of administrative power; the dominant role played by central government in the unitary state; and the simple majority electoral system favouring the major parties. This background is particularly apparent in the relationship between the regulatory authorities and industry, which until recently has been based on flexible negotiation behind closed doors. At the same time this 'chummy' and 'secretive' practice permits only limited public access to the authorization process. Because of the paucity of political arenas, the weak jurisdiction of the administrative courts, and the first-past-the-post voting system, political interest groups in particular have an uphill struggle in generating political initiative and government action.

Apart from these institutional factors, there are a number of specifically British factors that have an important influence on the type and form of state action on air pollution. The country's peripheral geographical position, highly varied structures of settlement and industrial density, and strong reliance on indigenous coal for energy supplies are prime factors in encouraging a quality-related approach to the issue. The focus of attention is essentially on local environmental quality. Government intervenes only when economic costs and ecological utility are in reasonable balance and there is scientific proof of a causal relation between the emission of harmful substances and environmental impairment. Until the mid-eighties, this British viewpoint tended to be comforted rather than called into question by more recent medium and short-run developments. The unfavourable economic situation; the absence of environment-related 'shocks'; a British environmental movement committed to traditional values; scientific uncertainty about causes; and the absence of a pollution protection equipment industry were all circumstances discouraging Britain from reacting to growing international pressure to change its regulatory concept. Conservative efforts at restructuring the state also tended to stabilize existing practice.

2.2.1 Problem Perception as the Basis for State Action

The British understanding of environmental pollution is concerned not with the quantity of harmful substances in the environment but with ambient quality. The focus is not on the simple question whether certain hazardous substances are present ('pollution as an undesirable material') but on the extent to which such substances impair the environment ('pollution as an effect'). 'From this, it follows that undesirable materials present in low concentrations, widely dispersed or transformed by natural processes may turn out to be quite harmless' (Boehmer-Christiansen/Skea 1991, 15f.); or as the chief inspector of the competent supervisory authority put it, 'There are no harmful substances, only harmful concentrations' (quoted in Knoepfel/ Weidner 1985, 24). Inherent to this quality-oriented concept is the premise that the environment can perfectly well absorb a certain emission load without suffering harm (Boehmer-Christiansen 1988, 7). This gives right of way to cost-benefit arguments. The objective is not to avoid emissions at (almost) any cost, but to define the 'least-cost' use of the environment, which can differ depending on local factors, the cost of avoidance technology, and the economic situation of the firm concerned. From this perspective there is no point in investing in emission-reduction technology as long as science has its doubts about the utility of such measures for the environment. 'Scientific uncertainty about the harmful effects of pollutants on the environment and human health is also used as an argument by the British government as to why abatement measures are therefore economically unreasonable as well as unnecessary from an environmental perspective' (Weidner 1987, 116). Science and not technology is thus decisive, which in view of the uncertainty and ambiguity of scientific knowledge, encourages reactive, 'wait-and-see' policy patterns.[68]

The predominant policy attitude to clean-air issues in Britain hence makes state intervention contingent on unequivocal scientific proof that pollutant emissions are at the root of the environmental impairment in question. This science-centred approach relates closely to a cost-benefit perspective, i.e., the careful weighing-up of demonstrable negative impacts from air pollution against the potential costs of avoidance measures. This stance can be explained in terms of various long, medium and short-term problem-perception factors.

Long-Term Factors in Problem Perception

There are three long-term variables affecting attitudes to be mentioned in this context that experience no or only gradual change over time: the geographical position of the country; settlement and industrial density structures; and the structure of the energy sector.

Geographical Position: Emissions and Ambient Air Quality
The singularity of Britain's geographical position plays an important role. As an island or rather island group with favourable prevailing wind conditions, the United Kingdom faces far fewer problems of transboundary air pollution than its continental neighbours. As tables 6 and 7 show, Britain exports large quantities of pollutants. About 65 per cent of British SO_2 emissions and almost 90 per cent of NO_x emissions are transported by long-range air currents to central, northern, and eastern Europe. The United Kingdom is hence one of the biggest pollution exporters in Europe. By contrast, Britain's share of total pollutant imports is comparatively small. The British thus enjoy a large 'trade surplus' in pollutant emissions. The dimensions involved are made clear by comparison with Germany (a country that also has high SO_2 and NO_x emissions). Imports and exports of SO_2, each at about 62 per cent of total emissions, are relatively in balance (EMEP study, quoted in Boehmer-Christiansen/Skea 1991, 5). The effect of Britain's peripheral geographical position is that emissions from domestic sources have far less impact on the British environment, especially when (as in the seventies) appropriately tall smokestacks ensured that emissions were widely dispersed.

These geographical peculiarities explain why the problem of air pollution did not become politicized despite the considerable extent of British pollutant emissions (see table 8). Important in this connection were the not insubstantial emission reductions achieved in the United Kingdom between 1975 and 1990. They are largely attributable to the increased use of low-sulphur fuels for domestic heating.[69] Over this period, emissions of particulate matter and SO_2 fell by almost thirty per cent. By contrast, CO_2 emissions fell only marginally (4 per cent), while NO_x discharged into the atmosphere increased by 24 per cent, mainly reflecting the growth in traffic volume.

Settlement and Industrial Density Structures
Although the structure of settlement and industrial density in the United Kingdom shows several industrial agglomerations (Greater London, Birmingham, Leeds, Manchester, Liverpool) and relatively high population density (ca 242 inhabitants per km²), this has by no means heightened awareness of the issue. This might be attributable to the substantial improvement in air quality that the introduction of smokeless areas has brought since the early sixties, and to the willingness in traditional mining areas to put up with air pollution for economic reasons[70] (interview with IEEP, December 1991). The uneven distribution of industrial regions within the country favours a quality-oriented approach that can take account of local environmental loading. Uniform emission standards throughout the country would fail to make adequate allowance for strong local variations in pollution levels.

The Structure of the Energy Sector
Although these factors quite plausibly explain the low level of issue awareness with regard to air pollution, this attitude is nevertheless surprising considering the structure of the British energy sector. Fossil fuels, especially British coal and North Sea oil, are of crucial importance for power generation in United Kingdom power stations. Unlike in France, where nuclear power meets the greater part of energy needs, this source plays only a subordinate role in Britain. Because of the high share of fossil fuels, power utilities are responsible for the lion's share of British SO_2 emissions.

This effect is heightened by an agreement between the largest power generator in Britain, the Central Electricity Generating Board (CEGB), and the British Coal Corporation (BCC), which substantially restricted the CEGB's room for manoeuvre. Under the agreement, the Board was required to fire a specified quantity of British coal in its stations. 'We had to have the British coal whatever price they charged' (interview with former CEGB-employee, September 1993). The share of British coal in total fuel fired by the CEGB is thus over 80 per cent (Boehmer-Christiansen/Skea 1991, 142f.). The structure of the energy industry is hence responsible for a high proportion of British air pollution, since despite high SO_2 exports, a relatively large amount of SO_2 remains in Britain (about 1600 kilotonnes compared to 1300 in Germany). In view of this circumstance, it might be expected that the high SO_2 emissions discharged by the utilities would at least to some degree politicize the issue.

Table 6:   British Balance on Sulphur Dioxide[1] — 100 tonnes of sulphur/year

Country	Imports			Exports		
	1985	1988	1991	1985	1988	1991
Belgium	54	33	53	125	124	90
Germany	132	64	175[2]	562	468	467[2]
France	85	34	107	608	558	429
Italy	2	1	2	92	101	65
Yugoslavia	3	0	1	89	61	37
Netherlands	41	20	27	202	184	164
Norway	2	1	1	208	273	261
Austria	1	0	1	125	67	36
Poland	48	18	33	176	107	151
Sweden	4	1	2	204	222	230
CSSR	45	10	31	116	76	56
USSR	16	2	11	384	249	273

1) Only the most important countries are taken into account.
2) Figures after German unification.

Source: EMEP/MSC-W Report 1/1992 and own calculations.

Table 7:   British Balance on Nitrogen Oxides[1] — 100 tonnes of nitrogen/year

Country	Imports			Exports		
	1985	1988	1991	1985	1988	1991
Belgium	16	14	20	52	59	48
Germany	100	54	98[2]	310	278	312[2]
France	54	54	89	311	286	258
Italy	2	1	2	58	62	44
Yugoslavia	1	0	0	53	35	27
Netherlands	40	22	36	86	83	87
Norway	5	2	3	148	206	227
Austria	1	0	1	62	45	26
Poland	17	7	12	115	147	110
Sweden	7	3	6	143	161	198
CSSR	15	3	10	75	55	41
USSR	6	1	8	272	189	273

1) Only the most important countries are taken into account.
2) Figures after German unification.

Source: EMEP/MSC-W Report 1/1992 and own calculations

Table 8:   Pollutant Emissions in Britain — 1,000 tonnes/year

Pollutant/year	1975	1980	1985	1990	1991
Dust	672	560	545	473	498
- Transport	116	123	146	211	212
- Domestic heating	426	316	285	151	177
- Power stations	32	29	28	27	25
- Industry	91	85	79	79	79
- Other sources	7	7	7	5	5
Sulphur dioxide SO_2	5,368	4,898	3,724	3,780	3,565
- Transport	150	117	102	128	124
- Domestic heating	301	226	202	118	133
- Power stations	2,941	3,007	2,627	2,722	2,534
- Industry	1,757	1,330	657	714	684
- Other sources	219	218	136	98	90
Nitrogen oxides NO_x	2,245	2,365	2,392	2,779	2,747
- Transport	885	976	1,136	1,558	1,578
- Domestic heating	62	68	72	68	76
- Power stations	837	880	807	777	718
- Industry	401	374	308	316	311
- Other Sources	60	67	69	60	64
Carbon dioxide CO_2	165,000	164,000	154,000	158,000	159,000
- Transport	22,000	24,000	28,000	34,000	34,000
- Domestic heating	23,000	23,000	24,000	22,000	24,000
- Power stations	57,000	58,000	52,000	54,000	53,000
- Industry	63,000	49,000	40,000	40,000	39,000
- Other sources	9,000	10,000	10,000	8,000	9,000

Source: Digest of Environmental Protection and Water Statistics (1992) and own calculations

Medium and Short-Term Problem-Perception Factors

Problem perception is also influenced by a range of medium and short-term factors, including the economic situation; environmentally relevant events; public environmental awareness and the environmental movement; international environmental policy; and advances in environmental technology and changes in the environmental technology market. For the United Kingdom there are also specifically British developments to consider, particularly the outcome of Conservative government efforts to restructure the state.

The Economic Situation
Whereas until 1973 Britain enjoyed relatively high rates of economic growth and comparatively low inflation and unemployment,[71] key economic indica-

tors deteriorated increasingly after the first oil price shock in the autumn of 1973. The consequences were high inflation, low growth and mass unemployment. The crisis hit the United Kingdom harder than any other western European country (Scharpf 1988, 11f.). This was also due to the British economy's structural disadvantages vis-à-vis countries like Germany and Japan. German industry, forced to reorganize completely in the aftermath of the Second World War, was able better to accommodate to technological developments, while the British continued to trust in the traditional structures of Empire and Commonwealth (White-Grove 1992, 102; interview with DoE, September 1993). 'At a time when Germany and the United States were innovating and investing in the new growth industries associated with chemicals, electronics and transport, Britain, lumbered with an outmoded technology inherited from the past, was unable to maintain productive efficiency' (Doherty 1989, 14). Moreover, little long-run capital investment occurred in Britain, owing both to relative 'short-termism' in corporate planning and to the nationalization of key industries, where high levels of indebtedness largely precluded comprehensive forward-looking investments (Boehmer-Christiansen/Skea 1991, 119ff.).

Between 1973 and 1979 average GDP growth was only 1.5 per cent, with inflation over the period averaging 16 per cent and unemployment 5 per cent (OECD Statistics 1960-1984). The situation deteriorated further after the second oil crisis in 1979. Only from 1982 did the slide appear to bottom out for the moment. Between 1973 and 1982, average growth was only 0.7 per cent (as compared to 1.5 per cent from 1973 to 1979 and 3.0 per cent from 1982 to 1987).[72] Under these poor economic conditions, environmental demands had less prospect of ranking high on the political agenda. In this phase, investment in improved control and avoidance technology was politically difficult to push through.

Table 9:   Outline Data on the Economic Situation in Britain — percentage change over the previous year

	1973	1975	1978	1981	1984	1987	1990
Economic growth[1]	7.7	-0.6	3.6	-1.3	2.2	4.8	0.8
Unemployment	2.2	3.2	5.9	9.0	11.4	10.4	5.5
Inflation	9.2	24.2	8.3	11.9	5.0	4.1	9.5

1) Real change in GDP.

Source: OECD Historical Statistics 1960-1990 (1992), 1960-1986 (1988), 1960-1982 (1984)

National Environment-Related Events
Not only was the economic situation unfavourable: Britain had also been spared the sort of environmental incident apt to generate public pressure for action to be taken in the environmental field. The geography of the British Isles renders them less vulnerable to transboundary air pollution than continental Europe. Whereas phenomena such as acid rain and forest dieback in Scandinavia and Germany very rapidly became highly charged political topics, no such politicization occurred in Britain. The advocates of more stringent environmental regulation were consequently unable to profit from 'external shocks' such as the 1952 smog disaster in London, which resulted in the introduction of smokeless areas under the 1956 Clean Air Act (CAA 1956) (Ashby/Anderson 1981, 103ff.).

Public Environmental Awareness and the Environmental Movement
In an inauspicious economic situation and without the provocation of external shocks, no perception of long-range air pollution as a problem had developed among the British public comparable to the environmental awareness that had been provoked by forest dieback in Germany. Nor were British environmental organizations initially able to achieve much in this direction. The main reason was that British environmental groups tend to operate 'silently', aiming to exercise political influence by playing the institutional game according to the rules of confidentiality and discretion. This obviated polarization in the political field of environmental protection, a strategy which — as we shall see — does not always augur well under the institutional conditions prevailing in the British governmental system (interview with *EURES*, August 1993; White-Grove 1992, 114ff.). What is more, the traditional British environmental movements operate with quite different value orientations. They focus not on clean air but on conserving nature and the countryside — traditional Victorian values, which find expression in such phrases as 'amenity of life' or 'pleasant countryside' (Vogel 1986, 46ff.).

Such traditionalist environmental groups take a more consensual, cooperative stance vis-à-vis government. They do not call the existing system into question. 'Less tinged with utopian ideals and emotional protest, traditional British groups were arguably less able on a cultural level, and institutionally ill prepared, to have an impact on the large-scale international issues of the 1970s' (Boehmer-Christiansen/Skea 1991, 77). As in other industrial countries, environmental issues became increasingly politicized in Britain in the sixties. New pressure groups arose that differed markedly from the traditional conservation movement in both style and fundamental positions. Organizations like Friends of the Earth (FoE) or Greenpeace, which set up in the United Kingdom respectively in 1970 and 1976, sought to assert their interests in a more radical manner by open confrontation and effective, publicity-seeking activities (ibid., 80). However, the new environmental groups' strategy of polarization failed initially to heighten public environmental awareness or to provoke greater environmental activity on the part of government. The institutional structures of the British governmental system bear a large share of the blame for this failure, being more amenable to consultation than confrontation.

International Environmental Policy
The underdeveloped politicization of environmental matters, the absence of environmental shocks, and the poor economic conditions cushioned international pressure on the British position. Particularly since the mid-seventies, the Scandinavian countries had stepped up their efforts within the framework of the UN Economic Commission for Europe (UNECE) to obtain a reduction in British SO_2 emissions. Although the United Kingdom put its signature to the Geneva Convention on Long Range Transboundary Air Pollution (LRTAP), it did so only because it had achieved a negotiated concession in coalition with other countries that there would be no binding reduction targets for SO_2 (Vogel 1986, 104).

Scientific and Technological Progress
A further factor in problem perception is the course taken by scientific and technological progress, which — at least initially — was no inducement for greater state intervention. This reflected the strong emphasis on scientific evidence as the prerequisite for action, and the pronounced cost-benefit attitude to problem solving taken by the British. The statements issued by Scandinavian countries and the findings of a 1972 OECD scientific study that identified Britain's SO_2 and NO_x emissions as a crucial cause of environmental damage in Scandinavia was not confirmed by British investigations (especially research conducted by the CEGB itself) (Blowers 1987, 289). Doubt was cast above all on the question of emission range, on the proportionality of emission and damage reduction, and on the causal relations between emissions and the environmental damage recorded. This scientific ambiguity was partially eliminated only by further studies conducted in the mid-eighties (interview with DoE, January 1993; Boehmer-Christiansen/Skea 1991, 43f.). Moreover, the CEGB was extremely sceptical about the use of flue-gas desulphurization (FGD) systems. Such equipment had been tested in the United Kingdom since as early as the thirties, but with little success.

Changes in the Environmental Technology Market
Scepticism about new avoidance technologies and scientific doubts meant that very little attention was paid to the development of the corresponding environmental technology market. This attitude was fostered by the CEGB assumption that the problem of emissions from coal-fired power stations would disappear in the nineties with the increasing use of nuclear energy (ibid., 146f.). Neither economic nor scientific grounds were thus seen for government action. What is more, Britain has no environmental protection equipment industry to speak of — the only sector of the economy that has a potential interest in more stringent environmental standards. 'The market for air pollution abatement technologies in the UK has remained relatively small' (Weidner 1987, 97).

Specific British Developments in Government
In contrast to France and Germany, particular developments in British government constitute a specific, internal policy factor affecting problem perception. The efforts to roll back the state that had been undertaken by the Conservative government since coming to power in the late seventies as part of a general strategy to change British perception of the state made more thoroughgoing regulation in the clean-air field unlikely (interview with DoE, November 1992). Deregulation, privatization, and cuts in public spending are the essential elements in Conservative philosophy, the aim of which is 'to roll back the frontiers of the state' (interview with AMA, January 1993). Since most SO_2 and NO_x emissions were caused by state enterprises (especially the CEGB), it would have contradicted Conservative austerity policies to impose higher costs on these enterprises by introducing more stringent regulation requirements (interview with DoE, January 1993).

Conclusion

The interaction between the various long, medium, and short-run factors foster a British attitude towards problem-solving that makes government intervention contingent on two necessary conditions: first the existence of scientific evidence and second economic proportionality with regard to the capacity of the environment to absorb pollution. Britain's geography alone would suggest taking this stance. Favourable wind conditions ensure that British emissions are exported in large quantities; since this prevented external events from generating internal political pressure, the government had nothing to gain from stricter environmental standards, especially since the environmental protection equipment industry — the only sector that would profit from it — plays only a marginal role in the United Kingdom.

The British accordingly had no incentive to take precautionary measures to reduce pollutant emissions without scientific proof, since, in view of the high level of pollutant exports, no major improvements in domestic environmental quality were to be expected. Moreover, the economic recession in the late seventies would hardly have rendered any such move acceptable to industry. It would also have been inconsistent with Conservative government objectives if a more stringent environmental regime were to result in higher government spending on control technologies in state-owned utilities. Growing international pressure could do little to change things — not least of all because the British were able successfully to plead the ambiguity of the scientific evidence available. The British perception of the issue thus nurtured the negative attitude taken by the United Kingdom towards stricter supranational regulations to abate transboundary air pollution.

2.2.2 Institutional and Instrumental Preconditions for State Action

On the hypothesis that the rationale of individual countries in supranational negotiations is to minimize the costs of domestic legal and institutional adjustments necessary to implement European measures, it can be interesting at this point to look beyond problem-solving traditions and review existing institutional arrangements in clean-air policy. It is only in this context that British behaviour in negotiations on directives can be understood. At the same time, this raises the question of the domestic political conditions for initiating possible modifications to the regulatory system, since the British negotiating stance at the European level correlates with the strength of internal political pressure for change.

Political Initiative and State Action: Institutional Prerequisites

We assume that the scope for launching political initiatives and inducing the government to act depend essentially on three characteristics of a political system: the number of political arenas, the electoral system, and the role of the courts. While the first two factors refer to the direct political opportunities for parties to express themselves and for interest groups to exert influence, the courts offer the possibility of indirectly forcing the government to take certain measures.

The Dominance of the Centre
Key characteristics of the British governmental system are the unitary form of the state and the supremacy of Parliament. 'The United Kingdom is a

unitary not a federal state. No territorial assembly inside its frontiers enjoys a coequal status to that of the Parliament of Westminster' (Dunleavy 1993, 5). This severely restricts the number of arenas in which political groupings can initiate governmental activity. Although there are elected assemblies at the local and regional levels (district and county councils), their competence and scope for action depends directly on enablement by Westminster. Unlike the Federal Republic of Germany, the United Kingdom knows no constitutionally entrenched guarantee of local self-government. Local authorities can at any time have their powers rescinded by central government.

This does not necessarily mean that local authorities play an unimportant part within the political system. 'The degree of centralization was considerable depending on the degree of discretion allowed to local authorities in Parliament legislation or mandates, the absence of close national government inspection and auditing as well as on relatively wide scope for revenue raising' (Lane/Ersson 1991, 215f.).

Depending on the room for manoeuvre allowed an authority and on the resources and powers at its disposal, the local government level can in principle set its own innovative emphasis even in a unitary state. This is evident in 'wide variations in standards and practices in different parts of the country' (Steel 1979, 34). At the local level there is hence a certain scope for influencing local government policy.

However, from 1979 the Thatcher government set about clipping local authority wings in its pursuit of centralization. The 'central government's financial noose for throttling local government spending' (Uppendahl et al. 1988, 44) was progressively tightened by restricting allocations and curbing local authority rights to levy taxes. Local government dependence on resources is well illustrated by the 'smokeless areas' instrument. If a local authority designates an area as such, it is obliged to reimburse private households thirty per cent of the costs for converting to smokeless fuels. The Conservative government's austerity measures entailed the risk of financial considerations prevailing over environmental aspects in the designation of smokeless zones (Knoepfel/Weidner 1985, 259). Downing Street's centralization endeavours reduced local authorities to little more than central government agents. 'Local authorities can do things, but they must do what [central government] tell them to do' (interview with AMA, January 1993). At the same time, this loss of local flexibility reduces the scope for interest groups to mobilize and articulate their interests in this field. They are forced to rely more heavily on informal contacts with representatives of the bureaucracy who prepare or implement political decisions.

One consequence of the Westminster model, with its closed circuit of responsibility, is that substantive policy decisions in Britain depend in the final resort on a handful of people in the executive, which occupies a very strong position within the political system. 'Policy decisions have to pass the *bottleneck of Whitehall* [emphasis added] ... , maybe ten or twelve persons, in each policy area, not even the House of Commons' (interview with AMA, January 1993). The chances in the United Kingdom of political initiatives succeeding in pushing through more stringent environmental demands are extremely limited because of the mono-centric nature of the political scene, the unstable competence and financial position at the local level, and the strong concentration of decision-making powers in a few hands within the executive. Moreover, environmental organizations have little opportunity to enhance their influence within the system by organizing in the form of political parties. A glance at the peculiarities of the British electoral system shows why.

The Electoral System and Political Party Access to the Political System
The 651 Members of Parliament[73] (i.e. members of the House of Commons) are elected by age-old tradition in accordance with the relative majority (or plurality) principle. The entire country is divided into as many constituencies as there are MPs to be elected. The single constituency seat goes to the candidate who polls the highest number of votes. All other votes are 'wasted' (first-past-the-post system). This necessarily reinforces the majority party. In this manner the electoral system produces relatively powerful and stable cabinet government (Dunleavy 1993, 3f.). At the same time, however, small parties (such as the Green Party) have scarcely any hope of gaining a seat in Parliament. 'We have a Green Party in Britain but it doesn't have any effect because of our electoral law' (interview with Greenpeace, January 1993). The vast majority of MPs elected belong to one of the two major parties, the Conservatives and Labour (Jesse 1992, 176). This effect of the electoral system reduces the opportunities for broad political groupings to gain access to the political system as parties. Small parties have to overcome formidable obstacles before they can establish a Parliamentary base. The only party to succeed (with the exception of small, strictly regional groupings on the 'Celtic Fringe', i.e. Northern Ireland, Scotland, and Wales) have been the Liberal Democrats, an amalgamation of the former Social Democratic and Liberal Parties, which have become the third force in the Lower House. They play an important role in environmental politics: 'It was the centre parties which argued most strongly for higher standards of environmental protection' (Boehmer-Christiansen/Skea 1991, 102; interview with DoE, January 1993). Thus, especially in the South of the country, where they constitute a serious threat to the Conservative heartland, the Liberal Democrats foster competition between the parties in regard to their environmental image (White-Grove 1992, 112).

The Role of the Courts

After the range of political arenas and electoral modalities, the courts are a third important factor influencing political decision-making. However, in contrast to Germany and France, there are a number of reasons why the chances of compelling government regulatory action in Britain through the courts are slim. The first is the principle of the supremacy of Parliament, vesting sole authority in Parliament to adopt or avoid laws. Britain knows no judicial review subjecting the laws passed by the legislature to scrutiny as to their constitutionality; the House of Lords acts in formal capacity as the highest court: 'Parliament is regarded as the highest court in the land — this is how lawyers explain why the law courts cannot query or set aside its duly enacted statutes; they argue that they are inferior courts to the "High Court of Parliament" ' (Finer 1970, 148). Interest groups thus have to rely more on consensual relations with the authorities: 'In countries where the road to the courts is absent, interest associations are more condemned to reach an understanding with the regulatory authorities' (van Waarden 1992, 15; see also Wilson 1989, 298).

A second reason is to be found in British legal tradition, which has developed neither a system of public-law principles to guide administrative action nor an administrative court system comparable to that existing in the continental European tradition (Dyson 1980, 42). Although there are over two thousand 'administrative tribunals' in the United Kingdom, they do not form an integral part of the general judicial system. They operate more like a type of arbitration board to which the citizen may turn if he or she feels badly treated by the administrative authorities.[74] Although these tribunals have no competence to elaborate their own legal principles, they do play an important role as first port of call in disputes between citizens and administration. However, if no settlement is reached, the matter is referred to the 'real' courts. The jurisdiction and powers of the tribunals are thus extremely limited (Budge/McKay 1988, 170f.; Mény 1993, 336f.).

Court rulings are concerned for the most part with procedural issues. Technocratic standards play no role (Damaska 1986, 25). Norms are interpreted not by professional judges in obedience to objective criteria but by lay judges guided by judicial precedent. Whereas in continental Europe administrative law has developed as an 'articulation of the state, as a distinctive, binding and enforceable system', Britain has retained a 'narrow, formal concept of expressing a procedural philosophy' (Dyson 1980, 42). The observance of certain procedural rules is more important than respecting material principles. This philosophy gives British regulatory authorities extensive discretionary powers. There are few clean-air standards amenable to testing in the courts. A plaintiff is therefore required not to prove to the court that certain limits have been exceeded, but that a plant has failed to apply the best available control technology economically feasible, and that this infringement has caused damage to the environment (McLoughlin 1982, 65). The difficulty of providing proof in the individual case and the low penalties imposed in the event of the court ruling against the defendant make legal action not very attractive. The courts are relegated to a subordinate role in clean-air politics.

Considering the overall institutional opportunities for public and private actors to generate political initiative and government action within the British governmental system, it is apparent that there are relatively high hurdles to be taken. The strong position of central government in a unitary state, leaving little scope for independent initiative on the part of local government, reduces the number of political arenas to one, namely the Westminster Parliament, which the British electoral system places under the domination of the government and where the opposition has little clout. Moreover, the electoral system renders it practically impossible for small parties to enter Parliament. Real opportunities to launch political initiatives and bring them to fruition are thus severely restricted in the United Kingdom. Recourse to the courts also offers no satisfactory approach, since there is no constitutional court in the United Kingdom to review the laws, nor a comprehensive, developed administrative court system to guide and scrutinize government action.

The Mode of State Action and the Style of Regulation

The hypothesis that it is in the interest of every member state of the European Union to minimize the costs of legal and institutional adjustments to meet European measures implies that a country will support supranational moves only if they largely harmonize with domestic institutional arrangements. And vice versa it will reject any measures requiring substantial modification of the domestic regulatory system. If we are to understand British behaviour with regard to various EU directives in the light of this hypothesis, we must first take a look at existing and former institutional arrangements. Of particular interest are the legal structure of regulation, the policy instruments and typical modes of implementation applied until very recently. It is only against the background of former practice that the behaviour of the United Kingdom in the European arena and the transformations caused by European regulation in British policy patterns can be understood.

Legal Regulatory Structures and Policy Instruments

Jurisdiction over the control of stationary air-pollution sources in the United Kingdom is shared by central and local authorities (Vogel 1986, 70). Processes requiring higher technical expertise for control — 'the more important and complex stationary pollution sources' (ibid.) — had until recently been subject to supervision by the Alkali Inspectorate (AI),[75] a central government body. The local authorities were competent with respect to all other processes. In particular, they monitored smoke and particle emissions by private households and industrial plants that did not fall within the purview of the AI. Whitehall implemented the various instruments in the form of regulations after hearing the interest groups concerned; approval by Parliament was required, which was generally forthcoming.

The most important statutory basis for AI activities was the Alkali etc. Works Regulation Act 1906. Like the French *'nomenclature'*, there was a schedule to the Act listing all plants required to register with the AI before beginning operation ('registered works'). To this extent it can be said that registration was equivalent to licensing (Knoepfel/Weidner 1985, 42). The schedule, which had steadily lengthened since 1906 to keep pace with technical and industrial developments, covered plants with operational processes that were technically difficult to monitor or which emitted particularly noxious and hazardous substances (Vogel 1986, 71). A schedule of the emissions subject to monitoring — 'noxious and offensive gases' (McLoughlin 1982, 73) — was also appended to the Act. There were practically no statutory quality or emission standards for AI regulatory activity.[76] This was justified on grounds of greater flexibility, which permitted individual reaction to given technological, operational and local situations (Vogel 1986, 76). A control principle, termed 'best practicable means (bpm)', formed the basis of regulation. AI supervision derived two general precepts: 'First of all, to prevent emissions, whether they come from chimneys or any other source in the factory. And, secondly, to ensure that any pollutants that are emitted do not cause a hazard or a nuisance' (Frankel 1978, 63).

Neither legislation nor the courts provided substantive specification of the bpm principle and its application (Knoepfel/Weidner 1985, 141), which gave AI inspectors wide discretionary powers: 'The term 'practicable' has never been clearly defined; the Alkali Inspectorate itself was considered the sole judge as to whether the plants under its jurisdiction are employing the best practicable means of controlling their emissions' (Vogel 1986, 79). In effect, however, the definition amounted to weighing local conditions against modern technology and the cost to the factory of control measures. Great importance was placed on the economic components.

I have often said, and been criticized for it, that if money were unlimited, there would be few problems of air pollution control which could not be solved technically. ... We have the technical knowledge to absorb gases, arrest grit, dust and fumes, and prevent smoke formation. The reason why we still permit the escape of these pollutants is because economics are an important part of the word 'practicable' (HM Alkali and Clean Air Inspectorate 1974, 12).

It is interesting to note that, because of excessive costs to industry, the AI regarded no avoidance technology as 'practicable' with regard to SO_2 and NO_x emissions. The British approach to regulation logically admitted only of the second alternative, namely 'to ensure that any pollutants that are emitted do not cause a hazard or a nuisance', which in effect means a 'high smokestack policy' (Weidner 1987, 76).

After consulting representatives from industry, the chief inspector set so-called 'presumptive standards' separately for each industrial sector and each process, which were published as 'notes on bpm'. These standards set limits for substances that were subject to AI monitoring. However, they could not be compared with statutory limit values in legal status and binding effect: 'Although a company's failure to comply with the limits [could] be used as evidence in various legal proceedings, the limits [were] not legally binding' (Vogel 1986, 77). They were more in the way of points of reference for operationalizing bpm. If the standards were not exceeded, it was assumed that the firms were applying the bpm principle correctly. The 'notes on bpm' gave the district inspector a sort of basis for negotiation. Depending on local conditions, he might deviate from the presumptive standards, but only upwards (McLoughlin 1982, 81-83). District inspectors had a great deal of leeway in appraising the local situation, background concentrations: 'The concrete definition of the degree of air pollution [was] left largely to the inspectors' (Weidner 1987, 77). The notes on bpm also specified emission measurement procedure, the scientific basis for determining the control technology and the appropriate measures for maintaining it. They were checked every ten to fifteen years and brought up to date in technical and scientific terms: 'Accordingly, bpm ... should be regarded as an elastic band that can be tightened as science develops and places greater facilities in the hands of the manufacturer' (Alkali Inspector quoted in Vogel 1986, 81).

Local authority supervision was the task of the district environmental health departments. The counties had no jurisdiction in clean-air matters. The most important local authority powers were under the Clean Air Acts of 1956 and 1968. Their purpose was primarily to control smoke and particulate matter emissions, an aim that had been pursued with increased vigour since the twenties (Ashby/Anderson 1981, 86ff.). The immediate motive for adopting the 1956 Act was the 1952 smog disaster in London (Knoepfel/Weidner 1985, 43). The principal control instrument under both Acts was the possibility of designating so-called smoke control areas, within which only smokeless fuels could be used (Scarrow 1972, 261). The government reimbursed seventy per cent of the costs incurred by private households in covering to such fuels. The remaining thirty per cent had to be borne by the local authorities (Weidner 1987, 83). Whereas the 1956 Clean Air Act left the use of this instrument to the discretion of local authorities, the 1968 Act gave central government the power to direct districts in individual cases to designate smokeless areas. However, the government made no use of this power — 'the British way to threaten' (interview with IEEP, December 1991).

The Public Health Acts 1936/1961/1969 give local authorities a subsidiary legal basis for taking action against deleterious and offensive emissions ('statutory nuisances') not covered by specific statutes (such as the Clean Air Acts) (Bennett 1979, 95). However, these provisions present a problem in so far as they admit of no preventive action; intervention is possible only if a nuisance is already apparent. Since the authority has a hard time proving the existence of nuisance in individual cases, these Acts are of subordinate practical utility (McLoughlin 1982, 28f.).

Implementation: Structure and Mode
Implementation style in clean-air policy was largely consistent with British traditions of the state and government. In Britain no ideological boundary developed between the state and society as it did in the continental European tradition. Little credence has been given to the idea of the state as an abstract entity bearing inherent responsibility for the performance of public functions or as a collective actor representing the nation as a whole (Dyson 1980, 43; Wilks/Wright 1987, 279). Britain has thus frequently been described as a 'stateless society' (Nettl 1968, 562) or in terms of 'government by civil society' (Badie/Birnbaum 1983, 121). The British understanding of the state, according to which public activity arises not from purposive government action but from competition between social groups bringing their interests to bear in the political process, indicates participation by private actors — such as industry in the field of air quality — in the making and implementation of policy. It is thus seen as wrong for the state to 'impose' measures unilaterally on society. 'Britain is best characterized as emphasizing consensus and a desire to avoid the imposition of solutions on sections of society. In that there is no particular priority accorded to anticipatory solutions. ... Underlying the consultative/negotiatory practice is a broad cultural norm that government should govern by consent' (Jordan/Richardson 1982, 81; 83).

The English legal system, with its emphasis on procedural regulation and a dearth of substantive criteria for reaching decisions, encourages this *modus operandi*. It often results in bargaining between regulators and the regulated. Consensus takes priority over coercion. The bpm principle, being based on negotiation of this type, is in keeping with the tradition, all the more so since it left no room for statutory standards. The low degree of codification in English law also fosters informal administrative action, as reflected in the relationship between the AI and industry. This effect was reinforced by British administrative authorities. The public had great faith in their neutrality and loyalty. They took advantage of this circumstance to act informally and behind closed doors: 'Thus, administrators [would] have greater leeway in dealing with organised interests. They were less fearful of cooperating with them and involving them formally in public policy ... , providing some with privileged access, or engaging in informal and secret relations' (van Waarden 1992, 18).[77] A further reason for the high degree of confidence placed in the administrative authorities may be found in the fact that the risk of them being 'taken over' by private interests was considered slight given the high rate of internal mobility in the public service. And it was precisely this 'acquired' confidence that allowed the authorities to build up informal network relationships. To this extent, the behaviour of the AI reflected general administrative practice in the United Kingdom. Although the AI inspectors with their specialist technical training did not necessarily conform to the ideal image of the British public servant (generalist, highly mobile within the service), they nevertheless profited from the overall faith the British public place in the authorities, which included the AI.

Implementation practice was also shaped by the institutional structures that determined the network elaboration phase, at the so-called 'institutional watershed' (Skowronek 1982), and which had stabilized in the course of time (Krasner 1988, 90). Since about 1830 the alkali industry in Britain had experienced enormous growth. It produced soda, used in the manufacture of soap, glass, and textiles. The production process caused high emissions of hydrochloric acid, which had a disastrous impact on agricultural land. This led to increased political pressure from the relatively influential landowners. The outcome of this development was the 1863 Alkali Act, which laid down that every works had to reduce hydrochloric acid emissions by 95 per cent. The type of abatement technology to be applied was not prescribed. A central authority was set up to implement the Act, which in keeping with the sector to be supervised, was named the Alkali Inspectorate. The first inspector, the chemist Robert Angus Smith and his four assistants took up their task under relatively uncertain conditions.

It is not difficult to imagine the obstacles Smith had to overcome. An isolated government official based in Manchester, with very little backing or guidance from his employers in Whitehall, ... empowered to control emissions from a great and flourishing industry. His only hope was to secure the confidence and cooperation of the factory owners. One tactless letter, one injudicious prosecution for infringement of the Alkali Act — and Smith would have had the whole alkali industry ganged up against him. ... He had no precedent for what he had to do (Ashby/Anderson 1981, 25f.).

Smith very soon had successes to chalk up, and managed to win the confidence of the industry, not least of all through his cooperative style of regulation, which was based on 'advice and friendly admonition' (ibid., 27), but also thanks to the fact that emission control by the industry actually brought economic benefits: Smith had developed a recycling method that made the reprocessed waste even more valuable for the industrialist than the actual product (Vogel 1986, 240). The 1874 Alkali Act, which placed a further range of processes under the control of the AI, introduced the bpm principle. Under Smith's successor, Alfred Fletcher, who continued to nurture the cooperative and trusting relationship with industry, the bpm principle became the prevailing control philosophy: 'For my part I feel it to be more binding than a definitive figure, even if that could be given, because it is an elastic band, and may be kept always tight as knowledge of the methods of suppressing the evils complained of increases' (Fletcher, quoted in Ashby/Anderson 1981, 40).

The foundations were thus laid for the style of implementation that was to determine British clean-air policy until the late eighties. The relationship between the AI and industry was also favoured by specific attitudes cultivated within British industry, largely shaped by the culture and tradition of Victorian society. Industry saw itself as bearing responsibility for society as a whole, and was correspondingly willing to accept certain social and political restrictions in the public interest: 'Over the last century, British business elites have proved far more willing to accept a whole range of social and political constraints than have their counterparts in the United States' (Vogel 1986, 248). The values cherished by British industry facilitated the development of a cooperative, informal style of regulation, since both public servants and industrialists were considered by the public to be 'gentlemen desirous of doing what is right' (Ashby/Anderson 1981, 28).

Furthermore, the incremental and stable development of British clean-air policy, which is manifested particularly in the long tradition of the AI and the gradual adaptation of the bpm principle, is reflected in the general continuity of government institutions, the evolution of which has proceeded relatively gradually and uniformly in the United Kingdom. This effect was favoured by the fact that the country has had no foreign institutions imposed on it since the Norman conquest. Through its reliance on precedent, the legal system, too, looks back on continuity of development (ibid., 21). Symptomatic for change in British institutions is, for example, the fact that one of the greatest advances in British clean-air policy, the 1956 Clean Air Act, was adopted only after extensive negotiations over a period of no less than thirty-five years (Vogel 1986, 271).

The general relationship between regulatory bodies and industry in the clean-air field could thus be characterized as very cooperative and trusting.

The AI characteristically used 'soft' means of control: 'British regulatory officials [pursued] a consistent policy of close cooperation with industry. They [continued] to rely more on persuasion and voluntary agreements and less on coercion than any other industrial democracy' (Vogel 1983, 89).

While industry placed a great deal of trust in the inspectors' technical expertise and judgment, the AI assumed that industry was itself interested in abating emissions, which made the use of coercion superfluous: 'The Alkali Inspectorate's relation to the manufacturer [was] more like that of a doctor getting the patient's cooperation in treating a disease than of a policeman apprehending a culprit' (ibid., 96). This confidence was encouraged by the flexibility of the AI's scope for action, which gave it wide discretion in interpreting bpm that could neither be monitored by superordinate authorities nor challenged in the courts — 'every inspector can make his own little policy' (interview with CBI, September 1991): 'Although the law does not appear to say so explicitly, "best practicable means" is generally interpreted as "best practicable means to the satisfaction of the Alkali Inspectorate" ' (RCEP 1976, 23).

The cooperative relationship between the AI and industry found expression in copious consultation in every phase of the regulatory process (from fixing presumptive standards to individual implementation in specific cases), although the AI was under no statutory obligation to proceed in this manner. The following statement by a chief inspector (quoted in Hill 1983, 90f.) describes how cooperation in fixing presumptive limits operated:

Working parties and discussion groups are set up, consisting of the representatives of industry, its research organization, if any, and the Inspectorate. ... The Chief Inspector makes the final decision on any standards ... but this only follows mutual discussions with industry representatives. ... Frequent, usually annual meetings are held between trade associations and the Inspectorate to note progress, discuss new technology, review research and development and generally reassess situations with the object of gaining further improvements and possibly getting together standard requirements for bpm'.

This regulation philosophy, which preferred to convince and cooperate than to confront, made legal action an extremely rare event: 'Prosecutions are used only when considered necessary, which is usually where there has been a flagrant breach of requirements' (McLoughlin 1982, 87). When penalties were imposed, they were relatively mild.

It was logical for both the AI and industry to implement bpm in this manner. Since there was little prospect of sanctions, the AI had to rely on industry's willingness to cooperate (Vogel 1986, 83). This was because the Inspectorate was in a relative weak position vis-à-vis industrial interests in the politico-administrative system. Thus the Health and Safety at Work Act 1974 (HSWA 1974) integrated the AI in the Health and Safety Executive, which maintained comparatively close relations with industry (Boehmer-Christiansen/Skea 1991, 260). Moreover, the Environment Directorate of the Department of the Environment (DoE), which still formally controlled the AI, had less political clout than other government departments, in particular the Department of Trade and Industry (DTI), '[which had] an interest in the costs of environmental controls and their impact on the competitiveness of British industry', and the Department of Energy (DEn), '[which] held responsibility for 'sponsoring' the nationalized energy industries' (ibid., 111). The CEGB, for instance, had a powerful advocate in the DEn.

What is more, the wide discretion in defining bpm in individual cases meant that hard and fast rules could be decided only within a bargaining process. AI inspectors were especially dependent on the information supplied by firms when it came to the financial components of the bpm principle. On the whole, this situation meant that the AI, if it wished to stabilize its position in the network and safeguard its influence on policy outcomes, needed the trust and voluntary cooperation of industry. It obtained it chiefly through comprehensive consultation and very restrained recourse to legal action in the event of offences:[78] 'An aggressive policy of confrontation, involving prosecution for every lapse, would destroy this basis of cooperation; it would harden attitudes and dispose industry to resist the imposition of costly programs for pollution abatement' (RCEP 1976, 72). In any event, frequent court cases would not only have put industry in a bad light but would also have demonstrated the AI's lack of success in its regulatory functions (Vogel 1986, 87).

On the other hand, it was in industry's interest to keep the stable give-and-take relationship with the AI going. The processes of consultation and negotiation made it possible to influence decisions in terms of individual cases. Although confrontation with the AI was likely to have relatively negligible legal consequences, it would have affected the duration of the procedure, which could involve industry — where it depended directly on the granting of a license[79] — in additional expense and possible put it at a competitive disadvantage. At the same time, by choosing to take a conflictual approach, industry would have run the risk of the flexible bpm principle being replaced in the long term by more detailed requirements (such as statutory emission standards) and by more restrictive administrative practice, which could have diminished its influence within the network. 'It is the ability to take responsibility and give quick decisions which pleases industry in its negotiations with the inspectorate. ... Industry will pay for time saved ... and is prepared to accept tougher decisions than it might otherwise gain from protracted argument' (industry representative quoted in Vogel 1986, 86).

The local authorities also preferred to cultivate a more cooperative style vis-à-vis industry. The local level was very willing to cooperate with industry to safeguard the local industrial base and jobs: 'Local authorities are unwilling to pressurise too much for fear that they will go elsewhere' (Gunningham 1974, 73). This attitude was reflected in the rarity of legal proceedings taken by local authorities against both industry and private individuals.[80] Many local authorities fail to exercise their legal scope for action to the full: 'Local authorities exercise considerable discretion in determining how strictly to use it' (Vogel 1986, 77). Proceedings under the PHA were also extremely rare, owing principally to the difficulty of providing proof in individual cases. The firm could exonerate itself if it could show that it had respected the bpm principle (McLoughlin 1982, 65).

In relation to central government, local authorities enjoyed a relatively high degree of autonomy in implementing statutory requirements: 'As far as day-to-day pollution control activities are concerned, central government merely gives local authority a frame-work within which there is considerable scope for creative implementation' (Hill 1983, 94). There is no central government institutional control or directive mechanisms affecting the local level. Central government had only informal control instruments at its disposal, such as circulars on calculating the required height for chimneys, which were in no way binding (Knoepfel/Weidner 1985, 181). Intervention was a last resort: 'The harshest measure which the central administration has at its disposal is the ability to "call in" the approval process, placing it under its immediate supervision' (Weidner 1987, 84). Local authorities had the problem, though, that they had relatively little power to control industrial emissions. They had no anticipatory regulatory instruments they could use. Judge-made law and the PHA required a nuisance to have occurred before proceedings could be instituted.

There were frequent coordination difficulties between the central and local levels when industries under AI supervision caused local nuisances. There was little local authorities could do in such cases. Although the Control of Pollution Act 1974 (CPA 1974) stipulated that the environmental health officers might enter such works and carry out measurements, only the data that was also received by the AI could be released to the public. Moreover, firms could appeal to the Secretary of State for the Environment if these measures gave rise to extraordinary costs, if data were not immediately available, if they were not in the public interest, or if trade secrets were affected. A further limiting factor was that the local authorities were obliged in such cases to set up a consultative committee on which local industry and the AI were represented. Because of these restrictive conditions, very few authorities made use of the legal possibilities (Knoepfel/Weidner 1985, 159f.; Frankel 1978, 68). All these factors show how little competence was vested in local authorities, which were consequently unable to react to complaints from the public — who generally turned to them first — by taking appropriate measures or providing the relevant information.

The close cooperation between the regulatory authorities and industry left little room for effective participation by other actors, such as environmental protection organizations. Implementation of British clean-air policy tended rather to take place in 'family-like ... close-knit groups of experts proud of their traditions and the trust placed in them by the public', writes O`Riordan (1979, 239):

Both the negotiations between the Alkali Inspectorate and trade association officials to determine the best practicable means for each industrial process and the actual enforcement of the presumptive limits that are established take place strictly in private. ... At the local level, environmental health departments are required to get the approval of committees of elected representatives before taking legal action against a company, but there is no requirement that these discussions take place in public. As a result, the opportunities that environmental pressure groups enjoy to influence the control of pollution in Britain are extremely limited (Vogel, 1986, 92).

This general informality and the insulation of the British regulatory system against the outside world is most clearly apparent in the extremely restrictive public information policy pursued by the AI. This policy, which was by no means a legal imposition, was given statutory status by the HSWA 1974. This Act forbids the Inspectorate to release emission data on registered works to the public without authorization from the works concerned (Frankel 1978, 67). The HSWA thus institutionalizes long established administrative practice in the AI, provoking the comment from journalist John Tinker: 'In Britain, today, data on environmental emissions are guarded more closely than military blueprints' (1972, 530)[81]. According to the AI, such data were not very informative for the public: 'I am a great believer in informing the public, but not in giving them figures they can't interpret. You would get amateur environmental experts and university scientists playing around with them. People can become scared of figures, they can get the wind up' (Chief Inspector, quoted in Tinker 1972, 533). The supervisory authorities' restrictive information policy necessarily diminished the possibilities for environmental organizations to exert pressure on the implementation of British clean-air policy. 'Everybody says [the inspectors are] very good, but how can we tell if they don't publish the figures?', asked the FoE director (quoted in Vogel 1986, 97).

Despite the opaqueness of the regulatory system, there was no broad public pressure of any significance to change existing arrangements 'There is in Britain today no significant domestic pressure to change the way British pollution-control policy is either made or enforced. Complaints about pollution tend to focus on particular sources, not on the system of regulation itself' (Vogel 1986, 101). The lack of conflict potential was ultimately attributable to two factors. First, the supervisory authorities were largely unknown to the public — 'most people in the United Kingdom have never heard of the Alkali Inspectorate' (ibid.); and second, successes had in fact been chalked up over the past thirty years, especially in abating emissions of 'black smoke': 'In general air pollution is not regarded as a serious problem by the public and there is little public anxiety about it, perhaps because the single most visible component of air pollution — smoke — has declined' (Perry 1981, 135).

Conclusion

If one looks at British clean-air policy in the early eighties as a whole, the picture that emerges is one of informal, in camera negotiatory practice, which tied the need for emission-avoidance not to generally applicable emission standards but to the local environmental situation and the economic feasibility of abatement measures for the firm concerned. Individual determination of the control technology to be used — the best practicable means — was a matter of cooperative and confidential dialogue between the regulatory authorities and industry, scarcely intelligible to the general public because of the restrictive information policy pursued by the authorities. This hampered effective intervention by environmental organizations, whose influence was in any case limited by the institutional peculiarities of the British governmental system. Given the predominance of 'soft' regulatory instruments, legal proceedings were an extremely rare event. Typically, every effort was made to avoid confrontation in order to sustain the prevailing atmosphere of confidence. In contrast to the AI, local authorities, which were responsible for all technically less complicated industrial processes, disposed of no preventive powers whatsoever. They could intervene only if damage or nuisance had already occurred. Only in the field of domestic heating did they have an instrument, the designation of smoke control zones, which gave them more extensive possibilities for abating private smoke emissions. Regulatory practice in the clean-air field was definitively shaped by the institutional background of the British governmental and administrative tradition, which encouraged consensual bargaining rather than classical top-down regulation by administrative law. The network establishment phase had also had a vital influence on practice. The bpm principle developed in the context of a more than century-old tradition to become the central instrument of British clean-air policy. Moreover, its components reflected the fundamental characteristics of British problem perception. Taking account of local environmental quality and cost-benefit considerations within the framework of the bpm principle were key elements of this philosophy.

Compared with the two principal pugilists in the European ring, Germany and Britain, with their widely differing regulatory systems, France takes a peculiarly aloof, albeit well-disposed attitude, the effect of the particular nature of French clean-air policy.

2.3 France: Regional Diversity and Variation in Policy Instruments

Clean-air policy in France goes back to the eighteenth century. Even at this early date, the French government issued decrees limiting emissions from industrial plants. At the beginning of the present century, this legislation culminated in the Industry and Trades Act[82] of December 1917 (Rest 1986, 39). To protect the population from the pollutants emitted by industrial plants, the Act defined distances to be maintained between industrial and residential areas in the framework of a three-class schema.

French clean-air policy continues to be characterized by area-related measures. How intervention by the state is determined in France by the specific French perspective on environmental issues and by the various institutional and instrumental preconditions is the topic of following sections.

2.3.1 Problem Perception as the Basis for State Action

Environmental pollution is regarded as a regional problem in France. Only in regions where disproportionate concentrations of pollutants occur are regulatory measures taken. Rural areas with little industry are largely exempted from environmental protection measures. This limitation of French regulatory activity to industrial agglomerations means that in the clean-air field, for example, there are no supraregional, binding emission standards as in Germany. Instead, as in Britain, permissible pollutant emission volumes are fixed at the regional level.

The French problem-solving philosophy, like the British, is concerned with harmful impacts on the environment. However, whereas the British want to see scientific evidence of causal links between environmental degradation and the release of certain substances before they are prepared to intervene, the French take environmental protective measures before the event. Although the precautionary principle is not enshrined in law and finds application almost only in agglomerations (see Olier/Jarrault 1988, 3), the French approach corresponds more strongly in this respect to German regulatory thinking. Another characteristic of the French problem-solving approach is based on the polluter-pays principle (*pollueur — payeur*). Although this principle has also not been embodied in law and does not apply to all environmental sectors as it does in the Federal Republic, it is applicable at least in water pollution control and in the clean-air field.

Leaving aside these far from insignificant parallels with British and German approaches, French environmental regulation has a distinctive feature: the instruments available for reducing emissions are more numerous and varied than in the other countries. The controlling authorities have a wide range of regulatory means at their disposal for taking the action on the spot deemed appropriate for combating an environmental problem or a given instance of pollution. The extent to which such an approach has been feasible and 'rational' in France can be assessed against the background of the following factors.

Long-Term Factors in Problem Perception

Among the major factors influencing French perception of clean-air issues are the country's geography, settlement and industrial density structures, and the structure of the energy sector.

Geographical Position: Emissions and Ambient Air Quality

France's geographical position makes it far less difficult to take regionally focused clean-air measures than in Germany encircled by neighbours and with poor 'air-conditioning'. In northern France, the west winds sweeping in from the Atlantic favour the rapid dispersion of pollutants. In the southeast, the famous 'Mistral' blowing from the north-west contributes to dissipating pollutant emissions (*IUAPPA* 1991, 135).

There are few areas in France subjected to heavy air pollution because of geography. High levels of pollution arise only in industrial areas situated in valleys. Lyon, for example, embedded between the Alps and the Massif Central and suffering thermal inversions in winter, and Strasbourg between the massifs of the Black Forest and the Vosges experience considerable environmental pollution. The Alsace region is situated in a basin, which acts as a pollution trap especially in unfavourable winter weather. Moreover, Alsace is the recipient of transboundary air pollution transported into the region by north-eastern air currents (interview *ASPA*, July 1993).

Taken as a whole, however, air-quality in France is a great deal better than in Germany. Not only is appreciably less pollution imported, but also smaller amounts of pollutants are emitted — which certainly cannot be said of Britain.

The emission situation in France has improved substantially in recent years. A 40 per cent reduction in SO_2 emissions was achieved between 1980 and 1985 alone, primarily attributable to the increased use of nuclear energy. Only NO_x emissions have risen slightly over the past 15 years due to the increase in traffic volumes. A glance at the 'trade' balance confirms what France's geographical position suggests, namely that France exports considerable more pollutants than it imports from its neighbours. This 'trade surplus' is diminished in the case of SO_2 only by the transborder pollution emanating from Germany and in the case of NO_x through British emissions. French emissions primarily affect Germany and the surrounding seas, albeit to a lesser degree.

Table 10: Pollutant Emissions in France — 1,000 tonnes/year

	1975	1980	1985	1990
Sulphur dioxide SO_2	3,300	3,348	1,451	1,200
—Power stations	950	1,222	408	313
—Industry	1,250	1,065	369	259
—Domestic heating/others	650	633	373	302
—Industrial processes	350	302	194	179
—Transport	100	126	107	145
Nitrogen oxides NO_x	1,400	1,646	1,400	1,490
—Power stations	230	287	134	105
—Industry	240	207	111	81
—Domestic heating/others	110	122	101	97
—Industrial processes	190	170	144	146
—Transport	830	860	910	1,060
Carbon dioxide CO_2	460,000	503,000	388,000	381,000
—Power stations	80,000	112,000	53,000	43,000
—Industry	175,000	95,000	70,000	62,000
—Domestic heating/others	125,000	139,000	113,000	103,000
—Industrial processes	n.a.	65,000	49,000	48,000
—Transport	80,000	93,000	103,000	125,000
Dust	485	435	295	233
—Power stations	75	104	50	14
—Industry	55	39	23	11
—Domestic heating/others	45	38	27	14
—Industrial processes	275	214	149	121
—Transport	35	40	46	74

Source: *CITEPA* (107)1992, (110)1993, *Ministère de l'Environnement* (1991a)

Table 11:   French Balance on Sulphur Dioxide[1] — in 100 tonnes sulphur/year

Country	Imports			Exports		
	1985	1988	1991	1985	1988	1990
Belgium	236	204	211	112	113	131
Germany	1,136	436	578	550	454	509
Britain	608	558	429	85	64	107
Italy	289	178	262	96	148	160
Mediterranean	0	0	0	548	365	392
North Sea	53	49	47	384	358	430
Austria	0	3	5	118	77	80
Poland	160	39	67	85	80	94
Switzerland	19	14	19	116	98	101
Spain	328	401	324	153	108	114
CSSR	185	71	128	69	56	63
USSR	4	7	11	134	98	103

1) Only the most important countries are taken into account.

Source: EMEP/MSC-W Report 1/1992

Table 12:   French Balance on Nitrogen Oxides[1] — 100 tonnes of nitrogen/year

Country	Imports			Exports		
	1985	1988	1991	1985	1988	1991
Atlantic	45	51	40	299	240	412
Germany	462	320	343	376	391	449
Britain	311	286	258	54	54	89
Italy	127	84	132	170	158	186
Mediterranean	0	0	0	407	306	352
North Sea	34	32	33	180	220	242
Poland	49	12	22	82	94	114
Spain	16	152	107	125	101	115
USSR	7	3	4	114	126	139

1) Only the most important countries are taken into account.

Source: EMEP/MSC-W Report 1/1992

Thus far, the picture for France appears favourable. The air-quality situation is relatively good. Strongly polluted areas occur only in isolation, so that regionally focused clean-air measures suffice to contain pollution. Little is imported from other countries. In contrast, relative large quantities of pollutants can be exported without the fear of neighbouring countries complaining, since most pollution is more or less 'lost at sea'. Since releases by

French industrial plants cause few problems abroad, France is under little pressure to take action from both the internal and foreign policy points of view.

Settlement and Industrial Density Structure
Population density in France, which at an average of 91 inhabitants to the square kilometre is very low, varies considerably from region to region. On a quarter of the total national territory there are only 20 inhabitants per square kilometre. Two thirds of all départements are even below the national average (*Statistisches Bundesamt, Länderbericht Frankreich* 1992, 29ff.). This uneven distribution is a consequence of the increasing rural exodus, which has urbanized 70 per cent of the French population. In order to slow down the exploding growth of Paris, which as *the* metropolis has always exerted the greatest attraction, the French government attempted on the initiative of de Gaulle to create '*métropoles d'équilibre*', 'equilibration metropolises' (Parodi 1971, 229; Laborie 1985). In the mid-sixties the regional planning authority *DATAR*[83] was set up to enhance the attractiveness of the 'province' vis-à-vis Paris with the aid of infrastructural measures and the establishment of industries. 'To a certain extent this strategy was thus inspired by the theory of unequilibrated growth, since it made a strict selection of a small number of cities that were to play the same role in relation to their vast hinterland as that taken by Paris vis-à-vis France as a whole'[84] (Parodi 1971, 229). The principal beneficiaries of these endeavours were Marseille, Lyon, Toulouse, Bordeaux, Strasbourg, Rennes, Clermont-Ferrand, Dijon and Nice and the conurbations of Lille-Tourcoing-Roubaix, Nancy-Metz and Nantes-Saint Nazaire.[85]

In contrast to the broad distribution of industry in Germany, industry in France in concentrated in a few large agglomerations. The main areas concerned are the region Nord-Pas-de-Calais, Greater Paris, the eastern areas of Alsace-Lorraine, the upper Rhône valley around Lyon and Saint-Etienne, and Bordeaux and Toulouse in the South-west (see *INSEE* 1988). Most French industry is located east of the line Le Havre — Marseille.

The Paris region, which generates 25 per cent of gross domestic product, is the site not only of banks, insurance companies and large groups but also of industrial production in electrical engineering, the chemical and motor industries, and aircraft construction. In the North and East, coal, iron and steel, and textiles are the principal industries. In the South, in the Rhône-Alpes region, the power, chemical and textile industries have established themselves (Große/Lüger 1989, 100). Apart from some agglomerations like Nantes, Toulouse and Bordeaux, the West and South-west have retained their agricultural structure despite the accelerated development of urban areas (*IUAPPA* 1991, 136).

In accordance with the geographical distribution of industrial agglomerations, air pollution problems are particularly serious to the east of the line Le Havre — Marseille. The industrial metropolises like Paris, Lille, Strasbourg, Lyon and Marseille are especially affected by environmental pollution. There is very little air pollution in the rest of France, and the issue is therefore not perceived as a serious one. It is hence no wonder that France has opted for regional concepts in combating air pollution. For, unlike in Germany, clean-air problems are not virulent throughout the country but are concentrated in particular areas. Here the gravity of the problem, the concern of the population, and the resulting public pressure is great enough to induce the authorities to take appropriate measures. Since the problems arise at certain points only and in concentrated form, the authorities can deploy a range of instruments adapted to local conditions.

The Structure of the Energy Sector
Over half the power consumed in France in the early sixties was produced by coal. One third was obtained from oil. With the fall in the oil prices at the beginning of the seventies, the ratio reversed. Since in comparison to German and British coal deposits, those in France are less abundant, difficult to exploit, and of low quality (Menysch/Uterwedde 1982, 50), the French government sought to satisfy the growing demand for power by increasing oil imports. When the first oil crisis broke in 1974, two-thirds of French energy consumption was being covered by oil (*Ministère de l'Industrie* 1991, 23).

The sudden rises in the price of oil in the mid-seventies were an unexpected and severe blow to the French economy because at that time France was importing almost 80 per cent of its entire energy raw materials (Lebas 1981, 92). Immediately following the first oil-price shock, the French government moved to avoid further exposure to the uncertainties of oil-price developments on the world market by concluding an unprecedented nuclear energy programme with the state utility *EDF* (*'Electricité de France'*), which envisaged constructing 50 nuclear power stations by the year 2000 (Leggewie/de Miller 1978, 16)[86] — a project that would have been unthinkable in Germany because of opposition from the Greens and the anti-nuclear movement.

The greater part of French power needs is still met by nuclear energy (75 per cent). Only 12 per cent of electricity is now generated by fossil fuels, while 13 per cent comes from hydroelectric power stations (see EDF 1990, quoted in Berg 1992, 9). Owing to these energy policy measures and the accelerated construction of nuclear power plants, France was able not only to lower its energy prices but also appreciably to diminish releases of the central environmental poison SO_2.

Medium and Short-Term Factors in Problem-Perception

Perception of the air-pollution issue is influenced by both long-run variables and medium and short-term factors. They include the economic situation, environmentally relevant events, public environmental awareness, international environmental policy, advances in environmental technology and changes in the environmental technology market.

The Economic Situation

Industrialization occurred very much later and developed much more slowly in France than in Britain and Germany. The aftermath of the Second World War — high public indebtedness, devastated industrial centres, and technological backwardness — was clearly apparent well into the sixties. To overcome these problems, top priority was given to opening up the economy (*'impératif industriel'*) and increasing international competitiveness (Hall 1986, 148). The government provided impetus and initiated 'state-led development' (Zysman 1983, 115c) with the aim of attaining a satisfactory economic level. 'As a player [the French government] pursues specific outcomes on a case-by-case basis, assembling packages of incentives which can be used to persuade or coerce. It discriminates among firms and applies administrative rules and regulations in order to achieve particular objectives' (Zysman 1983, 75).[87] The forces for modernization like Jean Monnet and Étienne Hirsch embraced the concept of a 'concerted economy', where the principal problem was 'how to lead an immense collective effort without being in control in decision-making in the state or in industry' (Jean Monnet quoted ibid. 1986, 27). The Ministry of Finance with its treasury and budget sections and the state banks (Hayward 1986, 22), which give the French economic policy community an overall pluralist and fragmented stamp, (ibid., 23) have intervened particularly actively in the economy.

Table 13:   Outline Data on the Economic Situation in France — percentage change over the previous year

	1973	1975	1978	1981	1984	1987	1990
Economic growth	5.4	0.2	3.4	1.2	1.5	2.2	2.6
Unemployment	2.6	4.1	5.2	7.4	9.7	10.5	9.6
Inflation	7.3	11.8	9.1	13.4	7.4	3.1	3.4

Source: OECD Historical Statistics 1960-1990 (1992), 1960-1988 (1988), 1960-1982 (1984)

On the basis of five-year plans, *'planification'*, the government set the modernization of economic and social structures in motion. The initiation of industrial merger and concentration processes produced new 'national champions' in the motor, steel, and chemical industries. Selective government subsidization of forward-looking industrial sectors such as aerospace and the nuclear and computer industries, promoted economic growth. Development was also aided by the bridging of the sacrosanct gap between private and public sectors with *grands corps* civil servants, especially from the *Corps des mines* moving to technologically advanced industries (Suleiman 1978, 209). 'This development has important implications for the economic and industrial policy process ... The entrepreneurial state and the statized enterprise concentrate their joint energies and resources upon industrial development, focused upon the firm. Under the remorseless stimulus of foreign competition, the mobile members of the grand corps became the heads of the "national champion" enterprises, dedicated not so much to compete inside France as with foreign firms....' (Hayward 1986, 30). Hence, in the sixties and seventies, France enjoyed the highest growth rates in the world and developed into one of the leading industrial countries in Europe (Braudel/Labrousse 1980, 1011f.).

Already in the mid-seventies, however, government planners expressed their misgivings about the feasibility of ensuring the international competitiveness of individual industries through focused support for one or two groups. It eventually became apparent that the steel and chemical industries, despite massive government subsidies, had not attained the efficiency needed to maintain their international position (Hall 1986, 149). The world-wide economic crisis triggered by the 1974 oil price shock revealed the weaknesses of this selective industrial policy, and brought on an economic downswing. Traditional industries like clothing and textiles were ousted from the marketplace by growing competition from newly industrialized countries such as Taiwan and South Korea (Hall 1986, 183). French inferiority to other suppliers in the world market became apparent particularly in the capital goods sector, which in comparison with the motor vehicles and electronics industries profited less from public contracts and research programmes (Dubois 1987, 13f.). The weaknesses in French industrial policy, with selective support for individual sectors and enterprises, now took effect. In a context of structural adjustment problems and falling growth rates, investment dropped rapidly — by 15 per cent between 1974 and 1981 — while unemployment and inflation rose sharply (see table 13).

Although the economic crisis imposed restrictions on French environmental policy, too, ecological issues did not entirely go under. A study on the long-term perspectives for the environment commissioned in the early seventies by the then Prime Minister Chaban-Delmas resulted in an environmental programme defining a 'hundred measures for the protection of the environment'. Six months later, the French government replaced the originally envisaged 'High Committee for Environmental Affairs' (*'Haut Comité de l'Environnement'*) and an 'Interdepartmental Action Committee on the Protection of Nature and the Environment' (*'Comité Interministériel d'Actions pour la Nature et l'Environnement'*) by an independent Ministry for the Protection of Nature and the Environment (*'Ministère de la Protection de la Nature et de l'Environnement'*) (Bungarten 1978, 40; Morand-Deviller 1987, 16). Immediately after having been established, the Ministry directed its attention to the revision of all existing decrees and regulations. The next task was to obtain amendment of the 1917 Industry and Trades Act. The aim of this amendment was 'to create a coherent legal basis for integrated clean-air policy concepts' (Knoepfel/Weidner 1985, 24). Although the Environment Ministry was in a relatively weak position vis-à-vis the Ministry of Agriculture and the Economics Ministry, it nevertheless achieved remarkable environmental policy successes, which found expression in the adoption of new decrees and regulations and in the updating of the Industry and Trades Act.

National Environment-Related Events

Whereas forest dieback gave very significant impetus to environmental policy in the Federal Republic, and the London smog mobilized policy activities in Britain especially in the clean-air field, France had experienced no notorious environmental situations that could have had a comparable effect. However, various accidents in the private sector induced the authorities to supplement existing provisions. Thus in the mid-sixties there was an explosion in an oil refinery in Feyzin that killed a number of people. The government thereupon decided to intensify the control of industrial plants, and for this purpose enlarged the powers of the controlling authority of the time, the *'Service des Mines'*, now *'DRIRE'*[88] (interview with *DRIRE*, November 1992).

Public Environmental Awareness and the Environmental Movement

Despite the absence of 'internal shocks', environmental awareness has grown markedly in France over recent years. Public sensitivity to environmental hazards has been enhanced especially by global environmental issues like the greenhouse effect, and by the rise of the green parties in the early eighties.

Increased public ecological awareness found expression primarily in activities at the regional level. Individual action groups formed to address local environmental problems. Whereas most industrial plants had previously obtained a permit without difficulty, there is now hardly a licensing application to escape opposition (interview with prefecture, June 1993). However, the extent to which environmental protection groups at the local level are able to exert decisive influence on clean-air policy depends in each case on the constellation of actors involved and on the degree to which the local population are affected and can be mobilized (Knoepfel/Larrue 1985, 53).

At the supraregional level a strong environmental movement arose in the mid-seventies in opposition to the government's nuclear power programme (see Leggewie/de Miller 1978). But no environmental movement addressing environmental pollution issues at the supraregional level emerged as in Germany and Britain. The reason is clear: environmental problems in France are not nation-wide phenomena but arise in single regions, where they are intense. In dealing with them, the authorities take measures restricted to the area concerned; as a consequence, the public apparently also tends to regard such problems as requiring local solution. It is therefore hardly surprising that only one environmental association is active at the national level, *'Les Amis de la Terre'*.

International Environmental Policy
In the international environmental policy arena France was neither a moving nor a braking force. On most issues the French were disinterested but well-disposed. The debate emerging in the late seventies on the 'acidification' of Scandinavian lakes, for example, did not affect France because in contrast to Britain it was not a major pollutant exporter, and, with its comprehensive nuclear power programme, had already substantially reduced SO_2 releases.

Whilst the topic of 'forest dieback' became a political issue in Germany in the early eighties and set an internal policy dynamics in motion, there was little scientific and political discussion in France. French forests were neither as seriously degraded nor culturally nearly as salient as German forests (Neumann/Uterwedde 1986, 56). And French environmental groups displayed little commitment to saving the forests, since their interests and resources were largely engaged in combating the nuclear power programme of the French government (Roqueplo 1986, 405).[89]

In 1984 the French government finally adopted a programme to combat 'acid rain', under which it committed itself to reducing SO_2 emissions by 50 per cent and hydrocarbons by 30 per cent over a period of 15 years. These figures went far beyond what had been required under the 1979 Geneva Convention on Long Range Transboundary Air Pollution (*Ministère de l'Environnement* 1991a, 10).

Although the attitude of the French government in environmental discussions at the international level tended to be dispassionate, it supported the agreements negotiated at the international environmental conferences. France thus put its signature without demur to the 1985 Helsinki protocol on the reduction of sulphur emissions, the 1987 Montreal protocol on carbon dioxide emission abatement, and the 1988 Sofia declaration on the reduction of nitrogen oxide emissions (Herz 1989, 1; Mousel/Herz 1990, 69).

Technological Progress
As mentioned above, the French government in the seventies intensified the development of individual industrial sectors into 'national champions'. Whilst the motor, steel and chemical industries profited especially from these development programmes, other industries were left by the wayside. This is one reason why France — unlike the Federal Republic in particular — has no well-developed environmental protection equipment industry apart from the field of waste disposal (Neumann/Uterwedde 1986, 158).

The regulatory instruments deployed in France for abating emissions offer a further explanation. Whereas Germany, with its best-available-technology approach prefers *plant-related* measures, France often takes action already at the product level. The *product-related* specification of what pollutant concentrations input materials may contain renders superfluous the development and construction of elaborate filter systems. Upstream regulation of this type makes less demands on the equipment of production facilities. Given the lower domestic demand in France, it is obvious that the manufacture of environmental technology for industrial plants is less widespread and less developed.

Changes in the Environmental Technology Market
From the early eighties, France sought to catch up technologically and economically in the environmental protection equipment sector. In 1981, the French government established the '*Agence de la Qualité de l'Air*' (*AQA*), with the task of coordinating and promoting environmental technology developments (see Pezet 1984). Although the *AQA* took important initiatives in introducing new technologies and provided financial backing for pertinent state and private research programmes, no market breakthrough was achieved. In the clean-air field, France continues to depend on foreign, especially German suppliers.[90] France has a leading position in Europe only in the field of waste disposal (interview with *Les Verts*, June 1993).

Conclusion

From the picture we have drawn it appears that the French government has had an easy time in clean-air policy. With relatively little effort it has proved possible to keep the air at least as clean as in Germany and Britain. This has been no legerdemain. A range of factors have contributed to mitigating matters.

To begin with, the French can consider themselves lucky with their geography. Thanks to the fresh wind whistling about their ears, the pollutants the French discharge at home are generously distributed beyond the borders. On arrival, pollution is mostly deposited not on annoyed neighbours but on uninhabited areas (at sea). Because of the territorial concentration of industry, the pollution that stays at home — relatively limited in quantity due to the government's extensive nuclear power programme — presents a nuisance only for a fraction of the population. And this pollution can be dealt with by means of selective, regional measures. Where regulation does not appear to bring the necessary results, environmental groups form, which articulate their objections within the context of permitting procedures etc. In contrast to Britain and Germany it is conspicuous that, although environmental awareness has increased in France over the years, there has been no environmental movement at the supraregional level. The influence of environmental associations on governmental policy decision-making has been correspondingly limited.

France fares hardly worse in the international arena. Since it neither causes significant problems nor suffers unduly from transboundary emissions, it has no occasion for excessive commitment. However, this does not mean that France is on the sidelines. The important agreements adopted at the international environmental conferences also bear France's signature. France would dearly love to play a more important role in another international arena. Since the eighties, the French government has been endeavouring to make the environmental protection equipment industry, which had been neglected in the past, internationally competitive. But developments in environmental technology overtook France unawares. With the selective promotion of the motor, steel, and chemical industries, and with primarily product-related emission abatement measures, investors had little incentive to take action in the environmental protection goods sector. France is nevertheless leading in waste disposal.

The interplay among these various factors offers an initial explanation for the French approach to air-pollution as a primarily regional problem. This perspective, which is taken not only by government but also by environmental associations, culminates in problem-processing and solving strategies that also relate strongly to local conditions.

2.3.2 Institutional and Instrumental Preconditions for State Action

State action by member countries in the European context is to be seen not only in the interplay of factors determining specific perception of environmental issues. In order to understand the negotiating positions of individual countries and how they influence one another and affect the state, it is also necessary to record and analyse the institutional and instrumental preconditions for regulative action.

Political Initiative and State Action: Institutional Prerequisites

The degree to which political energy is mobilized with respect to institutional prerequisites is, in the terms of our hypothesis, determined essentially by the number of political arenas, the electoral system, and thus the access of political parties to decision-making, and by the role of the courts. These factors are to be analysed in the following sections.

Regional Initiative in the Unitarian State

As in Britain, there are few political arenas in France providing a venue for the development of political initiative. The concentration of power in British central government finds its counterpart in the strong position of president and government in the French governmental system.

During the III and IV Republics, parliament was still the core of governmental power (*'régime d'assemblée'*), but parliamentary influence was strongly restricted under the V Republic[91] (Grosser/Goguel 1980, 260). The president of the republic profited most from the loss of power by parliament. The French president, in contrast to his western 'colleagues' became the most influential figure at the national level.[92] In the period between 1958 and 1986, no western head of state or government — not even the president of the United States of America — had so much power as the French president (Mény 1993, 232). Although the constitution lays down that internal affairs are the domain of the government, until 1986 they were *de facto* in the hands of the president, since he defined the essential scope and focus of policy (Mény 1993, 233; Große/Lüger 1989, 37). It was only the advent of 'cohabitation' following the 1986 general election, when the government was no longer formed by the party of the president but the large oppositional bloc of parties, that the president's influence diminished markedly, to concentrate thenceforth primarily on foreign affairs and security (Große/Lüger 1989, 141).

France's centralist constitution is a further hindrance to the generation of political activities, which long gave the regions and départements little margin for initiative. The idea behind the centralist structure of France's political system is a quasi-military organizational concept, the aim or 'illusion' of which is that instructions and information can be directly transmitted to central government by innumerable agents. However, the reality of the comprehensive links between the prefect, the representative of central government, and local elites was not in accord with this concept (Mény 1993, 246).

Unlike in Germany, there is no constitutionally guaranteed local self-government in France. However, Article 78 of the Constitution of the V Republic does guarantee the *'libre administration'* of local authorities. In a range of rulings, the Constitutional Court laid down that there are limits to intervention by central government that may not be exceeded even by individual statutory decisions.[93]

Until the late seventies, the indirectly elected regional councils and the directly elected *département* councils (*'conseil régional'* and *'conseil général'*)[94] had no executive powers at all. Although the *département* had the power to adopt its own budget, the implementation of its decisions was in the hands of the prefect (Grosser/Goguel 1980, 63). The entire power of decision was accordingly vested in the respective prefect, who still today acts as the representative of the state.[95]

The prefects in their areas like the prefectoral corps in their declared view insist that it is their vocation to coordinate, integrate, and mediate collective action in their *département*. They dispose of considerable legal means for the purpose. [Nevertheless], the prefects cannot rely solely on their regulatory resources to ensure the organizational coordination of the administrative apparatus; still less are they able to do so if they wish to bring influence to bear on the local environment. Intervention by the prefect, his capacity for arbitration, and his integrative potential depend directly on the structure of elite leadership in the *département* (Grémion 1976, 222).[96]

It was only the decentralization legislation adopted at the beginning of the eighties that enlarged political scope for action at the regional and local levels, thus coming closer to the model of local democracy (Fromont 1983, 339). The regional and *département* councils were transformed into genuinely representative bodies composed of members directly elected by the people. The councils in their turn elected their chief executive officer for the legislative period. In the *département* council this was the *'président du conseil général'*, given control of the administration, and thus a large proportion of functions previously in the hands of the prefect (Mény 1987, 253). Whereas the *départements* are responsible for social and health affairs and exercise certain regional planning functions,[97] the regions have been entrusted with coordination and planning functions. Regional economic development plays a particularly important role (Kukawka 1993, 20; Fromont 1983, 401).

The prefect now no longer had the right to override local authority decisions. Where he had reservations he could now only appeal to the administrative court or the *Conseil d'Etat* (Fromont 1983, 400). Although the prefect still had decisive powers of decision in many matters, the *département* councils now disposed — in at least some areas — of greater scope for action. The extent to which this scope is to be further enlarged or limited again depends on whether prefects or councils gain the upper hand in pending jurisdictional disputes (Kukawka 1993, 22).

Besides the additional powers the regions had won, the financial resources at the disposition of the *département* councils were increased (interview with Knoepfel, March 1993; interview with *ASPA*, July 1993). The changes set in motion by the decentralization legislation also affected clean-air policy. At the regional level there were now more resources available, for example, to intensify and ameliorate the monitoring of air quality[98] or to initiate pilot projects for research purposes.

The Electoral System and Political Party Access to the Political System
It is not the number of political arenas alone that decides what potential political parties have to exercise influence. The electoral system of a country also plays an essential role in determining the prospects of access to the political system for political groupings. Two different electoral systems come to bear in France. Elections to the National Assembly (*'Assemblée Nationale'*) obey the absolute majority principle, whereas the regional councils (*'conseils régionaux'*) were elected in 1986 as full *'collectivités territoriales'* for the first time in accordance with the relative proportional representation principle. The applicable system for elections to the general councils (*'conseils généraux'*) is the majority system with two ballots; only candidates winning 10 per cent of the votes on the first ballot are admitted to the second.

As a rule, no party obtains an absolute majority in elections to the National Assembly. For this reason, a second ballot is held, for which the larger parties within the various political camps form coalitions, for example the *'UDF'* (*Union pour le Démocratie Française*) with the *'RPR'* (*Rassemblement pour la République*) and the *'PS'* (*Parti Socialiste*) with the *'PC'* (*Parti Communiste*). The smaller parties bring up the rear of these coalitions of large parties. The small fry have practically no chance of winning a seat in the National Assembly. Access to the most important decision-making arena is thus denied them. The two green parties in France, *'Les Verts'* and *'Génération Ecologie'* accordingly have yet to take their place in the National Assembly and initiate environmental policy activities at the centre.

It is a different matter with elections to the regional and general councils, the *'conseils régionaux'* and *'conseils généraux'*. In the case of regional councils, the proportional representation system makes it easier for ecological candidates to gain access to regional assemblies. In the most recent regional elections, the green parties succeeded in winning a substantial number of seats (Beerwerth 1993, 347).[99] However, the greens' regional electoral successes vary widely. In areas with high pollution and pronounced public environmental awareness, the greens can hope to obtain good results. In Alsace greens won the same number of seats as the Socialists, and were therefore able to push through more stringent, 'unpopular' environmental protection measures than in other regions.[100]

However, at the decentralized level, where the area concerned is within manageable limits, it is also possible to bring environmental issues to bear even under the majority electoral system applicable for general councils. For where environmental problems are territorially concentrated in a context of relatively small constituencies, a single-member constituency can bring success. There are indeed ecologically-oriented general councils, and the personalization of the majority electoral system can limit attempts by Paris to influence general council elections substantively through party channels. This means that any weakness at the central political level, as exhibited by the green parties, need not necessarily affect first-past-the-post departmental elections.

Hence the French electoral system makes it difficult for green parties to gain access to central-government decision-making bodies. There are, however, two considerations that may relativize this thesis. On the one hand, the aggregation and mediation of societal interests take place in the political process in France not exclusively via the political parties. Ecological topics are transported in other decisional channels and raised to the national level, through, for example, the personal union of offices, the so-called '*cumul de mandats*', and through 'cross-linkage' between political representative decision-makers and their administrative counterparts. On the other hand, only very general, framework legislation is forthcoming at the national level, leaving actors on the spot with a great deal of decisional leeway. The difficulty caused by the electoral system for ecological parties to accede to national decision-making bodies is at least partly compensated by a sort of '*pouvoir périphérique*' or 'peripheral power' through influence in local 'implementation networks' (prefect, *DRIRE*, local authorities, associations, citizens' action groups) (Grémion 1976). This power depends essentially on the given constellation in local networks.

The Role of the Courts
The possibilities for political groups to influence central-government and regional clean-air policy depend not only on the form of the state and the electoral system, but also on the significance of the courts. In comparison to Germany, the courts in France are little of a check on politics and administration, since they frequently review only the legality of proceedings, and, although entitled to do so, do not base their rulings on material criteria.

As in the Federal Republic, civil law and administrative law jurisdictions have developed in France (Samuel 1978, 71). Civil law proceedings are instituted when a person claims damages against a company.[101] However, such action is extremely rare. Even where a causal connection between the injury suffered by the plaintiff and the emissions by the firm can be proved, there is no guarantee that the court will find for the plaintiff and award damages. Not only the meagre prospects for success but also the high costs such proceedings entail, far exceeding the financial capacity of a private person, act as a disincentive.

No costs accrue to the plaintiff in proceedings under administrative law, since for this type of action he may go to court without a lawyer. Within the framework of this form of proceedings, prefectoral orders, such as concrete duties imposed on plant operators, can be reviewed as to their legality. Besides duly confirming or setting aside prefectoral orders, the administrative courts are entitled to impose more stringent requirements. In the concrete formation of local clean-air policy and as a means for environmental groups to exert pressure, these courts have a potentially important role to play (Rest 1986, 14; interview with *Nord-Nature*, June 1993).

In France it is possible not only for affected parties to go to court. France is the only country of the three under review that gives associations a general right to take legal action,[102] an innovation introduced in 1976 with the Nature Protection Act (Morand-Deviller 1987, 25). Since then it has been possible for environmental groups to institute proceedings before an administrative court. With the so-called administrative courts of appeal (*'cours administratives d'appel'*), a new form of appeal court was set up that further enhanced the effectiveness of associations' right to sue (Karwiese 1987, *idem* 1989, 713). Environmental groups could now bring a suit before an administrative tribunal acting as 'counsel for Nature' without having to prove that they are an affected party. French environmental associations thus have much greater leverage in defending their interests than do British and German organizations (see Despax/Coulet 1982; Rest 1986, 14).

The Mode of State Action and the Style of Regulation

In order to identify the influences exerted by the European Union on national clean-air policy and to permit conclusions to be drawn on the respective negotiating positions of the countries under review, detailed knowledge is needed of the legal regulatory structures, policy instruments, and modes of implementation.

Legal Regulatory Structures
Since the early years of the last century there have been legal rules dealing with clean-air policy in France. As early as 1810 a decree was issued requiring the general registration and inspection of industrial plants. At the beginning of the twentieth century, in 1917, the Industry and Trades Act relating to 'hazardous, insalubrious and offensive establishments'[103] was passed, which was to remain valid in its original form for many decades until its amendment in 1976 (Igl 1976, 20).

The essential features of the French policy created by this Act persist to this day. For example, already at the beginning of the century factories were assigned to one of three categories in terms of their hazardousness and the pollution they caused. The more dangerous and polluting releases from the plant were judged to be, the further away it had to be located from the nearest residential area. Already at that time spatial aspects — proximity of the factory site to a residential area — played an important role (Igl 1976, 25).

The 1961 Air Pollution Act[104] contained the most important clean-air policy provisions. The Act was concerned not only with stationary sources but also with mobile sources of air pollution. It permitted the authorities to supervise non-registered plants as well, and provided for the selective utilization of low-pollutant fuels. However, it contained no concrete standards for ambient air quality or emissions (Rest 1986, 11). The revised Act on Classified Installations of July 1976[105] laid down only two classes. Class A (*'Autorisation'*) included hazardous and strongly polluting plants that were subject on principle to authorization. Class D (*'Déclaration'*) covered less hazardous or polluting establishments subject only to a general notification requirement (Morand-Deviller 1987, 98). Factories were assigned to a given class in terms of a constantly updated and expanded nomenclature listing all industrial activities[106] (Rengeling 1985, 116).

Concrete licensing procedure was laid down by Decree No. 77-1133 of 1977. After an application for authorization — including an *'étude d'impact'* and *'étude de dangers'*, studies on environmental impact and hazards[107] to be commissioned by the applicant — had been submitted to the prefect, the latter forwarded it to the factory inspectorate, the *'Direction Régionale de l'Industrie, de la Recherche et de l'Environnement'* (*DRIRE*) (Olier/Jarrault 1988, 4). The *DRIRE*, acting as a technical consulting body, checked whether the application documents were complete, formulated a catalogue of requirements, and returns the dossier to the prefect.

At the same time the prefect initiated the public enquiry procedure (*'enquête publique'*). This was conducted by the *'commissaire enquêteur'* or 'commissioner-investigator' appointed by the prefect. As soon as the public had been informed by the press and announcement bills of the licensing application, interested citizens and environmental groups had an opportunity to file objections within one month (Backhaus 1980, 226). To permit private persons and environmental groups to obtain prior information on the subject-matter of the permitting application, a general right to inspect records[108] was established in 1978. This permitted all documents pertinent to authorization to be inspected. The only exception were documents affecting trade secrets or commercial confidentiality (Winter 1990, 187). After conclusion of enquiry proceedings, the *commissaire enquêteur* presents the prefect with all the opinions collected, including his own assessment (Backhaus 1980, 225).

To make public hearings in the context of permitting procedure easier and more democratic, the Act on the Democratization of Public Enquiries[109] was adopted in 1983. Under this Act, the *commissaire enquêteur* was no longer appointed by the prefect but by the president of the administrative court. Moreover, it was now the task of the *commissaire enquêteur* to convene additional public meetings, to inspect sites, and to demand further documents from the applicant in order to present the public with a more comprehensive impression of the state of affairs: *'De mieux faire participer le public et de se faire une opinion plus complète sur le projet'* (*Ministère de l'Environnement* s.a., 8). Despite his enlarged competence, the *commissaire enquêteur*, whose function had yet to be precisely defined, still had very limited margin for action. If there was no environmental group in the area to inform and mobilize the public, he could set little in motion himself. He had little incentive for active commitment, anyway, since his pay was extremely low. It is therefore no wonder that the reports of some *commissaires* met only minimal standards. 'The documents are badly typed or quite frankly illegible,' complained one president of an administrative court (quoted in Le Monde, 28/29 April 1991).

After conclusion of the hearing procedure, the reports prepared by the *DRIRE* and the *commissaire enquêteur* are submitted to the Health Council, the *'conseil départemental d'hygiène'* (*CDH*). This body is composed of representatives from various occupational categories (pharmacists, physicians, architects, etc.), the competent administrative officials, and two elected representatives of the départmental council (Knoepfel/Weidner 1985, 23). If need be, this council can hear the applicant anew.

It is then the job of the prefect to make the final decision on commissioning the plant. His decision in its turn is based on the various opinions submitted by the *DRIRE*, the *commissaire enquêteur*, the Health Council, and the general public. As a rule, the prefect accepts the requirements elaborated by the *DRIRE* and the proposals formulated by the *CDH* (interview with prefecture, June 1993). On the other hand, the prefect is influenced in his decision by the Environment Ministry guidelines (Legrand et al. 1987, 58), which prescribe minimum technical requirements for certain industrial sectors, thus defining the 'state of the art' (Rengeling 1985, 22). The guidelines take the form of *'circulaires', 'instructions'* or *'commentaires'*, and consequently have no direct binding effect at law. They take effect only when the prefect incorporates them in an individual permit. This is issued by the prefect in the form of an *'arrêté préfectoral'*, a prefectoral order, comprehending all standards, safety regulations, and necessary inspections (Backhaus 1980, 227).[110]

If in the opinion of an individual or an environmental group the ecological aspects have not been sufficiently taken into account in the licensing procedure, an objection can be filed against the prefectoral order with the competent administrative court, thanks to the right granted associations in France to take legal action. 'We attack the prefectoral order before the administrative courts because it costs nothing and you don't have to have a lawyer' (interview with *Nord Nature*, June 1993). If the administrative court finds for the plaintiff, this can effect a modification of the prefectoral order, or lead concretely to the imposition of more stringent standards (Rest 1986, 12).

Policy Instruments
France disposes of a multiplicity of instruments serving to abate and control industrial emissions. Among the emission-abatement measures are the declaration of protected and alert zones, the adoption of parafiscal charges, and the conclusion of voluntary agreements.

In 1964 — at a time when France was still pursuing a tall-chimney policy, the instrument of so-called 'special protected zones' (*'zone de protection spéciale', ZPS*)[111] was introduced in Paris. Initially there was no legal basis for creating protected zones in the capital. Only with the setting up of the Environment Ministry in the early seventies was a decree (No. 74-415) adopted allowing the designation of a protected zone on application by the prefect and with the concurrence of the Environment, Energy, Construction, and Health Ministries (Knoepfel/Weidner 1985, 24). Within these special zones only low-sulphur fuels[112] could be used for domestic heating and industrial purposes. There were also special requirements regulated by prefectoral order. They included special qualification standards for personnel operating heating and combustion plants, as well as specific technical rules on the construction and operation of such installations (ibid., 25). Six protected zones have so far been designated in France.[113] They primarily cover densely populated agglomerations where air pollution emanates not only from individual large industrial plants but where domestic heating is also a major source.

The second instrument, which like the protected zone is area-related and designed to abate emissions, is the so-called 'alert zone' (*'zone d'alerte'*)

(Knoepfel/Weider 1985, 26; Morand-Deviller 1987, 115). In contrast to the protected zone, where special regulations apply regardless of prevailing air pollution, special rules come into force in alert zones in the event of valid air-quality standards being exceeded. The arrangements laid down in Decree No. 74-415 include rules on the use of low-pollutant fuels and on production curbs. It is incumbent on the respective prefect to decide what areas are to be declared alert zones and what air-quality standards are material (*Ministère de l'Environnement* 1991, 4). As a rule, however, alert zones are declared where air pollution emanates from a few large sources and is easy to regulate by selective measures (interview with *AREMA*, June 1993). There are now eleven such alert zones in France.[114]

In the mid-eighties, the French Ministry of the Environment braved the opposition of the Ministry of the Economy[115] in introducing a further emission-abatement instrument that had already proved its worth in water pollution control: the 'parafiscal charge', '*taxe parafiscale*'. Since July 1985, power stations with a capacity greater than 50 MW, and other industrial plants with annual SO_2 or NO_x emissions higher than 2500 tonnes are required to pay a charge at a rate of 130 francs per tonne of pollutant released. At the time, a total of about 480 plants were affected by the charge. It brought the Environment Ministry additional revenues of about 100 million francs (*Ministère de l'Environnement* 1991a, 7).

The original idea had been to impose the charge for only a five-year period. However, in May 1990 it was renewed for a further five years to December 1994, since it was then the most important economic incentive for reducing air-pollution in France. '*Ce dispositif constitue la principale mesure d'incitation économique actuellement en vigueur en France pour la réduction de la pollution atmosphérique*' (Olier/Jarrault 1988, 4). The May 1990 decree not only prolonged use of the charge but also expanded the schedule of liable industrial plants and pollutants, as well as increasing the amount to be paid. Power stations with as low a capacity as 20 MW and domestic refuse incineration plants with a capacity in excess of 3000 kcal per hour now have to pay the charge. The number of liable plants thus increased to about 870. The price per tonne of pollutant emitted was raised from 130 francs to 150 francs, so that some 180 million francs in revenue accrues per year (Mousel/Herz 1990, 71). Much of government revenue from the charge (about 87 per cent) flows back to industry — rapidly and unbureaucratically — in the form of investment aid (interview with *AREMA*, June 1993). However, a plant operator receives financial support only if the projected investment serves to reduce emissions.[116] The funds are distributed by an administrative committee chaired by the Environment Minister. A further ten per cent of the money is spent on technical development, and about three per cent on administration, going to the competent authority, the *ADEME* (*Agence de l'Environnement et de la Maîtrise de l'Energie*) (*Ministère de l'Environnement* 1991a, 7; Olier et al. 1989, 54). The money levied from plant operators does not, however, flow entirely into the public purse. Firms have the option of paying a proportion of their dues to the local air-quality monitoring association (Olier et al. 1989, 55). Although the effectiveness of the charge as a contribution to emission abatement is considered relatively insignificant even by the Ministry of the Environment itself, the charge is of crucial importance at least from a fiscal point of view. The charge is an important source of finance especially for technological research and development and subsidization purposes. 'There are two charges [water-pollution-control and clean-air charges] with the same purpose; they're not dissuasive because they aren't high enough to force industrialists to modify their production processes, they're to finance the agencies carrying out anti-pollution operations'[117] (interview with *Amis de la Terre*, June 1993).

A further instrument for reducing industrial emissions was created in the early seventies, the so-called *'contrat de branche'* or sectoral contract. Such agreements, which are limited in time, are concluded between the Environment Minister and representatives of industry.[118] The aim is to reduce industrial pollution of the environment on a voluntary basis. As a rule, the contracts cover more than one medium, addressing both air and water pollution. However, the focus of the measures is on water pollution control. In the clean-air field, contracts have so far been concluded only on suspended particles, so that the instrument does not play an outstanding role in curbing air pollution (Knoepfel/Weidner 1985, 28; Rengeling 1985, 124), and is regarded with some reservation by Environment Ministry officials (interview with *Ministère de l'Environnement*, June 1993).

The air-quality monitoring associations, the so-called *'associations de gestion de réseaux de mesure'* are responsible in France for monitoring harmful environmental impacts, which occur primarily in strongly polluted areas (Delandre 1991, 377; *Ministère de l'Environnement* 1991a, 27). The associations are generally initiated and set up by the local *DRIRE*. Although the first was established only in the mid-seventies, air-quality monitoring has a longer tradition in France. The first measurements of sulphur dioxide, suspended particulate matter, and carbon dioxide were motivated by the 1952 smog alert in London. However, measurements were taken only selectively in Paris, Lyon and the Region Nord-Pas-de-Calais.[119]

The *APPA*, *'Association pour la Prévention de la Pollution Atmosphérique'* was founded in 1958 to assure the continuous and nation-wide monitoring of ambient air quality. By establishing their own regional offices and with the support of local political personalities, such as the mayor of Marseille Gaston Defferre, the *APPA* succeeded in setting up monitoring stations in many cities against the will of industry (Samuel 1989, 1). Thus, by the beginning of the seventies, the *APPA* had built up a substantial monitoring network covering an area with an urban population of more than seven million.

One serious problem in operating the stations was to secure their funding, which came largely from local industry. In regions like Saint Etienne where many factories were closing down, the monitoring stations could not be maintained and had to cease operations. No financial support was forthcoming from the government, since air-pollution was superficially regarded as a problem caused by SO_2, which more or less 'disappeared into thin air' with the shutting down of the factories in question, so that continued monitoring appeared to be beside the point. This circumstance is one explanation for the uneven distribution of measurement networks in France. Some regions are fully covered by monitoring networks while others have not a single station (interview with *Les Verts*, June 1993). However, the number of monitoring associations has continued to grow. In 1988 there were only 23 of them, whereas there are now 28 with more being planned (Samuel 1989, 3; *Ministère de l'Environnement* 1991a, 28; interview with *DRIRE*, June 1993).

Despite the continuous expansion of the monitoring networks, there are too few measuring stations to record air-quality adequately. For this reason a ministerial administrative order in force since March 1988 required larger plants to monitor themselves (*'autosurveillance'*) and measure pollutant emissions on their own responsibility. Only the measurement methods and the intervals at which the data is to be communicated to the *DRIRE* were laid down (Mousel/Herz 1991, 70). Surveillance of ambient air-quality was thus placed not only in the hands of the air-quality monitoring associations, but also largely in those of industry itself. The government therefore has to rely on the willingness of industry to cooperate and on their sense of responsibility (interview with *CITEPA*, March 1993).

In 1973, an interministerial planning committee, *'Comité Interministériel pour l'Amenagement de la Nature et de l'Environnement'* CIANE, elaborated a programme for the setting up of so-called monitoring network associations to handle the administration of the control stations. In accordance with the *APPA* principle, the monitoring network associations are composed of all relevant clean-air policy actors: representatives of the government, of industrialists, local authorities, environmental associations, and of *APPA* itself (Delandre 1991, 377). The *'conseil d'administration'* is the authoritative decision-making body of an air-quality monitoring association. It is responsible for elaborating concrete guidelines and projects, like measuring additional pollutants and the planning of new monitoring stations etc. The administrative council also follows the *APPA* model in its multipartite composition.

In summing up, it may be said that France disposes of a wide range of instruments for combating air pollution. Whereas in the early sixties France, like Britain and Germany, espoused a tall-smokestack policy, in the course of the following twenty years it developed control instruments that involved various regulatory strategies. A regulative law approach is displayed by the policy instruments of plant licensing and of localized and product-related protected and alert zones. Examples of 'soft', economic control instruments are the pollution-related charge and the sectoral contract.

Implementation: Structure and Mode
In the implementation field, the highest governmental representative in the *département* is the prefect, the authoritative decision-maker. The Environment Minister intervenes in local clean-air policy only when industrial projects affect several *départements* (Hoppe 1984, 161). The prefect, who has the sole legal power of decision, consults several advisory bodies. The regional *DRIRE*, comparable to the German factory inspectorate (*Gewerbeaufsichtsamt*), plays the most important role. It is the authority that is in constant contact with industry, which carries out regular inspections on request, and which as technical expert in permitting proceedings seeks to sound out the room for manoeuvre between formal, legal requirements and technical feasibility. Where several départements are affected, the regional prefect is involved, that is to say the prefect with his seat in the regional capital.

The standards and conditions elaborated by the *DRIRE* and submitted to the prefect are settled in negotiation with the operator. The preparation of such requirements is embedded in a negotiatory process that can also be interpreted as a 'dialogue ... and the resulting licensing decision as a joint charter of permanent relations or as a quasi-contract between environment, administration, and industry' (Rengeling 1985, 121). The leitmotif of these negotiations is weighing environmental interests against available technologies and costs.[120] The governing principle is that of 'requirements adapted to the individual case' (ibid., 119). After the suitable technology for the greatest possible avoidance of air-pollution has been identified, the actual negotiating process begins. Now the available technologies have to be examined in relation to the individual case as to economic feasibility, taking account of the unit-of-production prices, maintenance costs, net earnings, and total investments over recent years (Rengeling 1985, 120). If the parties, the *DRIRE* and the operator, come to the conclusion that the standards required from the environmental policy perspective are too costly, this does not necessarily mean that they will be renounced. Subsequent negotiations are concerned rather with agreeing on a period within which the operator is required to conform with the standards, or, to quote a *DRIRE* engineer, 'the objective is clear; on the other hand what one *does* grant is time, that is to say not weakness but a degree of flexibility'[121] (interview with *DRIRE*, June 1993).

If one considers the positions taken in the decision-making process, it is clear that the *DRIRE* and the plant operator negotiate on an equal footing (Hoppe 1984, 165). The decision-making process takes the form of a dialogue, a 'transaction' or 'consultation' between the authority and the entrepreneur (*Fondation pour le Cadre de Vie*, quoted in Hoppe 1984, 163).[122] In sum, relations between regulator and industry can be described as cooperative. This is understandable when one considers that the *DRIRE* is subordinated to the Ministry of Industry and has only recently become accountable to the Environment Ministry (interview with *DRIRE*, November 1992).[123] Moreover, the engineers employed in firms and the *DRIRE* have had the same training in specialized tertiary institutions and thus have the same educational background. This is common ground that ought not to be underestimated, and the implications of which raise the question 'whether it was the industrialists manipulating the administration or the administration manipulating the industrialists?'[124] (interview with Roqueplo, June 1993).

Even if the relationship between industry and administration can be described as a 'system of negotiation' (or as Roqueplo puts it 'the administration plays more of an advisory than a police role'[125] — interview June 1993), local authorities exert no inconsiderable pressure to enforce regulations. The means range from fines to shutting down the plant, although the latter sanction is extremely rare. More frequently the '*procédure de consignation*' is used as a threat under which the prefect demands a sum from the firm corresponding approximately to the estimated investment for '*dépollution*'. This sum of money is reimbursed to the operator as soon as he has taken the necessary environmental protection measures.[126]

Like the relationship between the regulating authority and industry, relations between the authority and the public can also be said to be cooperative. Various mechanisms ensure that the public is integrated and can participate in state regulation of clean-air matters. In licensing proceedings, for example, the public enquiries, the right to inspect records, and the right of associations to take legal action[127] make sure that affected individuals and environmental groups are informed and are able to participate and raise objections. In the context of air-quality monitoring, the public has two options for access. Both the administrative council of the air-quality monitoring associations like *ASPA* or *AREMA* and the multipartite bodies like *SPPPI* follow the model of the *APPA* in bringing together a range of actors relevant to clean-air policy. The two institutions are therefore important as locales for concerted action. By including local authorities and environmental groups in various bodies, the state channels or prevents potential protest, and vests responsibility and powers of codecision in regional and local, private and public representatives in a 'strong state'. Moreover, the multipartite structure not only increases opportunities for public participation but, as a Ministry report has emphasized, also enhances 'the "transparency" and credibility of the information disseminated' on the government side (*Ministère de l'Environnement* 1991a, 27; interview with *AIRPARIF*, March 1993).

Another important institution for integrating the public and controlling environmental conflicts came into being in the early seventies: the so-called '*Secrétariat Permanent pour la Prévention des Pollutions Industrielles*', *SPPPI*, the Permanent Secretariat for the Prevention of Industrial Pollution (Knoepfel/Weidner 1985, 67). At the beginning of the seventies, conflicts flared up in Fos-Etang de Berre in the South of France between central government and the left-wing local council on the development of an industrial estate. In the hope of putting an end to the public dispute on the local air-quality issue, an 'interministerial permanent secretariat' was created, bringing together local and national actors. Since then, such secretariats have been set up especially in regions affected by heavy air pollution and where there is consequently a great deal of pressure for action to be taken. The principal task of these bodies, composed of representatives from the Ministry of Economics, local authorities, regions, and *départements*, as well as industrial associations and environmental protection organizations, is to prevent and address tensions caused by environmental issues through discussion and coordinated policy action.

Although multipartite institutions in France have hitherto doubtless contributed to restraining the potential for protest and to channelling issues, such bodies have their detractors. Critics see the risk of public discussion not being provoked but rather avoided if the impression is given that handing a problem over to a commission is to eliminate it. 'The Community of Strasbourg is very good at it; when there's a problem, you set up a commission, you get everyone together, you talk' (interview with *Alsace Nature*, July 1993).

Conclusion

Looking at the preconditions for state action in France from the institutional and instrumental point of view, a contradictory picture emerges on the possibilities of mobilizing political energy and activity. At first sight, political structures appear — at the national level — to be very monolithic and centralistic, and thus 'averse to activation', while— at the regional level — they seem very integrative and participative, and thus 'sympathetic to activation'. However, a closer look reveals that actors at the decentralized and centralized levels are interdependent, and that central government structures are segmented rather than monolithic, which is further intensified by professional corps rivalry (Suleiman 1987).

At the national level it is practically impossible for smaller political parties, as the vehicles of new initiatives, to gain access via elections to the crucial arenas of political power. The strong position of the president of the republic, even where diminished under the conditions of cohabitation, and the centralistic structure of the state contribute to consolidating the power of the political parties that are strong to begin with, and to hampering the rise of less well-established parties. This is predetermined not least of all by the French electoral system. Whilst elections to the National Assembly are by simple majority vote, only the regional councils are elected by the proportional representation system; *'conseils généraux'*, the departmental councils, also being elected by simple majority. Proportional representation, but also the limited territorial scope, gave smaller parties access to the representative bodies — even in the case of the departmental councils. The regions began to gain influence vis-à-vis central government only in the early eighties with the advent of the decentralization legislation. The regional and departmental councils also won scope for action, whilst the prefect as representative of central government lost some powers without, however, losing his still decisive power of decision. Despite repeated attempts to strengthen the integrative and coordinative role of the prefect (see the 1992 Act), territorial fragmentation has continued, not least of all in the course of decentralization. The numerous local agencies of central government, which are to some extent active at different territorial levels (arondissement, *département*, academies, etc.) maintain close horizontal ties with their clients, but also with the local authorities concerned, and vertical ties (professional-bureaucratic specialist networks) with their ministries. This also holds for the network embracing the *DRIRE*, which again limits prefectoral integrative power.

At the instrumental level, too, France displays more variety than one would initially expect of a unitary state. Neither the United Kingdom nor the Federal Republic of Germany has such a range of different regulatory mechanisms that so directly obey the rationale of industry and the interests of the general public. There is a choice of instruments for abating and controlling industrial emissions, which instruct industry in regulative law manner (protected and alert zones) or incite it in market economy manner (charges and sectoral contracts) to curb the release of pollutants. If the public finds it difficult at the national level to induce innovation through the ballot box, vertical interlinkage offers ways and means to influence 'Paris', and at the regional level there are various possibilities for integration and participation. With the right of associations to sue, the right to inspect records, and the multipartite composition of various regulative bodies, the public has possibilities to obtain information, to cooperate, and to oppose.

Choreography on the French and British political stages is in striking contrast to the overall picture in Germany. Whereas Germany opens up more possibilities for democratic codecision-making at various levels in the agenda-setting and policy-formulation phases than does France, the latter is more attentive to citizen demands in the phase of policy implementation, and is more open to objections and protest than Germany. While the fear in Germany is of the citizen as affected party and 'carper', in France the fear is of the citizen as 'naysayer' to the state as such.

2.4 Comparing the Countries Under Review

A multiplicity of policy interests vie in the European arena. Member states bring them into play in the supranational decision-making game in competition for the title of Community policy. The diversity in Europe is a product of national geography, economics, society, politics and culture. The aim of our study is to analyse this by taking the examples of the three large member states Germany, Britain, and France in order to assess its significance for European policy and its impacts on the nature of state activities in member countries.

2.4.1 Determinants of Problem Perception

Even the perception of what constitutes air pollution differs from country to country in terms of national qualifying factors. In Germany the prevailing view is that the existence of polluting substances in the air is sufficient legitimation for state intervention, even in the absence of absolute scientific certainty that the pollutant emissions cause certain harm to health and Nature. Precautionary action is propounded. Both prevention and reactive abatement of damage that has already occurred are governed primarily by the principle that the polluter should bear the cost of pollution; in other words, harmful impacts on the environment should be reduced at source. Compliance with emission limit values is prescribed to achieve this end. This can be done only if the protective mechanisms utilize state-of-art techniques. Not to put too fine a point on it, in Germany clean-air policy equals 'good technology'.

Britain, by contrast, views the issue more from a nature conservation and scientific perspective. It is taken for granted that meticulous scientific proof must be produced that there is a causal relationship between certain pollutant emissions and environmental degradation. 'Absolute removal of potential pollutants without clear reference to dose-effect relationships has long been resisted as a sensible goal' (Macrory 1988, 670). State intervention in the production process is regarded as legitimate only on this basis. The anticipatory approach, although it exists, is secondary. In contrast to the German approach, British environmental protection therefore seeks to assure a certain level of air quality; that is to say it is guided by pollution concentrations. Individual emissions are reduced when the prescribed air-quality limit has been exceeded. Even if intervention can be justified on grounds of scientific findings and a high level of environmental pollution, strict orientation on the best available technology is mitigated by the proviso that it may not constitute an unreasonable economic burden for the individual plant (DoE 1986).

France's clean-air policy presents a mixture of emission and quality orientation. State intervention is deemed necessary when pollutant emissions by industrial plants (*'établissements dangereux, incommodes et insalubres'*) are judged to be dangerous. However, emission limits are set in statutory form by central government, but are fixed or negotiated at the local level. The polluter-pays-principle is applied. This is most explicit where a 'parafiscal charge' (*taxe parafiscale*) is levied on certain emissions and from a certain volume of releases, the revenues being paid into a fund to finance environmental technology investment. Unlike German practice, the authorization process takes plant emissions to air, water and soil into integral account. This is in contrast to German, predominantly medium-specific procedure. On the other hand, consideration of regional air quality is on an equal footing with this emission orientation. The use of low-pollutant fuels is the primary means for ensuring compliance with quality limits. The precautionary principle finds expression in environmental impact assessment, which now has a relatively long history. France, like Germany, is strongly technology-oriented especially in controlling emissions from stationary sources, which is the job of the *DRIRE*. Their inspectors are almost without exception members of the *Corps des Mines* (Roqueplo 1988, 110), and, like their main interlocutors in industry, are graduates of the *Ecole des Mines*, so that the largely cooperative negotiations on environmental protection arrangements in industry take place chiefly among engineers (Roqueplo 1988, 150).

The differences in perspective and practical approach between the three countries can be explained in terms of long-run factors like geography, settlement patterns, industrial density distribution, and energy supplies. Germany is a large importer and exporter of pollutants mainly because of its geographical position at the heart of Europe and the wide-spread and strong concentration of industry. The dense settlement of the national territory commends stringent emission reduction. The technological bias in environmental protection has been favoured by a long tradition and a great deal of experience in the development and production of technology. Furthermore, the great importance of fossil fuels, especially coal with its high SO_2 emissions, explains why some regions have developed a strong awareness of palpable air pollution. In Britain, by contrast, where coal mining has been just as major a factor, the insular position of the country and the favourable prevailing winds have meant that the problem of SO_2 emissions has not been perceived as so dramatic. 'If practicable, reliance has been placed upon the natural mechanisms of the physical environment to disperse and dilute waste emissions, a policy encouraged by the country's geographical features — its insularity, fast-flowing rivers, estuaries [and] prevailing winds' (Macrory 1988, 670). The uneven distribution of industrial areas has recommended a primarily quality-oriented approach, concentrating on abating pollution in specific regions. French geography and climatic conditions — prevailing west winds — have meant that the country is affected by transboundary environmental pollution only in the northern and eastern regions, and exports more pollution than it imports. The suitability of a quality-based approach is suggested by the high concentration of population and industry in very few centres, leaving many sparsely populated areas. Population growth over the past 35 years (of almost 50 per cent) has been concentrated in seven southern and south-eastern *départements* (Mediterranean coast) and four *départements* in the Paris region (Pumain 1993, 31f.). Industrial air-pollution in France is considerably mitigated by the low industrial use of fossil fuels, made possible by a comprehensive nuclear-power programme.

Perception of the air-pollution issue is also strongly determined by medium and short-term developments like the economic situation, environmental incidents, and technological and scientific advances. A direct connection is to be observed in all three countries between the general economic situation and the willingness to invest in environmental protection, to give high priority to environmental issues, and consistently to carry out the measures adopted. In Germany, the economic crisis of the seventies triggered by the rise in oil prices was the direct cause of the slowdown in environmental policy activity. Industry's resistance to the new environmental policy departure that marked the late sixties and early seventies intensified. In the present period, too, the cost of German unification and the worldwide recession have slowed environmental policy. All initiatives imposing new costs on industry are critically scrutinized by government at a very early stage in the political process. In Britain the recession of the eighties also sharpened political and administrative awareness of cost-benefit questions in environmental protection, contributing to the vehement rejection of EU regulatory demands. The conciliatory relationship between the British authorities and industry would have suffered. As it proved, in the period of relative prosperity in the late eighties, the political will for a comprehensive reform of environmental legislation that developed was encouraged by other national and international factors. The general connection between economic prosperity and environmental commitment is also evident in France. Industrial enterprises invested far less in environmental protection technology (in water pollution control) between 1975 and 1980 (Faudry-Brenac 1981, 52). The competent ministries also advised subordinate authorities to go easy on industry in negotiations: 'stand firm on the objectives but be more flexible on the timing' (Faudry-Brenac 1981, 60f.).

Perception of environmental problems is also strongly determined by national (and international) events like environmental disasters or critical environmental developments. They give impetus to social and political forces that champion environmental protection innovations, and (temporarily) enlarge their scope for action. In Germany, the discussion on forest dieback accelerated political decision-making processes and contributed to political consensus formation on the Large Combustion Plant Regulation. With its insular geography and tall-chimney policy, Britain was less affected by such impacts; the debate on acid rain accordingly attracted little political attention. The British were not under the sort of pressure generated, for example, by the famous London smog of 1952, which led to adoption of the Clean Air Act four years after the event. Like Britain, France did not regard acid rain as an urgent problem, because the French nuclear-power programme had kept SO_2 emissions much lower than in Germany. More heed was given to the issue only when attention shifted to NO_x emissions from mobile sources. However, early impetus was given by environmental incidents such as Seveso, Feyzin, and Bhopal. These disasters attracted a great deal of publicity, and roused political interest in the deficiencies of environmental protection regimes. After the Feyzin disaster, the control of industrial installations was intensified (Roqueplo 1988, 111).

Highly developed environmental awareness and a high level of problem perception find expression in the number and activity of environmental organizations, and vice versa: environmental association activities are apt to intensify problem perception. With the founding of the political party 'The Greens', environmental policy activities took on particularly spectacular forms in Germany, a country where population density and industrial concentration render environmental problems more frequently and more palpably evident than elsewhere. In comparison, British environmental organizations are more concerned to cooperate informally with the authorities, adopting a mode of operation largely outside party political channels. Both Britain and France are regarded as latecomers in developing an awareness of air pollution. British concern with the issue — as we will be showing — developed only in the second half of the eighties, and was attributable to increasing international pressure in connection with an environmental movement campaign against 'the dirty man of Europe'. Internal political changes and a greening of party programmes occurred at the same period. In France, which acknowledged even in the 1990 *'Plan National de l'Environnement'* that it was 'fifteen years behind' its neighbours, ecological debate and environmental activities concentrated more on concrete, regional environmental issues, with the notable exception of nuclear power, which nurtured general environmental awareness. Notwithstanding the growth in environmental awareness, greens managed to gain a foothold in political decision-making bodies only at the regional and local levels.

Three significant factors for all three countries, but which have a very dissimilar impact on their respective perception of the air-pollution issue are international environmental policy, environmental technological progress, and changes in the environmental technology market. Although all three countries were subject to the same pressure from international treaty obligations, the repercussions for domestic problem perception in the eighties differed widely. Germany, together with Britain, initially showed strong opposition to the Geneva Convention and the Helsinki Protocol, but to the chagrin of the British finally come completely round. This change of opinion occurred as it became evident that forest dieback was also a German (and not only a Scandinavian) problem, and the government found itself under growing pressure to take action. To begin with, Britain remained steadfast in its opposition because it regarded itself as little affected by the problem, and because in the recessionary period of the eighties it had no desire to impose an additional burden on the economy in the form of internal regulations. For its part, France was able to put its signature to the agreements with a light heart because its nuclear-power programme allowed it to comply with requirements without difficulty.

The way in which an issue is perceived is also influenced by the ideas about how it can be resolved. In Germany, with its advanced environmental technology industry, a technology-oriented perspective prevailed. If the abatement of emissions throughout Europe or the world were to be imposed by law and the 'state-of-the-art' in technology were to be the means to this end, promising new markets would open up for this industry, which at present supplies about 30 per cent of total German exports. Britain, by contrast, has no important environmental technology industry. Furthermore, the British had had early experience (in the thirties) with the predominant techniques for reducing sulphur dioxide, namely flue-gas desulphurization, which left them very sceptical about such technology. France has an environmental industry (about 1.6 per cent of GDP in 1987) that is mainly concerned with water supply and purification and with waste recycling (Schreiber 1991, 88). However, according to the *CNPF*, industry has still not learned to regard 'the environment' as a promising market to be developed (interview with *CNPF*, June 1993).

These long and medium term factors are significant in explaining the differences in the ways the three countries perceive pollution of the environment or the air. Whether and how air pollution as a problem is tackled as a political and administrative task depends on the specific characteristics of the political system and its capacity for mobilizing political energy, for generating new policy initiatives, concentrating them, and giving them access to the policy-making agenda. The three countries take very different paths to this goal, guided by the specific institutional and instrumental structures of their political systems. Three institutional factors are particularly significant in determining how policy initiatives are generated: the number of available political arenas; the electoral system and the consequent access opportunities for ecologically minded parties to the policy-making process; and the role played by the courts.

2.4.2   Institutional and Instrumental Preconditions for State Action

The decentralized, indeed fragmented structure of the Federal Republic offers a large number of political arenas where environmental interests can be voiced and brought into the formal political decision-making process. The federal structure of the political system and the inclusion of the states via the *Bundesrat* in central government decision-making multiply the opportunities for debating new policy ideas, and the standpoints of the states have to be taken into account. Since the states are primarily responsible for implementing legislation (administrative federalism), they have the right to participate in the making of many federal laws in the Federal Council, the *Bundesrat*. The *Bundesrat* thus provides the opportunity for institutional expression at the federal level of the multiple viewpoints and policy initiatives developed at state level. This also holds for federal environmental legislation. In addition to their participatory rights in central government, the states dispose of their own environmental powers on the basis of their constitutional autonomy.

Hence, the multiplicity of political arenas both generates a policy dynamic and — owing to the strong attendant need to achieve consensus at the federal level — produces policy incrementalism. Agreement frequently proves possible only on the least common denominator. The Federal Minister of the Environment always faces protracted negotiations with the competent state ministries when regulations implementing the Federal Pollution Control Act are to be adopted. Under certain circumstances, decision-making processes can even be blocked. This can occur where different majorities in the two Houses, *Bundestag* and *Bundesrat*, produce gridlock in the party political contest between governing and opposition parties. In effect, the decision-making situation is then comparable to that under a grand coalition (1969-1982 and again since 1990) (Lehmbruch 1976).

In Germany there is also scope for policy initiative at local government level, because the local authorities have a constitutionally guaranteed right of self-government. Whilst in the seventies, local authority environmental initiatives were still modest (with the exception of water measures), the states transferred additional responsibilities to local government in the eighties in the performance of regulative, planning, operational, and consultative functions. In order to handle these tasks, many large cities and administrative counties set up their own organizational units for environmental protection. The local authorities have never played a significant part in stationary source emission control. In the planning and authorization of plants, they find a hearing only in their capacity of representatives of the public interest. However, they take a great deal of initiative and are becoming more and more important in the field of quality control and in the melioration of air quality through traffic policy measures, vegetation programmes, etc.

There is a further type of 'political' arena in Germany where important regulative decisions are made out of the public eye, and which reflects the German corporatist state tradition of cooperation between the state and associations. It is the bipartite or multipartite body in which private and state actors collaborate in making and carrying out policy. Such quasi-governmental institutions, like the professional and trade associations, the German Institute for Standardization, or the Technical Control Associations prepare important provisions in environmental policy.

The governmental system in the Federal Republic thus contains a wide range of political arenas. The institutional features of federalism, a chamber of the states at the federal level, administrative federalism, local government autonomy, and the corporatist, bipartite and multipartite bodies[128] give the state a fragmented and decentralized structure. The many arenas and the actors that play them are interlinked in a complicated network and are subject to great pressure to achieve agreement. Political solutions are therefore more likely to be found in negotiation than in majority decisions. The central coordination mechanism and sinews of this polycentric state is the party system. Cohesion is created by orientation towards federal elections and the desire to retain or win power at the centre (Lehmbruch 1994, 23).

The institutional preconditions for generating policy initiative in the United Kingdom are quite different. The Westminster model is based on a 'closed circuit of responsibility' within which a majority of votes produces a majority in the House of Commons. This in its turn produces a government, which adopts a government programme and the corresponding policies. Whether these policies are successful or not is decided by the voter in the next general election. Under the Westminster model, policy-making powers are concentrated in the hands of the prime minister and the cabinet, and in Whitehall. Although government departments do not have the same measure of autonomy as their counterparts in the Federal Republic, important policy initiatives have in the past originated from Whitehall, and policy proposals for a specific domain have to find the approval of a small circle of key figures in Whitehall and close to the prime minister if they are to find their way into Parliament. Environmental policy centres on the DoE, established in 1970, and which in comparison with the equivalent German and French ministries, bears responsibility for a very broad range of tasks. Its functions are closely meshed with those of the Departments of Transport, Trade and Industry, and Energy and require much coordination. Sources of policy initiative at the central level outside Her Majesty's government are committees of the House of Commons and House of Lords, which in their supra-party deliberations often submit government policy to sharp and constructive scrutiny, as was the case with clean-air policy in the mid-eighties. Although committee proposals also have to pass the Whitehall bottleneck if they are to be incorporated in government draft legislation and find Parliamentary majority, they can be extremely important, especially if they cross party lines.

Traditionally of great importance in British policy-making is the expert advice proffered by Royal Commissions. These bodies, which are composed of 'the Great and the Good' in British public life appointed on the proposal of the government, are independent. In the past, their policy recommendations have often strongly influenced legislation, in, for example, social welfare, in health and safety at work, and also environmental protection. The RCEP, established in 1970 on the initiative of the Prime Minister, and which differs from other Royal Commissions in that it is a standing body, is composed of leading figures from cultural life, the sciences, and industry, and acts as a permanent watchdog (Schreiber 1991, 33). Although the government is under no obligation to respond to Royal Commission reports, it regularly does so. The recommendations advanced by the Commission in 1976 were crucial in the 1990 revision of environmental legislation.

Unlike the states composing the Federal Republic of Germany, the British subnational political arenas, the local authorities, do not systematically participate institutionally in central government decision-making. Local authorities at district and county level are ruled 'from on high'. Their functions and financial resources are decided by central government. In the opposite direction there are no formal institutional channels to transport policy impulses upwards. This function is left largely to the parties and to the MPs' ties with their constituencies.

In the course of the 'reforms' undertaken in the Thatcher era, the statutory margin granted local authorities in policy making, their range of functions, and their financial resources were severely curtailed. Some of the services they had provided that were used by important voter clienteles were transferred to new quasi-governmental organizations or 'Quangos' to call a halt to the party political instrumentalization of these services by Labour county and district councils. In short, since the changes of the eighties, it has become a great deal more difficult for local political bodies to develop policy initiatives of their own, especially where they depart from the government line (Windhoff-Héritier 1992). However, in the administrative implementation of environmental legislation, local authorities enjoyed a relatively broad margin for action. In the field of water pollution control, for example, it was so broad that the water authorities were able not only to set emission limits for plant operators, but also to fix environmental quality standards for waters in their territory. The central authority restricted itself to advisory functions (Schreiber 1991, 69).

Policy-making powers in the United Kingdom are thus concentrated in the hands of the government and the parliamentary majority to whom they owe their office. In the context of the existing balance of power and the general legal setting, local political arenas remain severely restricted in developing new policy initiative. It is hence not surprising that local authorities, as we will be showing, are coming to regard the European institutions as interesting interlocutors and potential 'allies' against London.

In France as a centralist and semi-presidential democracy, policy-making powers are concentrated in the hands of the president and the government. However, in contrast to Britain, centralized power is 'shared', the greater part going to the president, because both president and government have their own — albeit interrelated — functions and powers, and each has its own democratic legitimation. This is less relevant when president and government have the same party political backing. Where majorities diverge (*'cohabitation'*) as was the case from 1986 to 1995, the president cannot find support from a parliamentary majority but is condemned to working together with a prime minister in opposition to him. In a period of cohabitation, the president therefore loses much of the scope for action he enjoys in a consensual period. He has the support neither of a government appointed by him nor of a majority in parliament (Mény 1993, 235). One of the expressions of the centralist structure of the French state is that laws are made only by the government and parliament. Subnational authorities, the regions and *départements*, do not participate institutionally in central government legislation as do the German states. The position of the government in legislation is enhanced by the relative generality of formulation. Parliament adopts only framework legislation. The government departments fill in the framework by means of implementing regulations (*décrets*) and ministerial circular orders (*circulaires*) and pass them on to the prefects as representatives of the state in the *départements* (*Ministère de l'Environnement* 1991a, 10).

In France, important environmental policy legislative initiatives thus originates at the national level, with the prime minister and government, but also with the president.[129] The Environment Ministry, which had been set up in the seventies at a time when there were wide-spread demands for national environmental policy to be coordinated, bears responsibility for this policy. Its position is, however, relatively weak (Romi 1990, 10). Although it has supervision of dangerous, unhygienic (*insalubres*) and polluting industrial plants, the competent regional authorities (*DRIRE*) are subordinated to the Ministry of Industry. *DRIRE* inspectors thus 'wear two hats: industry and environment' (Roqueplo 1988, 111). Notwithstanding repeated efforts to do so, the Environment Ministry, '*un petit perdu parmi les grands*' (Romi 1990, 10),[130] failed to gain control of the *DRIRE* inspectorates, but did finally manage to have them made answerable to itself as well. The only regional agencies the Ministry has are *Délégués Régionaux à l'Architecture et à l'Environnement* (*DRAE*). Their role is essentially consultative, 'a missionary activity of promotion, of consulting, and more generally of advising' (Romi 1990, 18). Ministry finances are also very modest. In 1990 it had a right to 0.058 per cent of government spending (*Ministère de l'Environnement, Plan National de l'Environnement* 1990).

Despite this centralization, local authorities have substantial weight in the implementation of framework legislation, as they did in Britain before the reforms of the eighties and to a lesser extent still do. Especially at the *département* level, secondary regulation within the local ambit is possible on the basis of ministerial orders, 'a possibility much exploited in the environmental sector' (Schreiber 1991, 19). The regionalization laws passed in 1982 also provide scope for the development of new political and administrative initiative in the environmental field. For the past ten years or so, a very interesting development has been taking place. Although, even after the reform, environmental affairs have in principle remained the domain of central government and its representative in the *département*, the prefect, who disposes of extensive licensing, control, and punitive powers, some '*collectivités locales*' have been making eager use of the principle of co-responsibility and the related rights of co-decision and the initiation of legislation, without central government having transferred significant powers to them.[131] Step by step, the elected authorities in the regions and *départements* (*conseils régionaux* and *conseils généraux*) have been evolving into political actors to be reckoned with. On the basis of their democratic legitimation, they are seeking to expand into new and different fields of action. 'Without really having sought to do so, the authors of decentralization have brought about a transition in France from a system of local administration to one of local government'[132] (Mueller 1992, 294).

As a consequence, relations between the prefect and the locally elected representatives have become increasingly competitive and conflictual. The coordination of subpolicies[133], previously in the charge of the prefect, has gradually passed to the elected authorities. The resources the rivals can throw in the balance differ widely. While the prefect can plead legal rules, the elected representatives resort willingly to credit allocation as a tool to gain their political ends (ibid., 293). The formal apportionment of functions in the contest proves secondary. Heedless of their modest formal powers, local authorities (region or *département*) happily tackle every new problem that turns up on the political agenda, and advance political solutions to them. They are 'new, fully competent political authorities, which base their legitimacy on their capacity to resolve all the problems occurring in their territory' (ibid.). Although, for example, the powers of the *conseils régionaux* and *conseils généraux* are very limited in environmental affairs, many *départements* pursue a committed environmental policy. A great deal is done especially in water matters, since the quantity and quality of water is the subject of intense public interest. The package of measures taken in individual *départements* to handle these problems and the policy patterns underlying them are highly varied in nature, and are determined by specific local problems and local demands (ibid.). But also in clean-air policy, such activities by '*élus locaux*' are to be found in various forms. The elected representatives of the regions and départements participate in the regional airquality monitoring measurement networks, which are being set up in increasing numbers. They take part in the development of these multipartite bodies (*associations de réseaux de mesure*), which as a rule are established under *DRIRE* administration. However, their administrative institutions include all environmental policy actors: representatives of the state, of regional and local authorities (*conseils régionaux, conseils généraux*), municipalities, industrial associations, and environmental organizations. Representatives of the local authorities

... have to win a role for themselves, because to begin with they basically had no power at all, no technical competence at all, but do have certain financial resources. ... For my part, I feel it at my level [air-quality monitoring networks] more and more ... That on the one hand there's the state, and on the other there's the region. ... The Ile de France Regional Council at present sets more in motion in the administrative council [of the network] than does the government, and potentially it's a more potent financier[134] (interview with *AIRPARIF*, March 1993).

In their endeavours to ensure more transparency on air quality, but also to prepare general environmental policy initiatives (Roqueplo 1988, 112), the actors in the multipartite monitoring networks act from different motives, and a peculiar mixture of regional and national ambitions can develop in these bodies. The local authorities see the importance of the networks in the improvement of regional environmental policy as an opportunity to enhance their profile as elected representatives. From the central government and *DRIRE* perspective, by contrast, they offer an opportunity 'to create credibility' (interview with *AIRPARIF*, March 1993), for the state to gain legitimacy and credibility in the eyes of the citizen. Employers, who pay towards these activities (sometimes picking up the whole bill), want to be informed about and have a say in environmental policy initiatives, especially because they can chose to pay their 'parafiscal charges' to these institutions. For the environmental organizations, by contrast, they are an arena in which environmental goals can be attained. The activities of some of these multipartite bodies (*SPPPI*) go far beyond monitoring activities to promote general action on reducing industrial air and water pollution.[135]

It is obvious that any such change in political activity at the local level must have an impact on the role of state institutions, especially on that of the prefect. They are in stark contradiction with conventional conceptions about the clear and hierarchical structure of government and administration in France, with the prefect, 'this strange animal from the French menagerie' (quoted in Mény 1993, 286), like a spider at the centre of the web controlling the external services at the behest of Paris (Mény 1993, 286).[136] This picture, already relativized in the past by the interlinking interests of and cooperation between the prefect and local *'notables'*, has been called into question since the centre of gravity in local politics has shifted from the prefect towards the elected representatives in the local authorities (Mueller 1992, 293). In this situation, state services have to seek new ways to maintain their presence in the local web. One possibility is to reinforce the rights of the prefect (as the 1992 Act on territorial administration aims to do), another is to increase the presence of the state in the region by institutionalizing cooperation with regional actors. The recent establishment of the *'directions régionales de l'environnment'* (*DIREN*) represents such a move (Mueller 1992, 294).

One consequence at the policy level of this regionalism, which is gaining increasing political momentum, is that a very diversified policy is developing from the bottom up — in counterbalance to Paris — resembling a patchwork of individually composed sets of measures, producing a variegated 'policy spread'. 'Everywhere you find local initiatives. ... A very strong cultural diversification representing the different "tribes and ethnic groups" has grown up, which ... is remarkable, in the administration as well' (interview with Knoepfel, May 1993). How the policy mix is put together and what the focus is depends on exchange processes in local and regional networks. 'This situation, compared to that existing before 1982, generates dynamism and diversity, but also uncertainty in handling political problems' (Mueller 1992, 295).[137] Hence, what is striking in France is the simultaneous existence of a strong centre and substantial political dynamism emanating from the regions.

Apart from the availability of political arenas, the electoral system is a crucial factor in deciding whether new political groupings have a chance of bringing their policy ideas into the political process. The German electoral system, based on the proportional representation principle (although the distribution of seats depends partly on the votes gained by individual candidates in individual constituencies), offers broad access to such groups. The only parties not represented are those that win less than 5 per cent of the overall vote. On the basis of this electoral system, it has thus been possible for a party like the Greens to win seats in the *Bundestag*.

In the United Kingdom, the chances for introducing policy initiatives into the political decision-making process through new political groupings are severely restricted by the electoral system. It is extremely difficult for a new party to win a seat. 'Unless some form of proportional representation comes to the United Kingdom, the British Green Party will never become a serious political force' (O'Riordan 1988, 12). It is therefore not surprising that environmental interests move in other channels than in Germany. Influence is exercised in informal cooperation with Whitehall and other public authorities rather than via the political parties.

Unlike the British relative majority ('first-past the-post') electoral system, the French relative majority system allows the two large blocs of governing majority and opposition minority to consolidate while maintaining a pluralist party system. In the first ballot the door is wide open (Mény 1993, 173), and all parties can put up candidates. The contest is often between five or six rival candidates hoping to win the necessary absolute majority. In the second ballot, where victory goes to the candidate winning an absolute or relative majority, the fight is generally between the bloc formed by the parties of the Right and the Socialists. The candidate with the best prospects is the one able to profit from the withdrawal of allies[138] who had been rivals in the first ballot. The system tends to filter out the influence of small parties in the second ballot. The small groupings therefore have better prospects of gaining access to decentralized political arenas, where, as in the case of the *conseils régionaux*, voting is on the proportional principle, or in the communes where the political situation is more manageable and environmental events make themselves felt locally. In 1989, greens won 1369 seats in communal assemblies (Schreiber 1991, 39f.). But, despite the growing importance of elected regional bodies, electoral successes at the local level (as in elections to the European Parliament, where greens gained eleven per cent of the vote) are of less consequence. If one's voice is to be heard to effect in a political system where, notwithstanding more powerful regions, the most important initiatives come from the centre, it is necessary to be on the political stage in Paris. This is achieved through the wide variety of existing vertical interlinkage mechanisms.

In addition to the number of available political arenas and the access thereto for political parties, controlled by the electoral system, the courts play an important role as an — indirect — source of policy initiative. Constitutional courts, administrative courts, and special-purpose courts may correct existing legislation and scrutinize administrative decisions.

The German Federal Constitutional Court or *Bundesverfassungsgericht* acts as a crucial corrective to politics and administration, before which every citizen who believes his basic rights have been violated by legislation can take action to test the constitutionality of the legislation at issue. In recent years there has been a growing tendency to 'pass the ball' to Karlsruhe in this manner on particularly controversial legislative issues, to change 'political arenas' in the hope of achieving a result favourable to one's own interests by having the Federal Constitutional Court make the decision.

The administrative courts, introduced in Germany in the late nineteenth century on the model of the French *Conseil d'Etat*, play an important role as clearing-houses for administrative decisions. In contrast to France, the federal principle finds expression in this area as well. There are administrative courts at the regional, state, and federal levels. Precisely in clean-air policy, the discrepancies in administrative court rulings from state to state were a crucial factor in the adoption of the 1982 Large Combustion Plant Regulation. Ecologically minded administrative court judges had taken steps towards more stringent regulation of emissions in some states. This induced the power utilities to urge uniform regulation of SO_2 emissions throughout the country to balance competitive opportunities. The outcome was the adoption of the Large Combustion Plant Regulation, shifting decision-making competence in the clean-air field to the federal level. Paradoxically, federalism, precisely because it produces new sources of policy initiative, sometimes tends to cancel itself out.

In Britain, the concentration of political power in the government is not counterbalanced in a well-defined division of powers with strong, independent, judicial review. The principle of the 'supremacy of Parliament' subordinates all other powers to Westminster. The House of Lords — also the highest court in the United Kingdom — is the supreme instance in the court system and exercises supervision over the system. However, the absence of a constitutional court does not mean that there is no system of checks and balances. Judges interpret legislation frequently and extensively, 'even to the point of "rewriting" parliamentary law or changing its meaning' (Mény 1993, 340). Many important legal rules — especially in the field of basic rights — are not embodied in statutes but derive from judge-made law (Mény 1993, 340). It nevertheless does make a difference whether government decisions are reviewed by a constitutional court fully independent of parliament or by a chamber of the legislating parliament.

A sort of quasi-administrative court (Budge/McKay 1988, 166), the so-called 'administrative tribunal', was introduced after the Second World War. Today there are numerous such arbitration bodies in Britain (Mény 1993, 336). But they are not important clearing-houses for environmental policy (Macrory 1985, 201ff.) like the German administrative courts, primarily because they are subordinated to the traditional court system and have no competence to develop independent administrative law.

The French *Conseil Constitutionnel*, set up in 1958, has a relatively weak position in comparison with the *Bundesverfassungsgericht* in Germany. Its rulings do not have to be accepted by the *Conseil d'Etat* and the *Conseil de Cassation*. However, over the years it has gradually assumed the role of a watchdog for constitutional basic rights and of a general instance for judicial review as to the compatibility with the constitution of substantive legislation, administrative decisions, and court rulings. However, it has no power to impose sanctions of any sort (Mény 1993, 344f.).

The time-honoured French administrative court system is strongly centralized, and is composed of 25 *'tribunaux administratifs'* (administrative courts) headed by the *Conseil d'Etat* established by Napoleon. The latter institution is the court of final appeal from administrative court decisions (in 1989 it transferred part of its functions to five further appeal courts), and thus also reviews the legality of government and administrative decisions. However, before a statute or regulation is passed, it also delivers an authoritative opinion on the proposed legislation (Mény 1993, 334).

The corrective function of French administrative courts is less pronounced than that of their German counterparts. One reason is that they work very slowly. '5000 cases are pending before the Strasbourg Administrative Court, which has jurisdiction over Haut Rhin, Bas Rhin and Moselle' (interview with *DRIRE*, November 1992). Another is that the courts are not entitled to give instructions to the administration (strict division of powers), and administrative court judgments cannot be enforced (Schreiber 1991, 15). Since laws adopted at the national level are usually only framework legislation, the administrative authorities, having no concrete rules on emission standards, have wide discretionary powers. The administrative court judges usually refuse to circumscribe administrative discretion by providing substantive interpretation of the law, restricting themselves to reviewing formal procedures (Schreiber 1991, 35).

The main features of the political system we have mentioned, namely the number of available political arenas, the electoral system and the related access to political institutions for parties, and the courts that can be called upon to revise political and administrative decisions, are crucial political system elements in determining the form and quantity of policy initiatives that develop and enter the decision-making process. The differences in institutional arrangements we have outlined explain why exchange processes in policy-making within national policy networks take such different paths.

If we turn to the conditions under which decisions in the area of clean-air policy come into being and to the institutional preconditions for generating policy, we must investigate the typical differences in legal regimes, policy instruments, and modes of implementation, which are an outcome of the policy-making process, and which member states are keen to see embodied at the European level so as to avoid institutional and legal adjustment costs and economic disadvantages.

The mode of regulation in German environmental policy and the instruments deployed are conspicuous for their strong reliance on the law. There is a copious and detailed body of legal rules in the spirit of a regulative-law intervention philosophy. But German arrangements also typically exhibit interlinkage between state and private rules. Legislation includes concepts like 'state of the art' or 'state of research' which require interpretation, which is entrusted to private regulatory bodies.

The relevant public law provisions are set forth predominantly in the Federal Pollution Control Act and in the regulations and administrative directives implementing the Act. The objectives of the statutory rules are environmental care, environmental protection, and the remedying of damage. They address stationary sources, products and specific territories. The major tools deployed are prescriptions/prohibitions or the refusal to authorize plants, processes and substances relating to a given pollutant emission. In order to reduce air pollution, limit values are set which presuppose the use of a certain level of technology in installing protective facilities.

This orientation towards a state-imposed regime does not, however, exclude the competent authorities from negotiating at an informal level with operators to attain their emission abatement goals. The lack of resources is the best recommendation for such a procedure. Public participation in such permitting processes is guaranteed only in the case of 'formal licensing procedure'. Within a certain period, objections to the authorization of a plant may be filed. In the case of 'simplified procedure', on the other hand, the applicant's plans are not made public.

Under the German system of administrative federalism, implementation of clean-air provisions is the task of the states. In most states of the Federation, the factory inspectorates (*Gewerbeaufsichtsämter*) as special-purpose authorities (or for very large plants the regional administrative authorities or *Regierungspräsidien*) are responsible for enforcing environmental legislation. Factory inspectorates are answerable to the respective state ministry of labour and not to the environment ministry, as is the case for competent authorities in the United Kingdom.

Britain, too, has a long tradition of environmental protection legislation, although it has not been as differentiated in extent and intensity, leaving the implementing authorities more margin for action. Until 1990, licensing decisions were guided not by statutory limit values but by 'presumptive limits' negotiated between regulator and operator, and which had guideline function for other operators of comparable plants. Relations between regulators and industry were cooperative and consensual, and, as in Germany, extremely sparing use was made of the legal sanctions provided for on the statute books. The concrete obligations imposed on emittents were decided on the basis of the 'best practicable means' principle, in other words they were determined by what was technically feasible and economically reasonable. Application of bpm was intended to prevent the emission of potentially hazardous substances in industrial processes to the greatest possible extent, in order to keep the dangers to human health and damage to the environment to a minimum. This precautionary approach, guided by the state of the art in technology and by the 'polluter pays principle' (PPP) could function only through constant cooperation and coordination between industry and controlling authorities (DoE 1986). However, until 1990, local authorities were unable to pursue precautionary emission control but had to wait until the courts ruled on the existence of a public nuisance.

France, too, has long had an Act on the abatement of emissions from industrial plants (since 1917). The decisive breakthrough in environmental legislation came between 1964 and 1976 with four key measures on water, waste, nature conservation, and industrial plants. However, it is constitutively characteristic of French environmental law that it is based on a small number of central, generally worded statutes to be adapted to regional conditions at the subnational level by means of secondary legislation (Schreiber 1991, 34).

The tools deployed in French clean-air policy include a wide range of control instruments, from the individual authorization of plants and processes via a quality-oriented policy of fuel substitution where air-quality limits are exceeded, to a form of tax levied in the event of non-compliance with emission standards, and the revenues from which are ear-marked to subsidize investment in environmental technology. Use is also made of industry agreements (*contrats de branche*), deals on long-term emission reductions in certain industrial sectors involving financial aid from the state for protective facilities (Schreiber 1991, 13). For a long time now, France has had a procedure for public participation in authorization proceedings (*enquête publique*). On applying for a permit, the operator is required to submit an environmental impact assessment covering all environmental media. However, until 1993, France set no fixed statutory limit values for plants and processes to guide authorization decisions (Knoepfel/Larrue 1985, 43; Roqueplo 1988, 103ff.). Permits were granted at the discretion of the competent *DRIRE* inspectors, whose stance was one of cooperative administrative action, although they disposed of what are from a German point of view draconian instruments like the *'procédure de consignation'*. 'You know, in France we have an extremely meticulous administration. The French administration, especially in the environmental field, are resolute people, very good, they're not at all lax about the environment' (interview with *CNPF*, June 1993). 'We're a country with relatively few laws, but those we have are complemented by a powerful, a very powerful administration' (interview with Roqueplo, June 1993). On balance, however, the consensual style appears to prevail (Roqueplo 1988). Even though some industrialists 'complain bitterly about the arrogant and curt manner of certain *directeurs régionaux*, the vast majority of them seem to profit from the way local negotiations are carried on (ibid., 110). This practice of amicable settlement and negotiation is facilitated by the circumstance that the top ministry officials (especially in the Ministry of Industry), the *DRIRE*, and the large enterprises are all 'old boys' of the same establishment, the *Ecole des Mines*. 'The regulator and the regulated come from the same college. This produces an old-boy solidarity that short-circuits intermediaries ... One negotiates directly with each other. It's a specifically French phenomenon with far-reaching consequences' (ibid., 114).

As we have seen, there are substantial differences between Germany, Britain, and France in the perception of issues as the basis for state action, and in the institutional and instrumental preconditions for processing air-pollution problems. In the context of European policy-making, the three leading member states seek to impose their specific policy approaches. We shall be investigating whether they succeed and, if so, why.

At the beginning of the eighties, it was the Federal Republic of Germany that imposed its problem-solving approach, putting its stamp on European policy in the regulation of industrial pollutants. In the regulatory contest it managed with the aid of the Commission to oblige the other member states to alter their practices.

2.5 European Union: Centralized Regulation and Subsidiarity

The way member states behave in supranational decision-making is determined not only by the properties and processes of national policy networks. The assertion of national interests in regulatory competition is also strongly conditioned by the characteristics of the supranational network. The institutional and legal setting for European politics and the patterns of interaction between national and supranational actors are crucial factors in delimiting member states' negotiating position. The strategic stance taken by the central actor in the supranational network, the Commission, also influences this position. In the course of the eighties, both the institutional and instrumental aspects of European environmental policy underwent important changes, which have had a lasting impact on the goals and strategies of individual member states of the Union.

2.5.1 The Beginnings of European Environmental Policy: Measures and Institutions

Until the adoption in 1986 of the SEA, European environmental policy was not a field of action for the EU established by treaty. The Commission was obliged to justify proposed action in this domain by reference to other objectives covered by the Treaties of Rome. Despite these limitations, an independent European environmental policy developed, the superordinate strategies of which were set forth in several action programmes. In the early eighties, the Commission began to espouse a quality-based approach to clean-air policy. However, concrete regulatory arrangements were hampered by the distinctive institutional features of the European decision-making process (especially the unanimity principle in the Council of Ministers), which usually permitted only lowest common denominator measures.

The Institutional and Legal Setting

In formally defining the tasks and powers of Community institutions, the Treaties of Rome provide the general institutional and legal setting for supranational decision-making. Certain distinctive features arising from the supranational nature of the Community also play an important part in shaping European legislation, especially important being the tendency to impose rather abstract framework requirements open to interpretation — within certain limits — by member states.

Institutions of the EU and their Functions before the SEA

Under the Treaties of Rome, member states transferred functions and rights to four principal institutions of the European Union as the entity vested with sovereign Community powers: the Commission, the Council of Ministers, the European Parliament (EP), and the European Court of Justice (ECJ). While the Commission and the Council of Ministers have legislative and executive functions, and the European Parliament advisory and supervisory powers, it is the task of the European Court of Justice to exercise judicial review with respect to the interpretation and application of Community law.

Commission competence covers three complexes: legislative, control, and executive functions (Weidenfeld/Wessels 1992, 247). Among the legislative tasks is the planning and elaboration of proposals for Council decisions. Without a proposal from the Commission, the Council of Ministers cannot (with few exceptions) adopt any legal measures; the Commission has sole right of initiative. The enforcement powers of the Commission are concerned with upholding primary and secondary Community law. The former consists of the treaties on which the Communities are based, while secondary Community law is law created by the institutions of the Community. In exercising its control functions, the Commission may demand all necessary information and conduct investigations in member states. Member states have far-reaching duties to provide information to the Commission. If the Commission discovers infringement of Community provisions, it may institute legal proceedings against member states or other Community institutions before the European Court of Justice on grounds of non-compliance with the Treaties. Finally, the executive functions include representing the Community externally in negotiations under both private and international law, and management of the EU budgets and the various funds (Ebers 1989, 358f.; Woyke 1984, 365).

At the time the Commission consisted of seventeen[139] commissioners appointed by the common accord of member state governments. Below the commissioner level an independent administrative infrastructure had developed, comprising several directorates general and special services, which in their turn were subdivided into various directorates and divisions (Harbrecht 1978, 75-77). In preparing legislative initiatives, the Commission was assisted by over a thousand standing and temporary committees with coordinative and advisory functions. They were composed mostly of European and national civil servants, private experts, and representatives of various interest groups (Schmitt von Sydow 1980, 131ff.).

Members of the Council of Ministers were representatives of member state governments. Depending on the matter in hand, governments were represented by different ministers. The office of President of the Council was held in turn by each member state for a term of six months. The Councils of

Ministers were under the direction of the European Council, which twice a year assembled the heads of state and government of member states and the President of the Commission, accompanied by the foreign ministers and a member of the Commission, to discuss matters relating to the further development of the EU and, where necessary, to reach fundamental decisions or 'conclusions' (Peters 1992). The most important task of the Council of Ministers was to legislate within the framework of secondary Community law. The Council adopted the legal measures proposed by the Commission. As a rule, it was not bound by specific deadlines. In addition, the Council had a right to initiate legislation in certain cases under Article 152 of the EEC Treaty, which permitted it to require the Commission to submit a proposal (Hrbek 1984, 393). Finally, the Council had rights of control vis-à-vis the Commission. It could call on the ECJ to determine the legality of Commission actions (Ebers 1989, 360). As provided by the EEC Treaty, decisions were taken in the Council of Ministers by simple or qualified majority, or by a unanimous vote. However, voting subject to a particular majority requirement (qualified majority or unanimity) were the rule (Weidenfeld/Wessels 1992, 278f.). Decisions on environmental policy matters based on Article 235 or Article 100 of the EEC Treaty had to be adopted unanimously.

The Committee of Permanent Representatives (COREPER) prepared Council meetings and decisions in the Council of Ministers. COREPER I was composed of deputy permanent representatives. This body was concerned with subjects that were politically less controversial. COREPER II, comprising the ambassadors (Permanent Representatives) of the member states, dealt with politically more significant matters. Members of COREPER were bound by instructions from both the Council and member states (European Communities 1991, 23). Before certain measures were discussed in the COREPER, technical details were settled in various working groups, predominantly composed of national civil servants. The fact that the national officials in these working groups were usually identical with those sitting in bodies consulted by the Commission in preparing their decisions indicates the close meshing of national and European bureaucracies (Weidenfeld/Wessels 1992, 280). Moreover, all meetings at the Council of Ministers level were also attended by representatives of the Commission.

The legal form taken by Council and Commission resolutions can differ. Whereas directives were binding only with respect to the goals to be attained, leaving it to the member States to decide in what form and by what means objectives were to be met, regulations had binding effect with respect to both means and ends. As opposed to the abstract, general nature of directives and regulations, decisions regulate specific cases; they have individual application and are binding only on those to whom they are addressed.

However, the distinction between directive and regulation is hazy. Some directives in the clean-air field contain very detailed requirements, for example on measurement. It may also be the case that so-called framework directives define very general scope for action to be filled in by comparatively specific directives that can come relatively close in type to regulations (Rehbinder/Stewart 1985, 35).

Compared with national parliaments, the European Parliament had few powers. In the early eighties, the EP had only marginal possibilities for co-operating in lawmaking. The EP can only debate projected decisions; action is the reserve of the Commission and the Council (Harbrecht 1978, 100). The focus of Parliamentary competence was on controlling the other Community institutions, especially the Commission. This control function operated essential through questions, debate and consultation (Bieber 1984, 340).

The European Court of Justice, as the judicial organ of the Community, ruled on the interpretation and application of the Treaties in disputes between member states, between the EU and member states, and between the EU and individuals. Furthermore, national courts in member states had recourse to the ECJ, which in cases submitted to it by the national courts could deliver 'preliminary rulings' on the interpretation and application of Community law (Shapiro 1992).

Formal Decision-Making Procedure
Preparation of an EU directive or regulation begins at the Commission level. The Commission has an unrestricted right of initiative. It nevertheless often occurs that in proposing legislation, the Commission reacts to proposals put forward by individual member states (interview with EU Commission DG XI, March 1993) or the European Council. As a rule, the competent directorate general passes the proposition to an appropriate working group, which elaborates the details of the measure and assess its political feasibility and practicability. The outcome of the process of consultation, in which national officials, experts, and interest groups participate, is an official draft directive, which is published in the Official Journal of the EU. The proposal is then examined by the Council of Ministers. In certain cases the Council has first to submit the draft to the EP or the European Economic and Social Committee for their views. The Council of Ministers then passes the proposed measure to an appropriate working group for further processing and examination. At this level, the first official discussions and negotiations between the civil servants of the various member countries take place. The outcome of these consultations provided the basis for negotiation at the COREPER level. Final details are settled and the fully negotiated proposal is submitted to the Council of Ministers for adoption without further discussion taking place at this level.

However, depending on how politically delicate the measure is, further negotiation is frequently needed at ministerial level. If agreement still proves elusive, renewed discussion takes place in the working groups. A considerable time might thus pass between the original Commission proposal and final adoption of a measure by the Council of Ministers (Gregory 1983, 25f.; European Communities 1991, 32).

The Legal System

Law-making by the European Union is essentially in the continental European Roman law tradition prevailing in the six founding countries France, Italy, Germany, Belgium, the Netherlands, and Luxembourg (Bridge 1981, 357). Typical of this tradition are codified, abstract rules intended to guide future action ('law as "leading" society') rather than to settle particular disputes like English common law. Practical application of these general principles is in the hands of professional judges, who interpret the law by objective criteria (van Waarden 1992, 13ff.). The predominant mode of interpretation is teleological, seeking the sense and purpose of norms. Although judicial precedents have a role to play within this tradition, they do not constitute a source of law in themselves as they do in the inductive and pragmatic common-law system (Bridge 1981, 357). EU law with its extremely abstract and vague formulation is hence in the tradition of Roman law.

This tendency for vague formulation is encouraged by another factor. It should be remembered that law-making at the European level is always the result of negotiations and compromises, necessarily reflected in more abstract and less precise language. 'Political compromises are often attained by the use of ambiguous words' (ibid., 360). Relatively broad phrasing gives individual member states leeway in implementing EU law at the national level. However, it also causes great uncertainty in interpretation, especially in countries with a different legal tradition. Interpretation of EU law under common law, primarily oriented on judicial precedent, bears little scrutiny by the ECJ, which uses teleological criteria in applying rules to specific cases (ibid., 375).

Actors and Processes

There are two faces to the European Union. It is both a supranational entity (embodied in the Commission, the European Parliament, the European Economic and Social Committee, the European Court of Justice) and an intergovernmental system (embodied in the European Council, the Council of Ministers, and COREPER). The respective significance of these two elements cannot be defined with any exactitude. Fluctuations are apparent between both individual policy areas and between various policies within the same policy area (Kohler-Koch 1993, 100). Nor is it as simple a matter as might appear at first glance to categorize an actor as either supranational or intergovernmental. Each exhibits a varying mix of specifically constitutive supranational and intergovernmental characteristics. Assignment to one or the other category is possible only on the basis of relative weighting.

Problem Definition and Agenda-Setting
Within the supranational network, it is possible to characterize the Commission as a corporative actor, because rights and resources have been transferred to it by the member states, endowing it with its own powers to act. Unlike an intergovernmental actor, the interests of which correspond to the aggregation of individual interests, the Commission pursues not only Community-wide goals, but also institutional interests of its own, which can be contrary to the notions of individual member states (Kenis/Schneider 1987, 438ff.). The relatively limited room for financial manoeuvre, with the lion's share of financial resources going on agriculture and development, means that the Commission will tend to safeguard its vested institutional interests by expanding its regulatory powers (Eichener 1992, 51). This development is clearly apparent in the steady growth of EU environmental regulation, a policy area that the Commission has done much to open up (Majone 1989, 167). The Commission's principal resources consist in its power to initiate EU legislation and hence to influence decision-making, and in its role as mediator between conflicting national interests. 'The main resources of the Commission in dealing with the member states are its functions as a process manager, initiator and policy-broker' (interview with EU Commission DG XI, March 1993; see also Eichener 1992, 52ff.). Although the Commission with its monopoly on proposals can be regarded as the initiator of Community law (Schumann 1991, 50), this does not necessarily mean that all EU measures can be attributed to the Commission. Individual member states, interest groups, or 'quangos' also approach the Commission, requesting certain policies to be elaborated for reference to the Council of Ministers (Siedentopf/Hauschild 1990, 448). Precisely in the clean-air field, the Commission, with its meagre personnel resources, is evidently more willing to take up the initiatives of individual countries than to lead the way itself (interview with EU Commission DG XI, March 1993).[140]

By delegating national civil servants to the Commission, member states have the opportunity to influence problem definition and agenda-setting at the European level (interview with DoE, Nov. 1992). The relative paucity of personnel resources renders this factor all the more important. The Commission expressly welcomes the delegation of such experts, since they are a valuable source of information (interview with EU Commission DG

XI, March 1993). Thus the Industrial Emissions Division in Directorate General XI has a staff of sixteen, of whom half are delegated from the member states and quit the Division after a maximum period of three years. 'We address member states and ask for people who have expertise in a specific field. Personnel all in all is very scarce' (interview with EU Commission, DG XI, March 1993). 'The Commission is what might be called an "adolescent bureaucracy". It is still very dependent upon national experts and groups for detailed information about diverse technical standards, legislation and organizational structures throughout the EC' (Mazey/Richardson 1993a, 5).

EU activity is also governed by external factors. When treaties and international agreements are signed, the EU must take appropriate steps to implement them. One example is the 1979 Geneva Convention on Long Range Transboundary Air Pollution, which requires SO_2 emissions to be avoided as far as possible or gradually reduced (Hohmann 1989, 38). It appears that both supranational and national interests can influence both decisions on whether a social problem needs to be dealt with politically at the European level, and how the problem is perceived.

Programme Formulation
During programme development there is close interaction between national and supranational elements. Drafting by the Commission is already characterized by close cooperation with national bureaucracies. This takes place primarily in the numerous working groups, which are composed of national civil servants, Commission experts, representatives of interest groups, and external experts (Siedentopf/Hauschild 1990, 448). While these groups constitute an important source of information for the Commission, they offer member states the possibility of influencing policy-making through their own experts: 'National civil servants try to influence the Commission towards their own position' (Wessels 1990, 238). There are nevertheless some indications that the general approach at this level is not 'bargaining' but 'problem-solving' (Scharpf 1985). The working group activity is not directed towards resolving politically sensitive issues. Their job is rather to settle legal and technical difficulties that could make agreement in the Council of Ministers and subsequent implementation more difficult. It is thus the task of the experts 'to highlight and then iron out those elements in a proposal which will render its implementation or application in the member states difficult' (Weiler 1988, 35).

As soon as the proposed measure leaves the informal forum of the working groups at Commission level and is referred to the Council of Ministers as an official draft proposal, the picture changes. National interests come more strongly to the fore, bargaining in the sense of safeguarding positions becomes the principal problem-solving mode. Trade-offs between the interests of member states determine the climate of negotiation (Siedentopf/ Hauschild 1990, 448).

The phase of negotiation in the Council of Ministers is, however, not exempt from supranational influence. Especially important in this respect is the Commission's right of proposal, which largely determines the agenda and the substantive discussion in the Council of Ministers. Drafting input provides a rough setting for decision-making that member states can only revise completely if they reach agreement among themselves, an improbable event in view of conflicting national interests (Eichener 1992, 45). An alternative is for the Council of Ministers to block a decision, but because of the political pressure the problem exerts, this is unlikely to happen frequently. More often than not, it may therefore be said that 'once a directive has been proposed by the Commission it should go through and be discussed, either rejected or accepted in roughly the same form as in which it was proposed. ... You can't change the nature of it on the Council table' (interview with the Office of the EU Commission London, January 1993). The Commission can furthermore threaten to withdraw a proposed alternative. The Council of Ministers is then under greater pressure to reach agreement, because in this event the political decision-making process would be at an end: 'Once the Commission takes it off the table, that's the end of the process' (ibid.).

Finally, the Commission, together with the country holding the Presidency, has an important role to play in this phase as mediator between member states with diverging interests (Eichener 1992, 45). There is close cooperation between the Presidency of the Council and the Commission when it comes to setting the agenda for working groups, COREPER, and the Council of Ministers. Within the framework of the timetable laid down by the Commission, the Presidency has the possibility of setting certain priorities and elaborating its own proposals and compromises. The ways and means of discussion, and the timing of decisions can also be influenced by the Presidency (Kirchner 1992, 90). At the same time, the incumbent country has the opportunity to introduce its own interests in the policy-making process, although the six month period is relatively brief for this purpose (Gregory 1983, 144f.).

A trade-off often occurs between the Presidency and the Commission in setting the agenda, within the context of which the Presidency is conceded the opportunity to promote matters close to its heart if in exchange it gives the Commission's proposals for directives corresponding priority (interview with DoE, September 1993). National interests and Community-wide goals thus meet in the Presidency (consensus-finding in the Council of Ministers). The factor linking the two elements is political responsibility and political prestige, which play a part for a national government in giving form to its period in office. The necessary conditions for the success of a presidential term is often the willingness of a country to desist in some degree from placing its own interests first (Kirchner 1992, 109).

Implementation

In the implementation phase, the emphasis in European policy shifts from the supranational to the national level.[141] The EU has to rely on national governments to carry out its policies, which in their turn can involve regional or local government or non-governmental actors (Schumann 1991, 240). The Commission itself has no executive competence in the field of environmental policy. It may neither issue direct instructions to national authorities nor is it empowered to set up authorities at the national level (Rehbinder/Stewart 1985, 137). It is the task of member states to take the organizational steps necessary to implement EU directives. They include the setting up of authorities, the provision of financial and manpower resources, and the 'translation' of European law into domestic law, which in the clean-air field, for example, can include the specification of emittents' rights and duties, supervisory and monitoring procedures, and measures for legal control (ibid.). In this context, implementation depends very much on specific national and sectoral administrative styles and legal traditions (Siedentopf/ Hauschild 1990, 453). EU law is, so to speak, spelled out in national practice. The degree of influence exercised by national political and administrative systems depends, however, on how much scope for action EU law allows. If little policy and administrative leeway is given, this can lead to changes in national patterns.

The powers of the Commission in the implementation phase are limited more or less to supervision and information. Member states are responsible to the Commission for the due and proper implementation of EU law. In this regard the Commission has far-reaching rights of information and control. If a country fails to honour its obligations, the Commission may institute proceedings against the delinquent country before the ECJ. However, the Commission has limited scope for imposing sanctions. National actors (such as environmental organizations) may also bring an action before the ECJ to review compliance by their government with European directives (Rehbinder/Stewart 1985, 143ff.). Such actors' radius of action is thus enlarged by the interlinkage of supranational and national policy networks.

EU Clean-Air Policy up to 1980

From the political and legal perspectives, EU environmental policy differs from other policy fields such as trade, agriculture, or transport primarily in that it is not explicitly mentioned in the 1957 Treaties of Rome that established the Community. This was certainly due to the almost complete lack of awareness of the issue at the time, likely to have been particularly extreme in the case of a community like the EU designed primarily to promote economic cooperation (Johnson/Corcelle 1989, 1). It was only fifteen years after the signing of the Treaties of Rome that the formation and development of a European environmental policy began to become apparent. Since the mid-sixties, numerous environmental disasters have provoked greater politicization of environmental problems. The increasing pressure exerted by these issues resulted in first, wary steps being taken towards a Community-wide environmental policy. At the summit meeting in Paris in October 1972, the EU heads of state and government assigned the Commission the task of elaborating an environmental action programme. The Commission formed a 'task-force group' for this purpose, which was later to develop into the Directorate General XI Environmental and Consumer Protection and Nuclear Safety (Bongaerts 1989, 579).

Legal Basis

At that time, however, it was quite unclear what should constitute the legal basis for European environmental policy. 'It was a question of discovering hidden possibilities in the Treaty' (Scheuing 1989, 154). Subsequently, the legal basis usually taken was Articles 100 and 235 of the EEC Treaty. Article 100 provides the authority to issue directives for the purpose of harmonizing legal and administrative provisions with a view to completing the Common Market. The adoption of environmental policy directives could thus be justified on grounds that, from an economic point of view, divergent environmental requirements in member states constituted barriers to trade (Johnson/Corcelle 1989, 4; Kelemen 1995, 7). Article 235 of the EEC Treaty permits the Community to take 'appropriate measures' to attain the goals of the internal market, even though the Treaty does not explicitly provide the necessary powers (Scheuing 1989, 156). The two articles are usually taken as the enabling basis for EU directives. Both article 100 and article 235 stipulate unanimity in the Council of Ministers (Johnson/Corcelle 1989, 4).

It is remarkable that both provisions were originally intended to grant the institutions of the EU the necessary powers to achieve and secure economic integration. They were by no means designed to achieve protection of the environment as a goal in itself. Nevertheless, in a certain sense European environmental policy — encouraged by the appropriately broad ECJ interpretation of the Treaties — took on a life of its own beyond these purely economic motives. Improving the quality of the environment is regarded as an aim in itself, not a mere appendix to issues of economic integration (Rehbinder/Stewart 1985, 16; 27). The programmes and measures undertaken in the clean-air field up to 1980 will be described in brief in the following section.

Actions Programmes and Directives
The environmental action programmes of the European Union ultimately have two objectives. They specify the focal areas of EU legislation for the coming years and determine the fresh strategic directions in environmental policy. They constitute a policy framework to be filled in by appropriate directives (Haigh 1990, 11).

The adoption of the First Action Programme (1973-1976) marked the beginning of independent EU environmental policy. It set very comprehensive, demanding goals, which can be regarded as guiding principles for European environmental policy (Bongaerts 1989, 579). The superordinate objective of environmental policy was:

to bring [economic] expansion into the service of mankind by procuring for mankind an environment providing the best conditions of like and to reconcile this expansion with the increasingly imperative necessity of preserving the natural environment (European Communities 1973, 5).

Besides implementing the precautionary and polluter-pays principles, the programme also states as an objective the long-term harmonization of European and national environmental policies. A concrete clean-air measure adopted in the context of the 1975 Action Programme was a directive on the sulphur content of gas oil. It laid down product limits for the sulphur content of fuel oil and diesel fuel. The aim was to reduce both trade barriers and SO_2 air pollution (Haigh 1990, 177). In the same year, the Commission proposed a directive prescribing quality standards for the atmospheric lead pollution. This was to be adopted by the Council only in 1982.

The Second Action Programme from 1977-1981 was, as regards its strategic orientation, a continuation of the First Action Programme (Bongaerts 1989, 580). The setting of quality standards for the ambient air was required, covering especially concentrations of the pollutants lead, NO_x, CO, SO_2, and suspended particulates, photochemical oxidation agents, asbestos, and hydrocarbons (European Communities 1977, 13). Within the time frame of this action programme, however, this air-quality oriented strategy was pursued only with respect to SO_2 and suspended particulates. A corresponding directive laying down quality standards and limit values for these substances was adopted by the Council in 1980 (Boehmer-Christiansen/Skea 1991, 232). In the context of the Third Action Programme (1982-1986) the EU changed direction (Bongaerts 1989, 580). Apart from stressing the precautionary principle, emission-related strategies for combating air pollution were now the order of the day, seeking to abate the emission of pollutants at source (European Communities 1983, 2; Johnson/Corcelle 1989, 17). These principles are accordingly reflected in the directives on the combating of air pollution from industrial plants and on the abatement of pollutant emissions from large combustion plants, both of which are based on the Third Action Programme.

2.5.2　The Establishment of a European Environmental Policy: Institutional and Instrumental Developments

The importance of environmental policy relative to other Community fields of action, concerned primarily with economic integration, increased markedly as the eighties advanced. This development has become particularly apparent in recent institutional innovations at the European level: the SEA and the Maastricht Treaty. Both broaden the scope for action in the environmental field (and hence in clean-air policy). These institutional changes are accompanied by a strategic reorientation on the part of the Commission, partly revising and supplementing existing EU approaches to clean-air regulation.

Institutional Changes: the Single European Act and the Treaty of Maastricht

The SEA and the Maastricht Treaty initiated important innovations in the general decision-making processes at the supranational level, and thus in European environmental policy. Central elements in each are concerned with strengthening supranational institutions in Community decision-making, the explicit embodiment of environmental protection as a task of the Community, and the underlining of subsidiarity as the maxim for European regulation.

The decisive motive for adoption of the SEA, which came into force on 1 July 1987, was to speed up economic integration of the Community in order to implement the internal market. Stagnating growth, high unemployment, and fears of losing ground to the United States and Japan encouraged member states to renounce further national sovereign rights in order to reinforce supranational powers. Moreover, it was predicted that, with the accession of Spain and Portugal in 1986, integration could suffer further delay. It had already proved difficult enough to obtain unanimity in the Council of Ministers with ten member states, so that the prospect of additional accessions promised to aggravate the situation even more (Merkel 1993, 18).

The SEA therefore stipulated qualified majority voting instead of unanimity for all decisions pertinent to the internal market. The votes of member states were weighted according to size. Germany, France, the United Kingdom and Italy were accorded ten votes each; Spain eight; Belgium, the Netherlands, Portugal, and Greece have five each, Denmark and Ireland three; and Luxembourg two. A qualified majority was to be obtained with at least fifty-four of the total seventy-six votes. The weighting of votes is designed to prevent 'small' countries being outvoted by a bloc of 'big' countries. Vice versa, it is not possible for two 'big' countries alone to form a vetoing minority.[142] Where a qualified majority is required, the Single European Act increases the influence of the European Parliament. In these cases, consultation of the EP is supplemented by the so-called cooperation procedure. In a second reading,[143] the EP may by an absolute majority reject Council of Ministers measures or propose amendments thereto. These proposals can then be rejected by the Council of Ministers only in unanimity, whereas their acceptance requires only a qualified majority. However, the fundamental precondition is that the Commission accept Parliament's proposed amendments (Corbett 1990, 26f.).

Although these new decision-making procedures at first found no application in environmental protection (they applied only when the measure concerned also served to complete the internal market),[144] the SEA nevertheless brought a degree of order and purposiveness to European environmental policy (interview with EU Commission, DG XI, September 1993). Within the framework of the SEA, environmental policy was for the first time explicitly embodied in the EU Treaties. The significance of the SEA is that it recognizes the need not only to link the goals of free trade with a 'high level' of environmental protection (Art. 100a), but also to pursue environmental objectives as a legitimate goal in itself (Art. 130r,s,t). The SEA, which defines environmental protection as a cross-sectional task — 'environmental protection requirements must be integrated into the definition and implementation of other Community policies' (Art. 130r) — thus gives environmental policy a special status (Scheuing 1989, 176).

The Maastricht 'Treaty on European Union', which has been in force since 1 November 1993, continues the developments initiated by the SEA. The European Union (EU) forms a new overall institutional setting for the European unification process, which rests on three essential pillars, the EC, the Common Foreign and Security Policy (CFSP), and cooperation in the fields of Justice and Home Affairs (CJHA). In this context, the EC plays a decisive role. In addition to introducing a European economic and monetary union by a 1999 deadline, and giving tangible form to European civil rights (Union citizenship), the Treaty provides for new and expanded powers for the Community in individual policy fields. The areas covered are not only consumer protection, health, research, technology, education, and culture, but also environmental protection[145] (European Communities 1992, 8).

The enhanced standing of environmental policy is especially evident in the changes the Treaty introduces in Council decision-making procedures. In principle, voting is now by qualified majority. However, the unanimity requirement is retained for important areas, in particular tax matters, regional planning, land use (with the exception of waste management), as well as water management and measures substantially affecting the choice of energy sources by member states and the general structure of national energy policy (Art. 130s). In a certain sense, member states' concurrence was 'bought' by excluding a number of important subjects. In addition, southern member states demanded financial support for environmental protection and infrastructural measures within the framework of the Cohesion Fund[146] (Strübel 1992, 147).

Nevertheless, reformed decision-making procedures in the Council of Ministers have given new impetus to European environmental policy. A member state can no longer rely on blocking negotiations in the Council. Majorities now have to be 'sounded out' and coalitions formed if national interests are to be pushed through: 'Decisionmaking in the Community therefore relies more on coalition building than on national vetoes for protection' (Peters 1992, 83). At the same time, the willingness on the part of individual countries to compromise has grown, since it is now far more difficult for an isolated government to obtain concessions from others (ibid., 84). The Commission, too, has new options open to it. It plays no passive role in the formation of coalitions. From a early stage in programme development it can attempt to forge alliances (see Peters 1992, 83; Eichener 1992).

All in all, greater innovation that is not reduced to the smallest common denominator has now become possible (Strübel 1992, 147). Whereas the joint decision trap (Scharpf 1985) ultimately left a choice only between accepting a minimum solution or blocking negotiations, member states' rationale for action must now adapt to changed conditions. A government will now better safeguard national interests by taking the initiative in European decision-making with innovative proposals of its own rather than relying on vetoing other countries' initiatives at cross purposes to its own projects. This constellation will increasingly induce member states to outbid one another in their regulatory proposals, and, in the context of this regulatory competition, to adopt appropriate prior domestic rules so as to be able to enjoy the benefits of having set the pace in the supranational arena.

The expansion of EP powers over and above those provided under the SEA add another aspect to decision-making procedure. If differences of opinion with the Council arise, it is now possible in many areas to negotiate a compromise in a conciliation committee, where decisions are taken on an equal footing with the Council. While the Council has the final say in the cooperation procedure, the new powers granted the EP give it a veto and the right to block decisions.[147] This 'co-decision' procedure applies with respect to all rules relevant to the internal market and to the adoption of multiannual action programmes[148] (European Communities 1992, 26f.).

In addition to these institutional changes, both the SEA and the Union Treaty lay down an important fundamental condition for Community activity, the so-called subsidiarity principle. This principle is embodied in the Treaty of Maastricht as a superordinate requirement:

In areas which do not fall within its exclusive competence, the Community shall take action, in accordance with the principle of subsidiarity, only if and in so far as the objectives of the proposed action cannot be sufficiently achieved by the Member states and can therefore, by reason of the scale or effects of the proposed action, be better achieved by the Community (ibid., 11).

This principle had already been introduced in the area of environmental policy by the SEA (Scheuing 1989, 164). However, concrete definition of the subsidiarity concept is still controversial among and within member countries. While the states of the Federal Republic of Germany regard the subsidiarity as a guarantee of the powers vested in them by the German Basic Law, the British government sees subsidiarity mainly as a means of limiting the further transfer of national sovereign powers to Brussels (Wilke/Wallace 1992, 4ff.). British local government for its part interprets the principle as an opportunity for them to attain greater autonomy vis-à-vis central government. They point to the greater weight given the regions under the Union Treaty (European Communities 1992, 11).

Instrumental Changes: New Commission Strategies

Parallel to the changes introduced by the SEA and the Treaty of Maastricht, the Commission revised its problem-solving approach in the field of clean-air policy. The emission-oriented strategy, guided strictly by technological capabilities, has now given way to greater emphasis on ambient air quality, as was the case in the early eighties. However, there is an important difference: far-reaching informational rights for the public are intended to generate 'pressure from below', ensuring the proper implementation of quality standards by individual member states. Member states are required only to comply with certain clean-air standards. How they go about it is their affair.

In this way the Commission takes account of the subsidiarity principle embodied in the SEA and the Maastricht Treaty. However, member states are obliged to reveal the results of their regulatory arrangements — i.e. their fulfilment of EU requirements — and to release air-quality data to the public. The Commission assumes that infringements of standards — about which the public is informed — will give rise to pressure from below for directives to be duly and properly implemented. 'The system works only if pressure comes from below' (interview with EU Commission, DG XI, March 1993). A framework directive on the monitoring and evaluation of air quality is being negotiated at the European level to ensure uniform measurement and supervision of air quality throughout the EU (interview with DoE, September 1993). In keeping with this new, subsidiarity-based approach, the new European Environment Agency that began work in Copenhagen in 1994 is not designed to act as a European 'environment police'. Its primary task is to ensure the flow of information among the public, national regulators, and the Commission (interview with EU Commission, DG XI, September 1993; see also Haigh 1990, 380).[149]

With its shift in strategy, the Commission drew the logical conclusion from the problems that earlier approaches had provoked. The strategy based on ambient air quality was very poorly implemented, since member states had a great deal of leeway and the Commission has inadequate executive competence at the national level for carrying out EU directives. It has to rely on national actors (Schumann 1991, 240; Rehbinder/Stewart 1985, 137). Emission-related procedures, which allowed easier monitoring of implementation at the individual source, were also problematic. Since such an approach demanded high investment in control technology for many countries, and benefited the environmental technology industries of individual states (Héritier 1993, 38), it proved extremely difficult to reach decisions and agreement in the Council of Ministers (as amply illustrated by the Large Combustion Plant Directive). Moreover, the subsidiarity principle made legitimation difficult.

The shift in Commission strategy was already apparent in the Fourth Action Programme (1987-1992), which — although no departure from the emission-oriented strategy was yet on the books — placed more emphasis on improving public access to information (European Communities 1987, 15ff.). However, the real turning point in strategic thinking came only with the Fifth Action Programme in 1992. Besides stressing comprehensive 'transparency' and public information, the Programme sets out to combine the subsidiarity principle with closer involvement of the relevant actors in dialogue among partners (European Communities 1992a, 48f.). To bring about this dialogue, the Commission wants in future to achieve the abatement of pollutant emissions by means of voluntary private-law contracts with industry. The move was intended to bridge the gap between the frequent reproach of over-regulation and the call for subsidiarity. However, since such contracts would displace national legislation, a degree of opposition can be expected from member states (interview with EU Commission, DG XI, September 1993).

At the supranational level, three phases and modes of problem-solving can thus be distinguished. In the early eighties, EU clean-air policy was initially characterized by a consistent orientation on ambient air quality, which, under growing pressure from Germany, was replaced by an emission-related strategy based on technological capabilities. This change sharply contradicted the prevailing problem-solving mode in Britain, which is based on consideration of local environmental quality. British resistance was correspondingly stiff, as became abundantly clear in the negotiations on the Large Combustion Plant Directive. With its Fifth Action Programme, the EU performed an about-face, returning to a strategy based on ambient air quality, which, 'enriched' with further public information rights, was nevertheless intended to ensure improved implementation through pressure from below.

Notes

[1] Abteilung Umwelt.
[2] Umweltkabinett.
[3] Ständiger Abteilungsleiterausschuß für Umweltfragen
[4] Article 74.
[5] Umweltbundesamt, UBA.
[6] Bundesministerium für Umwelt, Naturschutz und Reaktorsicherheit.
[7] Bundes-Immissionsschutzgesetz (BImSchG).
[8] BImSchG, Section 1: 'Zweck dieses Gesetzes ist es, Menschen, Tiere und Pflanzen, den Boden, das Wasser, die Atmosphäre sowie Kultur- und sonstige Sachgüter vor schädlichen Umwelteinwirkungen ... zu schützen und dem Enstehen schädlicher Umwelteinwirkungen vorzubeugen'.
[9] This Federal government view is not shared by the Advisory Council on Environmental Questions (Sachverständigenrat für Umweltfragen [SRU]).In their 1987 environmental expertise the SRU emphasized the need for environmental quality standards to be defined explicitly, which they suggested should be based on two evaluations, a conservation value profile and a vulnerability profile. The environmental quality sought by the government, the Council criticised, is frequently determined not positively but ex negativo or indirectly, environmental quality being understood as the inverse of pollutant potential to be eliminated: 'Environmental quality under the conditions of industrial production techniques is primarily a result or implication of the structure and level of emissions, which enter the media air,

water and soil, and which have a lasting effect on their quality' (SRU 1988, 55). Although there is no uniform, accepted research framework within which environmental quality can be analysed, the Council asserted that it was nevertheless necessary to arrive at a political definition of quality objectives for environmental media and for individual components (SRU 1988, 54).

10 BImSchG, Part Three.

11 This is not to deny that Germans have a pronounced faith in science. In the politico-administrative system 'facts are first collected, analysed, and evaluated before a decision is made' (Hey/Brendle, 1992, 73). On the government side, the bodies that are to provide scientific backup and legitimacy for political decisions include the Federal Environment Office, the Advisory Council on Environmental Questions, and the Radiation Protection Commission (Strahlenschutzkommission).

12 Sub-paragraphs 5(1),(2).

13 Koordinierungsstelle Umwelt der Stadt Duisburg.

14 See section 1.2.

15 Given this situation, Dirk Maxeimer, former editor in chief of the periodical Natur, feels it is now 'absurd to refer to transport by car as "private transport". In the Federal Republic, the share of cars in passenger traffic is six time that of all other means of transport taken together. Travelling by tram is thus perfectly private, going on foot is almost elitist. Whoever wishes to join the masses is at home in the traffic jam' (Maxeimer 1993, 93f.).

16 Articles from the daily press in the fifties and sixties provide graphic documentation of the drastic daily and direct confrontation of the population in the Ruhr District with polluted air. Victims' complaints ranged from child respiratory disease to soot-covered window sills and washing.

17 BGBl. I781

18 GV. NW 225

19 BGBl 721.

20 For more detail see Section 1.2.2.

21 Enquête-Kommission 'Vorsorge zum Schutz der Erdatmosphäre'.

22 The major emphasis was on efficient energy use, including the use of cogeneration and industrial waste heat for district heating purposes (see UBA 1990, 130ff.).

23 The topic of the environment had already been used for tactical purposes in elections before the environmental policy actually began to emerge in the early seventies. In the 1961 general election campaign, the SPD called for 'blue sky over the Ruhr'. Once this demand appeared to have been met in about the mid-seventies, an 'all-clear effect' set in (Jänicke 1978, 155). In the eyes of the public, the environment issue lost in urgency, since it had been possible to abate particularly visible environmental pollution (dust and soot in the air, foam and dead fish in rivers and lakes) (Posse 1986, 48f.).

24 Only a few years later it was recognized that these effects are not important.

25 Großfeuerungsanlagen-Verordnung (GFAVo).

26 Gesetz zur Erleichterung von Investititionen und der Ausweisung und Bereitstellung von Wohnbauland, known in short as the 'Investitionserleichterungs- und Wohnbaulandgesetz', 22 April 1993, BGBl. 16, 466, Bonn 1993. The Act came into force on 1 May 1993.

27 Umweltdezernat Bielefeld.

28 Unabhängiges Institut für Umweltfragen.

29 Critical voices were raised in industry, as well, but criticism was from the opposite perspective, castigating economic (and thus also environmental-policy) ineffective-

ness. It was complained that the Investment facilitation and residential development land act was 'a typical example of "management by helicopter": swirl in, raise a lot of dust, and swirl out. What's been the result? Has it shortened construction phases? Of course not. We'll see where it gets us' (Interview with steel producer, March 1993).

30 In the 1994 budget, however, funding for the Federal Ministry for Environmental protection, Nature Conservation and Reactor Safety was once again increased by 5,5 per cent over the previous year.

31 1st Regulation Implementing the Federal Pollution Control Act (1. BImSchV)

32 13th BImSchV

33 Germans' strong affective relationship with the forest is reflected in numerous songs that are highly popular among those broad sections of the population who appreciate so-called 'folkloristic' music. High in the 'charts' of forest numbers, to mention only a few, are evergreens like 'Ein Jäger aus Kurpfalz' ('A Hunter from Kurpfalz'), 'Ein Männlein steht im Walde' ('There is a Wee Man in the Woods'), and 'Oh Tannenbaum'.

34 See also section. 2.2 in this chapter.

35 Landesanstalt für Immissionsschutz.

36 However, the share of expenditure on environmental protection measured against gross output is extremely low (in manufacturing industry in 1989 only 0.7 per cent) (DIW 1993, 201).

37 According to more recent studies, however, the locational quality of German industry is not impaired by the relatively stringent environmental protection requirements (DIW 1993).

38 In Germany, technical rules are compiled by public technical committees, such as the Nuclear Technology Committee, by trade associations in the form of accident prevention regulations, by the legislature as administrative directives, e.g. TA Luft, and by about 150 private organizations in Germany, such as the German Institute for Standardization (DIN) and the Association of German Engineers (VDI), as generally applicable technical rules.

39 The progressive Recycling Act (Kreislaufwirtschaftsgesetz) drafted by the Federal Ministry of the Environment, for example, 'sat on the Minister's desk' for a long time (interview with steel producer, March 1993), since the recession and the tense situation in eastern Germany made it more difficult to impose further environmental regulations.

40 See Section 2.2.1 for greater detail.

41 See von Prittwitz' 'capacity thesis', which covers both political, economic, and technological aspects (1990, 107ff.).

42 See section 2.1.

43 The wide-spread use of the loan word 'Land', pl. 'Länder' in English when referring to Germany is unjustified. The German 'Land' is essentially comparable to the entity termed 'state' in countries such as the United States, Australia and India, and there seems no point in using a superfluous loan word that in the singular is a homograph of an English word and in the plural contains a letter unknown to the English alphabet (translator's note).

44 The administrative purview of Kreis and Gemeinde covers areas where local authorities can act in their own right (so-called 'autonomous functions') and areas where the local authorities discharge functions mandatorily as directed by government ('delegated functions'). With regard to autonomous functions, the

counties and municipalities decide whether and in what form they will take action. State supervisory powers over local affairs in this category do not go beyond monitoring the legality of measures taken. Matters relating to air quality that fall in this administrative category include construction planning and the construction and maintenance local roads. In their exercise of delegated functions, which include statutory duties assigned to local authorities by the Federal or state governments, counties and municipalities are subject to supervision by state authorities with respect to the legality and expediency of the measures taken. The performance of mandatory functions, especially administrative regulatory tasks (subordinate regional and building authorities) is subject to government direction. Among the mandatory functions not subject to government instruction are town and country planning and waste disposal.

45 Local authorities, motivated by greater familiarity with the problems and with the people affected by them, often have an important function in political mobilization (greater detail in section 2.2).

46 On the theory and empirical recording of political interlinkage see Scharpf et al. (1976); Hesse (1978); Läufer (1985).

47 Each voter has two votes: the first vote influences the personal composition of the Bundestag, being cast to elect deputies by a relative majority in the constituencies (plurality system). The second vote for a party list at state level decides the relative strength of the political parties in the Bundestag, since the 662 seats are distributed among the parties in proportion to the number of second votes won. Only parties that have managed to gain at least five per cent of the second votes or which have won a direct seat in at least three constituencies are admitted to the Bundestag (so-called restrictive clause).

48 Jo Leinen from the Federal Association of Citizens' Groups for Environmental Protection Bundesverband Bürgerinitiativen Umweltschutz (BBU) in the Saarland and Monika Griefahn from Greenpeace in Lower Saxony (see Czada 1993).

49 Important examples of the 'anticipatory obedience of the Bundestag' are the decisions on the 1976 Act on Co-determination, the debate on the 1977 Military Service Amendment and the issue of extremists in the public service (von Beyme 1991, 375).

50 In the Federal Republic of Germany technical regulatory regimes are elaborated by public committees, such as the nuclear technology committee, by the trade associations in the form of accident prevention regulations, by the legislature as administrative directives (for example the technical guidelines on air quality, noise and waste), as well as by a range of about 150 private organizations like the German Institute for Standardization and the Association of German Engineers. Recognized technical regimes are almost all very close to the legal order. They can themselves become directly applicable in law through strict or dynamic reference in legal provisions. If technical rules are incorporated in administrative directives, the executive binds itself, and the standards or guidelines become in effect legally binding (see Grefen s.a.).

51 'Wesentlichkeitsprinzip', See BVerfGe 58, 257.

52 The Federal Pollution Control Act lays down the following arrangements for hearing interested parties: 'Where authority to issue regulations and administrative directives prescribe consultation of the parties in interest, a circle of representatives to be selected in each case from among the scientific community, the affected parties [i.e. potentially affected sections of the public, such as home and land owners], the

industries concerned, the transport systems concerned, and the highest state authorities with responsibility for pollution protection shall be heard' (Section 51 BImSchG).

53 This is indicated, as the BUND representative graphically illustrated in an interview, by the sequence of speakers: 'Thirty to forty different organizations are invited to a hearing. At times cards have to be distributed because of the huge number of people attending. And the sequence of speakers, regardless of who has a card, is always first the Confederation of German Industry, with Mr. Sander. Then, depending on the matter being discussed, come the Federal Association of the Chemical Industry and afterwards the Federal Association of the German Coal Industry or the German Energy Industry. And then as an alibi for the very democratic procedure comes a representative of an alternative view, usually BUND. And after that it's back to representatives of the industrial associations. But always in the same order: first Mr. Sander from the BDI, then the next biggest federal association, and then in third or fourth place some environmental organization' (interview with BUND, March 1993). According to the Federal Ministry of the Environment, by contrast, 'The order of speakers is determined by the rule: 'the first to ask leave to speak will be the first to speak' (interview with BMU, January 1994).

54 Although many representatives of industry do not deny their good relations with government, they do complain — and in this they are at one with the environmental organizations — about the sometimes substantial modification of ministerial bills in the Bundesrat committees, which act beyond the reach and control of the lobby organizations. 'Important changes happen there at the level between the committees and in the Bundesrat debates, which are carried out in the working groups of state departmental heads the associations can't get at' (interview with steel producer, February 1992).

55 Of German environmental associations, 'BUND is almost the only group concerned with pollution control law that is represented at the federal level' (interview with BUND, March 1993) and which considers itself expressly to be a lobby for Nature at all levels of society and politics

56 Belastungsgebiet, Smoggebiet, Schongebiet. In North Rhine-Westphalia the southern and central Rhine Corridor, and the western, central and eastern Ruhr District, were designated polluted zones. These areas total 3,167 km^2 (9.3% of the total state territory). In the late seventies, SO_2 emissions in these areas was almost nineteen times higher than the federal average. Both the eastern and western Ruhr Districts were also declared smog zones. As early as 1964, work began here on installing an automatic smog-warning system with eleven monitoring stations (see Knoepfel/ Weidner 1985, 368ff.).

57 Untersuchungsgebiete. In North Rhine-Westphalia the following towns and cities were designated as inspection zones: Bonn, Aachen, Düren, Wuppertal, Solingen, Remscheid, Mönchengladbach, Velbert, Krefeld, Wesel, Münster, Bielefeld, Paderborn, Hamm, Siegen, Hagen and Iserlohn (interview with the State Institute for Pollution Control, February 1993).

58 The possibilities of computer-aided remote emission monitoring are at present under consideration at both the political and administrative levels, which would permit the authorities 'to run emission data on-line from every plant to a central desk.' Industry has reacted guardedly to this project, since 'no supplementary effect is expected for the environment' and there are fears that 'in the event of infringement they'll come at us at once with the public prosecutor and all the trimmings. And then on the

slightest "beep" automatic routines will be triggered before it can be discovered whether there's something in it or if it's a false alarm' (interview with a chemical company, February 1993).

59 According to the Federal Ministry for the Environment, the precautionary principle provides for state intervention to combat environmental dangers that are already apparent ("Gefahrenabwehr"), to minimize anticipated environmental risks ("Risikovorsorge") and to promote environmentally compatible technical developments ("Zukunftsvorsorge") (BMU 1990, 15ff.). Although the precautionary principle was already postulated as a key precept for environmental protection at the outset of Federal environmental policy and in the First Environmental Programme adopted by the Federal government in 1972, the Federal Environment Office estimates that, 'in the final resort, something in the way of reparatory environmental policy or damage abatement was pursued on a large scale, and only in the last two or three years has an attempt been made to stop damage occurring in the first place' (interview with the Federal Environment Office, November 1992.

60 Administrative authority of the Regierungsbezirk (translator's note).

61 Administrative unit covering a number of counties and county boroughs over which it has general supervisory power; in size a Regierungsbezirk many often be equated with a large English administrative county (translator's note).

62 Ministerium für Arbeit, Gesundheit und Soziales (MAGS).

63 Trade supervisory activities in the field of pollution control in the period 1963 to 1976 can be categorized as follows: imposition of restrictions on fuels, constructive and procedural measures, exhaust gas sanitation, chimneys, operational measures, monitoring and maintenance, settlement of locational issues, and licensing (Arbeits- und Sozialminister des Landes Nordrhein-Westfalen 1969, 47f.). Since 1976 the now much more comprehensive supervision of industrial plants has been classified differently: individual case regulation concerning commercial and industrial emission sources (e.g. prohibition of operation, monitoring), monitoring compliance with materials and products requirements (e.g. sulphur content of light fuel oil), risk-avoidance measures in the event of low-exchange weather conditions (e.g. changes in operation), measures to prevent accidents with air pollutants (e.g. reporting accidents) (Minister für Umwelt, Raumordnung and Landwirtschaft des Landes Nordrhein-Westfalen 1986, 24ff.).

64 Especially projects under Sections 31, 33, 34 and 35 of the Building Code.

65 If it turns out that the local authority has no legal basis for withholding its consent, the decision may be overridden by the legal supervisory body, and the municipality can be held liable for compensation.

66 This unintelligibility or impracticability of rules can give rise to situations where, in relation to new or amended laws, 'the authorities are not the ones to approach firms and demand implementation: firms take the initiative. They jump the gun and fulfil the new requirements.' Because of the eagerness with which plant operators meet their statutory requirements in such cases, the authorities are handed a 'standard' that they can thenceforth apply as a yardstick in their demands for implementation. 'In this way they find out all about the matter and then, when they take action themselves, they say OK, listen you guys, this is what we've got from company X, and that's what we now want from you, but the same way' (interview with environmental consultancy firm, September 1993). However, the brakes are applied to such implementation fervour as soon as a firm 'puts forward a proposal and has its wrist slapped by the industry' (interview with environmental consultancy firm, September 1993).

67 The new type of environmental organizations created in the course of the eighties at the local government level can be classified in terms of centralization/decentralization as follows: 1. Decentralized functions without 'functional supplementation'. As in the seventies, environmental protection is dealt with not by specific organizational units but by the appropriate departments within the framework of their given functions. 2. Decentralized functions with 'functional supplementation'. Environmental affairs continue to be within the purview of the appropriate departments, and are merely supported by additional, interdepartmental institutions (environmental protection working groups; environmental protection project groups; environment commissioners). 3. Centralized functions. Environmental protection functions are carried out by one institution: a directorate for environmental protection, which comprises various offices dealing with environment-related functions (trading standards, health, town planning, etc.) or an environmental protection office with specific competence in environmental protection matters (see KGST 1985; MURL 1986, 9f.; Jaedicke et al. 1990, 45; 49).

68 Boehmer-Christiansen (1988) gives an interesting description of how divergent cultural values and points of view in German and British problem-solving approaches find linguistic expression. The meaning of certain German words like 'Schadstoff' (literally harmful or damaging substance) or 'Immissionsschutz' (literally protection against pollutants — where Immission 'refers both to the presence of pollutants and to their effects') induce a far greater sense of threat against which nature has to be protected ('we guard and protect our environment'), whereas British parlance expresses merely the need to organize and control the environment ('we manage our environment').

69 Since 1956, so-called smokeless fuels can be prescribed by law by local authorities for domestic heating purposes in certain areas within their jurisdiction. Both this instrument and societal changes have reduced the share of coal heating in private households (interview with IEEP, December 1991).

70 Miners received cut-price coal and saw this privilege of cheap heating endangered by the possible prohibition of coal for domestic heating purposes.

71 In 1973 GDP rose 7.6 per cent inflation at 7.1 per cent and unemployment at 3.0 per cent (OECD Historical Statistics 1960-1984).

72 Source: Annual Abstract of Statistics, London, HMSO (quoted in Boehmer-Christiansen/Skea 1991, 117) and OECD Statistics 1960-1984.

73 The number of seats is due to increase from 651 to 659 (see Economist, June 10th 1995, vol.335 No.7918, 37).

74 These tribunals were set up in the course of establishing the Welfare State after the Second World War. The intention was to provide citizens with the possibility of asserting claims to government services before these 'quasi-courts' (Budge/McKay 1988, 170).

75 The AI was responsible for supervisory activities in England and Wales and in Northern Ireland. Scotland has a similar body, the Industrial Pollution Inspectorate, which has the same material functions. It should be noted at this point that reference throughout the book is to the laws and institutions of England and Wales, since arrangements in Scotland differ only marginally and are in substance directly comparable.

76 Limit values are laid down only for salt and sulphuric acid, now irrelevant relics of earlier alkali legislation.

77 Although from a superficial point of view there is no direct incentive for administrators to exhaust their room for manoeuvre within the framework of

informal and 'in camera' relations, a glance at the poorly developed administrative law in Britain shows that in the absence of sanctions, the administrative authorities ultimately have little other choice (Vogel 1986, 83).

78 Between 1920 and 1966, for example, the AI took only two firms to court (Vogel 1986, 88).

79 Delay in the approval process could on the contrary be in the interest of a firm precisely if it did not depend on prior authorization to operate a plant, since additional conditions might emerge from the approval process.

80 Between 1970 and 1974, the annual number of cases brought by local authorities against industry ranged from 50 to 133 (Vogel 1986, 88).

81 The AI's policy of secrecy went as far as omitting the names of firms that had obviously infringed regulations in the AI annual report. Only the sector to which such firms belonged was stated (e.g., 'a metal processing company') (Vogel 1986, 93).

82 *La loi du 19 décembre 1917, relative aux établissements dangereux, insalubres ou incommodes'*.

83 'Délégation à l'Aménagement du Territoire et à l'Action Régionale' (Parodi 1971, 229).

84 *'Cette stratégie s'inspirait donc dans une certaine mesure de la théorie de la croissance déséquilibrée puisqu'elle sélectionnait sévèrement un petit nombre de villes qui devaient jouer vis-à-vis de leur vaste "Hinterland" le rôle détenu naguère par Paris vis-à-vis de la France entière.''*

85 The promotion of these agglomerations intensified the disparities between large and medium-sized urban places and rural areas (Hadesbeck 1991, 34).

86 France is one of the few countries in western Europe with exploitable uranium deposits (see INSEE 1988). After the Second World War, the military use of nuclear power was to the fore rather than civil uses (Berg 1992, 26).

87 This markedly slower economic development has been attributed to 'economic Malthusianism' (Sauvy 1965), which, as was typical for French entrepreneurship at the time, was concerned with stability rather than expansion and with protection rather than competition (Neumann/Uterwedde 1986, 37).

88 See section 2.2.2 in this chapter.

89 The phenomenon of forest dieback was discussed in France only after the Germans had come to consider road transport as the major culprit of acid rain and had demanded the introduction of the catalytic converter throughout Europe (interview with Roqueplo, June 1993). This affected the economic interests of the French motor industry. Fitting catalytic converters would have been a money-losing proposition because it paid off only for large cars of the type constructed in Germany (Larrue/Prud'homme 1990, 10). The French government complained — without effect — that French industry was having 'to pick up the bill for an internal policy turn-around in the FRG' (Roqueplo 1986, 413).

90 Their success is acknowledged: 'The Germans have had the intelligence and wisdom to consider the environment as an economic weapon like any other, it's a market like any other' (interview with *CITEPA*, March 1993).

91 The political struggle for power between parliament and government paralysed government policy, so that frequent changes of government were the order of the day, and a continuous policy became impossible (Wright 1983, 128).

92 Among the powers vested in the President in the policy decision-making process is the appointment of the Prime Minister, the dissolution of the National Assembly, the

issue of emergency powers and the calling of referenda. Moreover, the President decides on foreign policy and defence issues (Duhamel 1991, 58ff.).

93   We are grateful to Burkhard Eberlein for this information.

94   Half the members on the regional council were members of the National Assembly and the other half representatives of *département* and communal councils (Fromont 1983, 399).

95   Because of the interlinking interests of the prefect and local *notables*, the prefect's power of decision was to a certain extent accommodated to local conditions (Grémion 1976, 222).

96   *Les préfets dans leur circonscriptions comme le corps préfectoral dans ses prises de position ne cessent de réitérer leur vocation de coordination, de synthèse et d'arbitrage de l'action collective dans le département. Ils disposent pour ce faire de moyens juridiques considérables...[Cependant] les préfets ne peuvent pas s'appuyer sur leur seules ressources réglementaires pour assurer la coordination organisationnelle de l'appareil administratif, ils peuvent encore moins le faire si ils veulent agir sur l'environnement local. L'intervention du préfet, sa capacité d'arbitrage et ses possibilités d'intégration dépendent directement de la structure du leadership notabiliaire du département.'*

97   Including road building, supralocal public transport, and water and power supplies.

98   After the reactor accident at Chernobyl and the subsequent protests by French nuclear power opponents, the first monitoring stations independent of the state were set up on the initiative and with the financial backing of the Alsace Regional Council. Before adoption of the decentralization statutes this would have been impossible against the declared will of the prefects (interview with *Les Verts*, Sept. 1993).

99   In towns or urban boroughs greens won between five and ten per cent of the vote, while in Paris and Alsace the figure was just under fifteen per cent.

100 The thesis already advanced by Knoepfel/Larrue (1985, 64) that precisely individual political personalities in decisive positions with their ecological commitment exert the most influence on clean-air policy is confirmed by the case of the Alsace region (interview with *Les Verts*, July 1993; interview with *ASPA*, July 1993).

101 In the Region Nord-Pas-de-Calais, for example, a farmer took legal action against a major concern because its emissions had caused the death of his productive livestock (interview with *Nord Nature*, June 1993).

102 Before an environmental group can sue, it must have been recognized as an association for at least three years ('*agrément*') (Rest 1986,14).

103 *La loi du 19 décembre 1917 relative aux établissements dangereux, insalubres et incommodes.'*

104 *La loi No. 61-842 du 2 août 1961, relative à la lutte contre les pollutions atmosphériques et les odeurs et portant modification de la loi du 19 décembre.*

105 *La loi No.76-663 du 19 juillet 1976 relative aux installations classées pour la protection de l'environnement.*

106 The nomenclature is expanded in accordance with expert opinions from two bodies, the '*Conseil Supérieur des installations classées*' and the '*Conseil d'Etat*' in decree form (Rengeling 1985, 116).

107 France's hazard study was supplemented by the 'Seveso' Directive adopted by the EU, so that firms were now subject to more stringent rules, resulting, for example, in some plants that were now 'Seveso'-classified, converting to different, less hazardous production methods (interview with Alsace Nature, July 1993).

108 *La loi No.78-753 du 17 juillet 1978 portant diverses mesures d'amélioration des relations entre l'administration et le public et diverses dispositions d'ordre administratif, social et fiscal.*

109 *La loi No 83-630 au 12 juillet 1983 relative à la démocratisation des enquêtes publiques et à la protection de l'environnement'* is also referred to as the '*Loi Bouchardeau*' because it was initiated by the then Secretary of State for Environmental Affairs, Huguette Bouchardeau (Jegouzo 1990, 269; interview with prefecture, June 1993).

110 An '*arrêté*' can be issued by both a prefect or a mayor or a ministry. '*Arrêtés*' are legally directly applicable in contrast to the '*circulaire*' or '*instruction*', which correspond more close to the German ministerial guidelines or '*Runderlaß*' ('circular order'), and are binding only internally on the administration (Constantinesco/ Hübner 1974, 7; Rengeling 1985, 121).

111 The regulatory instrument 'special protected zone' is comparable to the 'smokeless areas' deployed in Britain to abate SO_2 emissions from domestic heating.

112 For the most part the fuels in question are classified *TBTS* ('*Très Basse Teneur en Soufre*') (interview with *DRIRE*, November 1992).

113 Paris, Lyon et Ville Urbane, Marseille, Petite Couronne de Paris, Agglomération Lille-Roubaix-Tourcoing and Strasbourg (*Ministère de l'Environnement* 1991a, 3).

114 Dunkirk, Le Havre, Rouen, Paris, Strasbourg, Montbéliard, Nantes, Lyon, Grenoble, Fos and Marseille (*Ministère de l'Environnement* 1991a, 149).

115 The introduction of this charge was pushed particularly strongly be the environmental association '*Amis de la Terre*'. Their representative Brice Lalonde, later to become Minister of the Environment, conducted talks on the issue with the then Prime Minister Laurent Fabius (interview with *Amis de la Terre*, June 1993).

116 There are precise rules on the financial contribution the government will make to clean-air measures. If the measure contributes to a more than 40 per cent desulphurization of operating equipment, 10 per cent of investment costs are assumed by the government, and where pollutant reduction exceeds 60 per cent, the government pays 15 per cent of the costs (Olier et al. 1989, 56).

117 '*Ce sont deux taxes dont le but est le même; ce ne sont pas des taxes dissuasives, parce qu'elles ne sont pas assez élevées pour forcer les industriels à modifier leur processus de production, C'est pour alimenter les Agences à faire des opérations antipollution.*'

118 Contracts have been concluded with, for example, sugar factories, distilleries, with the cement industry and various refineries, iron works and asbestos and steel works (Rengeling 1985, 124).

119 The *Institut Pasteur de Lille* began taking air-quality readings as early as the mid-fifties (Delandre 1991, 376)

120 Art.17 of Decree No.77-1133.

121 *L'objectif est clair, par contre, ce qu'on donne, C'est les délais, C'est-à-dire pas de faiblesse, mais un peu de souplesse.*'

122 Operators frequently seek information on standards, conditions, and possibilities from the competent *DRIRE* prior to submitting an official licensing application (Hoppe 1984, 163).

123 One of the last official acts of Brice Lalonde, the then Minister of the Environment was to place the DRIR (without final E) under so-called '*double tutelle*', 'double guardianship'. Only since 1992 has the *DRIRE* ('*Direction Régionale de l'Industrie*

et de la Recherche et de l'Environnement') also been accountable to the Ministry of the Environment (interview with DRIRE, November 1992).

124 *Si c'étaient les industriels qui manoeuvraient l'administration ou l'administration qui manoeuvrait les industriels?*

125 *L'administration joue plus un rôle conseil que de police.*

126 The severity with which the authorities intervene differs from region to region. In Alsace, for example, only the threat of a *'consignation'* is issued, whereas in Nord-Pas-de-Calais an average of 15 fines per year are imposed (interview with DRIRE, November 1992; interview with prefecture, June 1993).

127 The willingness to take legal action and sensitivity to environmental issues have grown enormously in recent years. For example, Nord-Pas-de-Calais registered increased interest in participation in public enquiry proceedings and greater readiness to go to court. A prefecture official remarked that the licensing of a pig-breeding concern used to be a formality, whereas today all sorts of objections were raised (interview with prefecture, June 1993).

128 Not to forget the pronounced departmental autonomy of federal ministries and the independence of the courts and the *Bundesbank*, the German central bank.

129 Mitterrand, for example urged measures to be taken against depletion of the ozone layer in 1989, when France held the presidency of the Council of Ministers.

130 'A small [ministry] lost among the big ones', namely Industry, Infrastructure, and Culture.

131 With the new transfers of powers under the 1982 and 1983 decentralization acts, the regions are now responsible for planning in the development of the local economy. The *départements*, for example, are empowered to draw up their own plans for financing construction works benefiting agriculture, and municipalities have been granted the power to prepare both the general development plan for the community (*'plan d'occupation des sols'*) and the land use plan (*'schéma directeur'*) in associations of local authorities (Fromont 1983, 401).

132 *Sans qu'ils l'aient vraiment souhaité, les auteurs de la décentralisation ont fait passer la France d'un système d'administration locale à un système de gouvernement local.*

133 On closer inspection, the different administrative functions in the environmental field prove to be very fragmented. 'The "local administrative concert" in environmental affairs is in reality a real organic cacophony where all levels and sectors of the administration interfere with one another without real coordination and without areas of responsibility being clearly defined' (Romi 1990, 34).

134 *[Ils] ... ont un rôle à capturer parce qu'au départ ils n'avaient au fond aucun pouvoir, aucune compétence technique et qui ont d'une certaine façon des moyens financiers. ... Moi, je le ressens à mon niveau [des réseaux pour la surveillance de l'ai] de plus en plus .... Qu'il y a l'Etat d'une part, mais qu'il y a la région également. ... Le Conseil Régional de l'Ile de France est plus moteur à l'heure actuelle dans le conseil d'administration [du réseau] que l'Etat, et est potentiellement un financier plus fort.'*

135 Like the *Secrétariat Permanent pour la Prévention des Pollutions Industrielles* in *Basse-Seine* or the *Conseil Régional* of *Nord-Pas-de-Calais*, where with help from the Socialists a Green politician was appointed president and promptly initiated new environmental policy measures.

136 'Examination of the registered plants regime shows ... that, in matters to do with the industrial and urban environment, the prefect is perhaps the administrative authority best equipped with juridical means of intervention' (Romi 1990, 55).

137 *Cette situation, par rapport à celle qui prévalait avant 1982, est génératrice de dynamisme, de diversité, mais aussi d'incertitude dans le traitement des problèmes politiques.'* The impressive extent and multiplicity of activities results in available information on air quality being dispersed. Each monitoring network operates within its own context and pursues its own policies. It is only very recently that institutional steps have been taken for the *Institut National pour l'Environnement* to coordinate and concentrate information systematically (Larrue 1992, 298).

138 Important in this connection is the rule that candidates who fail to win 12.5 per cent of the vote in the first ballot must withdraw (Mény 1993, 173).

139 Since 1995 and the accession of Austria, Finland, and Sweden, there have been 21 Commissioners.

140 Although the Commission is often accused of being too eager to regulate, 90% of the proposals relating to industrial emissions discussed in recent years came from member states (interview with EU Commission, DG XI, March 1993).

141 However, this occurs not as a sudden break but as the direct prolongation of the prior negotiating phase. Continuity is often provided by the same civil service experts who negotiated the directive in Brussels being responsible for national implementation. Britain is the exception, since higher officials are regularly relieved every three years or so.

142 However, the SEA did not eliminate the so-called 'Luxembourg Compromise' of 1966. According to this principle, voting is not by qualified majority if a member state asserts that a crucial national interest is at stake. The compromise — an agreement to disagree — was accepted on the stubborn insistence of France, which had boycotted negotiations in the Council of Ministers — *'politique de la chaise vide'* (Ehlermann 1990, 139).

143 In the first reading, the EP still has only the right to be heard. The cooperation procedure introduces the renewed discussion of a Council decision in Parliament within the first reading, and the possibility of amending it in accordance with the procedure described.

144 This was, for example, the case of the 1989 directive regulating exhaust standards for small cars. In this instance the EP availed itself for the first time of its new powers, pushing through adoption of the more stringent American exhaust-gas standards against Council reluctance (Ehlermann 1990; Corbett 1990).

145 The greater importance of other policy fields vis-à-vis economic integration is reflected in changed terminology. Thus the most comprehensive of the three Communities, the EEC, is now officially to be referred to as the European Community (European Communities 1992, 8).

146 The Cohesion Fund is designed to secure balanced social and economic development in the various regions of the Community (European Communities 1992, 43).

147 The precondition is a absolute majority in the EP.

148 In all other areas where voting is by qualified majority, the cooperation procedure applies, and where unanimity is required, the consultation procedure.

149 The regulation setting up this Environment Agency had already been adopted by the Commission in 1990. However, it could not come into force until member states had reached agreement on the seat for the new authority, which occurred only in 1993.

3 Changing Roles in the European Negotiating Game: Initiative and Blockade

The Italian painter Ambrogio Lorenzetti had a very precise idea of what deeds and effects distinguish 'Good and Bad Government'.[1] Fourteenth century Siena was another world, and to draw a distinction between a good government and a bad one would doubtless be a both impossible and presumptuous undertaking. There is nevertheless an undeniable parallel to the subject matter of the sections to come. Our principal concern is what characterizes British, French, and German behaviour in European Union clean-air policy. In which areas, topics, and measures do these countries set the pace ('good' government), drag their heels ('bad' government), or play the well-disposed but disinterested spectator ('neutral' government).

Our assumption, which we shall be examining in relation to concrete EU directives and regulations, is that Germany sets the pace in substantive measures, while the United Kingdom does so in procedural matters. France plays a special role, in that neither vehement opposition nor outstanding initiative and support are in evidence.

3.1 The Federal Republic of Germany as Pace-setter: Substantive Measures

In Germany both the perception and handling of clean-air issues are characterized by a technical understanding of environmental protection. The risks and deleterious impacts of air pollution are countered by plant-related measures. Emissions are abated directly at source by means of highly developed environmental protection technology. Industrial plants are authorized to operate only if they meet state-of-the-art standards. The limit values stipulated in the Technical Guidelines for the Maintenance of Clean Air pursuant to the Pollution Control Act are to be respected in all cases, irrespective of whether the plant operates in a strongly polluted area or an unpolluted one.

Since the early eighties at the latest, this emission-related approach with its orientation on the best available technical solutions, has been the indisputable hallmark of German clean-air policy. Probably its most important outcome has been the Large Combustion Plant Regulation, which not only cost German industry billions of marks, but also brought appreciable improvements in air quality.

However, industry's prime concern is not clean air, but making a profit. This might sound banal, but it is a strong reason why the Federal Republic wishes 'to push through more exigent environmental regulations in the EC as well' (interview with the Federal Ministry for the Environment, July 1993). If stringent clean-air measures are to be obtained without abandoning the necessary negotiating position, the gap between industry and the EC 'must be kept as small as possible for competitive reasons. The threat we hear is that *Germany as a location for industry* [emphasis added] is at risk because of higher environmental costs, and that business will go abroad' (ibid.).

Another argument, also economic, in favour of German efforts to raise its national rules to the status of supranational standards has to do with sales opportunities for German environmental technology. The stricter environmental requirements become for Germany's European neighbours, the more demand there will be for environmental technology products, such as filter installations. The particular beneficiary of this increase in demand is the German environmental protection equipment industry, the number one in Europe (interview with steel producer, March 1993; interview with environmental protection equipment industry, July 1993).

Germany's commitment within the EU, and its endeavours to impose its emission-related, best-available-technology approach at the supranational level is also 'rational' as a strategy for minimizing legal adjustment costs. The characteristic German style of environmental regulation finds expression not only in regulatory legislation, but also in formal and informal administrative structures, implementation styles, and interaction patterns that have grown and consolidated over decades. Changing this legal, institutional, and instrumental pattern in submission, for example, to the European ambient medium quality approach would doubtless provoke opposition from many sides, and prove considerably expensive.

In view of the complex economic, legal, and political dimensions of state action and influence at the supranational level, it is understandable that Germany plays the pace-setter, leading the way 'in regulating emissions from air-polluting plants' (interview with Federal Environment Office, November 1993), in 'setting limits and technical standards' (interview with the German Nature Conservation Ring, February 1993). Especially in the eighties, the Federal Republic was reputedly top of the class, claiming that 'we don't need to be told what to do or preached at by international commissions. If you want to do something, take our regulations as a model and do it the same way' (interview with Federal Environment Office, November 1993).

Although the then Environment Minister Töpfer had 'a very good reputation in Europe and is one of the most eminent environmental politicians, who is always a bit of a locomotive in the EC' (interview with environmental consultancy firm, January 1993), the Germans also managed to arouse antipathy. Because 'they always felt they had to get environmental protection on the German model going at the European level' (interview with FDP, March 1993), they were soon denounced as 'arrogant crackpots and perfectionists who set their goals so high that countries like Portugal wouldn't even try to get there' (interview with environmental consultancy firm, September 1993).

Germany's model role is put in perspective, however, if one looks beyond legislation to the country's implementation record. Closer scrutiny reveals that 'reservations certainly have to be made with regard to pace-setter role' (interview with the Advisory Council on Environmental Questions, March 1993). When it comes to concrete implementation of EU directives, 'if its a matter of "butter or fish", then [the Germans] tend to trail the field.' The number of actions pending against the Federal Republic before the European Court of Justice for contravention of obligations under the Treaties show that Germany's implementation record is 'in blatant contradiction to the ostensible policy of the Federal government to set the pace in the EC' (interview with Independent Institute For Environmental Affairs (*UIU*), November 1992). EU directives are often implemented only 'when there is sufficient local pressure from an environmental movement' (interview with Greenpeace, November 1992). For this reason a representative of the Association of German Engineers also expressed the wish 'that legislation be really guided by what is happening throughout the EC, that for the moment one doesn't invent anything new as pacemaker, ... so that legislation doesn't get even further ahead, leaving engineers and implementation limping along way behind' (interview with VDI, January 1993).

In the case of Germany, too, it seems one must abandon the idea 'that there can be general pace-setters. If it's a matter of technical environmental protection, the Germans spearhead developments. But when it comes to cost-effective, non-technical environmental protection involving changes in behaviour, it's suddenly a quite different kettle of fish' (interview with EURES, August 1993). The late and inadequate implementation of the directives on environmental impact assessment and environmental information shows that, although Germany is 'materially a clear pace-setter this is not the case *with regard to methods* [emphasis added] and procedures' (interview with Association of German Chambers of Industry and Commerce [*DIHT*], March 1993; interview with Federal Environment Office, November 1992; interview with Advisory Council on Environmental Questions, March 1993).

A further factor qualifying Germany's pacemaking role is the primacy given industrial interests over environmental interests. This is especially apparent in times of recession. 'In the depths of a recession, Environment Minister Töpfer has a hard time of it and can't afford to play the pace-setter' (interview with European Environment Office, March 1993). Recently, there has been a renewed joint effort by the Germans *and* the British to scale back environmental regulatory efforts. The 'Molitor deregulation committee' seeks to alleviate the legislative and administrative burden on industry for reasons of competitiveness (The European, 15 June 1995, 1).

The following sections examine the influence the Federal Republic has exercised on EU clean-air policy by looking at actual directives. It becomes clear how successful Germany has been in its dealing with the Commission in installing its problem-solving approaches at the supranational level, and imposing them as the predominant regulatory philosophy.

3.1.1 From Air Quality to Emission Control

In the early eighties, the European Union adopted an initial two directives on air quality limits and guide values for sulphur dioxide and suspended particulates,[2] and on air quality standards for nitrogen dioxide,[3] both of which followed a strategy based on ambient air quality. The directive on standards for SO_2 and suspended particulates sets limit and guide values. The limit values are binding, guide values serve as points of reference for the long-term abatement of air pollution. However, amelioration of ambient air quality in strongly polluted areas was not to be allowed at the cost of shifting problems to relatively unburdened areas. This does not mean that the construction of industrial plants producing SO_2 and suspended particulate emissions were generally forbidden is such areas. It was merely stipulated that such plants were not to give rise to any significant deterioration in air quality, without the term 'significant' being clearly defining (Knoepfel/ Weidner 1985, 255). The directive also required member states to set up measuring stations to monitor air quality, especially in areas where limits were likely to be exceeded.

The directive on air-quality standards for NO_2 lays down limit and guide values for atmospheric NO_2. The choice of measures to implement the standards was left up to the individual countries. Where measuring stations record that limit values have been exceeded, the Commission was to be informed.

The orientation on air quality attracted increasing criticism as the debate on acid rain intensified and international pressure grew on the EU to abate pollutant emissions. Seeking a more effective policy, the Commission turned in 1993 to a more strongly emission and modern-technology oriented *modus operandi* on the model of the German Large Combustion Plant Regulation.

3.1.2   Primacy of Emission Control and State-of-the-Art Abatement Technology

Once the Federal Republic had revised and modernized its own legislation in 1982 with the adoption of the Large Combustion Plant Regulation in 1982, it seemed perfectly logical from a German point of view to impose German standards on other member states of the European Union. Germany had two aims: to safeguard the competitiveness of domestic industry, which was now subject to more stringent requirements than its foreign competitors, and to expand markets for the highly developed German environmental technology sector. At the same time, uniform European arrangements consolidated the negotiating position of domestic administrative authorities vis-à-vis industry, which was no longer able to advance the argument that its less rule-bound competitors were at an advantage. As we have shown, German intervention at the supranational level encouraged a strategic re-orientation on the part of the Commission from quality-based policy to an emission-related approach. The new strategy corresponded largely with German policy, which gave high priority to the precautionary principle and to abating emissions by using the best available technology.

The Commission's strategic turn around found expression in two measures in particular: the Directive on the combating of air pollution from industrial plants[4] and the Directive on the limitation of certain pollutants into the air from large combustion plants,[5] the latter was to a large extent modelled on the German regulation. Since we may assume that, in supranational negotiations, every member state seeks to minimize the adjustments its regulatory system has to undergo, the other countries could hardly be expected to acclaim the Commission proposals. Countries with problem-solving strategies and institutional arrangements strongly at variance with the purport of proposed directives had little interest in adopting such measures. Britain in particular could be expected to refuse the jump, since its regulatory philosophy was in complete contradiction to measures shaped by the classical German approach of administrative-law intervention. Negotiations on the Large Combustion Plant Directive were extremely protracted and onerous, reflecting the radical discrepancies between German and British thinking.

Combating Air Pollution from Industrial Plants: the New Orientation

Although this directive initiated the Commission's shift in strategy towards an emission-based policy, the decision-making process proved surprisingly unproblematic. The measure was adopted after only a year of negotiations — a very short period by EU standards. One reason for this was that the directive itself did not lay down special emission limits for pollutants, restricting itself to procedural arrangements for the licensing of industrial plants. Member states were thus not put under such great pressure to adapt their institutions, since most countries already had something in the way of an authorization system. Moreover, the absence of statutory limits possibly led some countries at the time of the negotiations to misjudge the full consequences of the remaining provisions. 'We were astounded ourselves at how quickly the thing was dealt with. We'd never expected that [the directive] would be adopted in less than a year. For me there's only one explanation: the member states didn't realize at the time what they were signing' (interview with the EU Commission, DG XI, September 1993).

The Content of the Directive

The Directive on industrial air pollution is the first important reaction from the EU to the problem of forest dieback and acid rain. It lays down fundamental principles that are to be observed in authorizing industrial plants. The directive does not itself fix emissions limits, but specifies various industrial sectors for which limits for certain — also listed — pollutants are subsequently to be set by so-called 'daughter directives'.[6] It is thus referred to as a 'framework directive': It's aim was 'to provide for further measures and procedures designed to prevent or reduce air pollution from industrial plants ... in the Community' (European Communities, 1984, 21). As regards authorization, the provisions of the directive fall under two headings: licensing requirement, and preconditions for licensing. All industrial plants listed in the schedule require authorization to operate. The categories covered are energy, metal production and processing, chemicals, paper and packaging, waste disposal, and non-metalliferous minerals (building materials). The licensing requirement means that no plant may operate without prior authorization by the competent authority. This also applies to existing plants that are altered to a significant extent. Before issuing an authorization, the authorities have to ensure that the following conditions have been met (Article 4):

1. The plant operator must have taken 'all preventive measures against air pollution ... including the application of the best available technology,

provided that the application of such measures does not entail excessive costs' (BATNEEC).
2. The use of plant will not cause significant air pollution.
3. None of the emission limit values applicable will be exceeded.
4. All the air quality limit values applicable will be taken into account.

In addition, both the application for authorization and the decision of the competent authority must be open to public inspection. When and for how long this should take place is, however, left to the discretion of member states. On the basis of the 'framework directive', the Council of Ministers is empowered to fix emission limits[7] for individual substances as proposed by the Commission by means of daughter directives. Apart from the BATNEEC principle, the precondition is the unanimous concurrence of all member states. Regardless of this, individual countries are required gradually to adapt existing plant to meet the new requirements set out in the directive. Particular attention is to be given to technical characteristics of the plant in question, its rate of utilization and length of remaining life, the nature and volume of pollutants emitted by it, and the economic situation of the undertaking concerned (Haigh 1990, 224).

The Decision-Making Process
The Commission proposal of April 1983 for the adoption of a framework directive to control industrial emissions was made under pressure from two directions, from within the Community and from without. The Scandinavian countries were seeking, in the UNECE context, to push through comprehensive measures to combat transborder air pollution; and Germany — under the impression of worsening forest dieback — was calling for the EU to take remedial action. Being dissatisfied with the outcome of the 1979 Geneva negotiations, the Scandinavian countries organized a new meeting in 1982, the tenth anniversary of the Stockholm Conference, in an attempt to increase the pressure on the countries that had been dragging their feet, namely Britain, Germany, and the United States. Scientific papers and discussions among experts drew attention to the necessity radically to reduce emissions of SO_2 and NO_x in order to guard against further environmental damage from acid rain. The final declaration stated, '[that] deterioration of soil and water will continue and may increase unless additional control measures are implemented and existing control policies are strengthened' (quoted in Wetstone/Rosencranz 1983, 148). As a first step, the Scandinavian countries suggested reducing SO_2 emissions by 30 per cent by 1993 (base year 1980). Whilst this measure failed, as was to be expected, to find the support of Eastern Bloc countries, the United States, and the United Kingdom, Germany surprised the other obstructionist countries by joining

the '30% Club' (Strübel 1992, 194; Boehmer-Christiansen/Skea 1991, 28). At the same time, the German government submitted a memorandum to the EU Council of Ministers calling for the Commission to elaborate a general strategy for abating air pollution from industrial plants as rapidly as possible. Further discussion in the Council of Ministers in June and December 1982 served to underline the urgency of the German request. The Commission finally reacted in April 1983 to the initiative of the Federal Republic by submitting a proposal for a directive.

France, which had long had such formalized authorization procedures (albeit without statutory limits) with extensive opportunities for public participation (*'enquête publique'*), faced few problems of institutional adaptation to meet the requirements of the proposed directive. Whereas the French assumed a relatively neutral stance, Britain's attitude was marked by a degree of ambivalence. Although the British regarded their national licensing practice as largely in conformity with the requirements of the framework directive, they did not welcome the idea of introducing emission standards in the various sectors of industry. The adoption of emission limits was in complete contradiction to British practice, with its emphasis on ambient environmental quality, which operated without any statutory standards at all. If agreement was to be reached in the Council of Ministers, some concessions to the British were therefore unavoidable.

The United Kingdom successfully insisted that emission limits be agreed only by a unanimous vote in the Council of Ministers instead of by a qualified majority as had previously been the case. The British thus secured a right of veto, enabling them to block decisions on emission standards (Boehmer-Christiansen/Skea 1991, 233). The EU had paid a high price for a British Yes to the framework directive. The unanimity principle had drawn its teeth (interview with HMIP, September 1991). Although Britain otherwise considered its regulatory regime largely to conform with the principles put forward by the directive, it obtained yet more concessions in the negotiations. Thus, in view of the strong emphasis on economic components within the British bpm principle, the BAT (Best Available Technology) concept originally proposed by the Commission was diluted by the appendage NEEC (Not Entailing Excessive Cost) to secure greater congruence between the two principles. Moreover, on British insistence, the implementation deadline was shifted by 30 months (from 1 January 1985 to 1 July 1987) (Haigh 1990, 225f.).

This period of grace was necessary mainly because many plants requiring authorization under the provisions of the directive were subject to supervision by local authorities, which had no means at their disposal for preventive control. Although the directive demanded no substantial changes with regard to the bpm principle, it did necessitate extending this concept to fur-

ther categories of industrial plants — mostly small and medium-sized undertakings. The fact that the United Kingdom gave way to this pressure for institutional change without much ado is attributable essentially to two factors.

First, there had for some time been moves in Britain itself towards reforming the regulatory system, which, among other things, envisaged precisely the changes implicit in the directive. These plans had been elaborated
by the Royal Commission on Environmental Pollution (RCEP) — an independent standing body advising Queen, Parliament and the public on environmental protection matters — in its fifth report published in 1976.[8] Between 1976 and 1988, the ideas of the RCEP were taken up by neither the
Labour government nor the Conservatives in power from 1979. However,
since it is the usual practice in Britain for the government to give a written
report to Parliament on whether and how it intends to implement RCEP
proposals, the fact that this had not yet been done generated a degree of internal political pressure, which was intensified by EU requirements. In its
reply to the 1982 RCEP report, the DoE accordingly announced a general
revision of the existing control system (DoE 1982, 7). Given this coincidence of internal and European pressure for change, the best solution from
the British point of view appeared to be to postpone the implementation
deadline for the framework directive in order to gain time for discussion and
consultation at the internal political level.[9]

The second and possibly more pertinent factor explaining Britain's willingness to agree to the directive is that the proposal contained a further
provision, much more problematic for the United Kingdom, which largely
upstaged the problem of the necessary expansion of authorization processes,
namely the fixing of emission limits by a qualified majority in the Council
of Ministers. The British thus concentrated on opposing this arrangement,
and succeeding in pushing through the rule that such limit values could be
adopted only by a unanimous vote. This behaviour by the British clearly
shows that, depending on the given conflict structure of the decision-making
situation, there are certain limits to member states' obstructionist behaviour.
Saying No to everything finds no political legitimacy; what occurs is rather
'conflict among conflicts' (Héritier 1993a, 444): 'Individuals are able to be
engaged in lesser conflicts ... only if they are not involved in conflicts
which, for them, are more important' (Kellow 1988, 720).

The rapid adoption of the framework directive was possible principally
because the pressure to adapt national licensing procedures to comply with
procedural arrangements under the directive was low. Furthermore, the directive set no concrete emission limits for specific substances, which was
welcomed especially by countries like Britain that took a quality-oriented
approach to air-pollution control. However, final British agreement came
only once their demand had been met to set emission limits within the

framework of daughter directives in unanimity rather than by qualified majority voting. Through this trade-off the directive lost its drastic effect, but in exchange the British voted for it, thus accepting certain institutional changes to their regulatory system.

Large Combustion Plants: Regulation of Emissions by the State

Whilst the decision-making process on the framework directive proved relatively fast and unproblematic, negotiations on the Large Combustion Plant Directive were to be extremely protracted and complicated. The main reason was ultimately the profound discrepancy between the British viewpoint and the 'German-influenced' stance adopted by the Commission. Two fundamentally different rationales were on collision course. Whereas the Germans found it logical to force through their national regime at the European level, thus safeguarding their domestic room for manoeuvre, minimizing legal adaptation costs, and enhancing market opportunities for German technology, it was logical from a British perspective to block this attempt by the Federal Republic. After all, the proposed emission limits, which were to be defined in terms of the most up-to-date technological capabilities, were in radical contradiction to the British view of the issue, guided by ambient air quality, strict scientific causality, and economic proportionality. Resistance from the United Kingdom was hence hardly surprising. It was only when the British found themselves under pressure from within the country as well as internationally and in the European context that they showed greater willingness to compromise in the supranational decision-making process.

The Decision-Making Process
Negotiations were initially shaped by the strategy of the Commission to isolate the British as the main opponents of the directive with the aim of forcing them to compromise. This plan was doomed to failure with the accession to the European Union of Spain, which formed a coalition with the United Kingdom against the directive. The initiative in giving form to the measure henceforth passed more and more from the Commission to the succeeding presidencies of the Council of Ministers, where agreement was finally reached in bilateral negotiations and by once again isolating the British government, which had now also been under pressure at home.
The Commission Proposal. The Large Combustion Plant Directive represents the first and as yet only daughter directive pursuant to the Industrial Air Pollution Directive.[10] The proposed LCP Directive and the development of the framework directive are closely related in time and content. In a certain sense it was even the perceived need for a special directive on large combustion plants that drew attention to the need for general framework

legislation. And it was the Federal Republic that insisted on a European regime, and which set the pace with its 1982 Large Combustion Plant Regulation (Boehmer-Christiansen/Skea 1991, 234f.).

The draft directive submitted by the Commission in December 1983 was accordingly based to a large extent on the German regulation. It provided for limits on SO_2, NO_x, and suspended particulates, which were to apply to all new power stations (i.e., those authorized after 1 January 1985). Moreover, the proposal provided for emissions from all plants (new and old alike) to be reduced by specific amounts by 1995. SO_2 was to be cut back by 60 per cent, and the values for NO_x and particulate matter was to be reduced by 40 per cent in each case (base year 1980). These reductions were to be achieved at the national level, that is to say each member state had to reduce total emissions of pollutants in its territory by the specific amount ('national bubbles'). In practical terms, these standards required the fitting of flue-gas desulpherization (FGD) equipment or comparable technology to reduce SO_2. The use of low-sulphur fuels sufficed only for plants under 300 MW. To comply with NO_x emission limits, it was necessary to use low-NO_x technology (ibid., 238).

The alignment of the draft directive on the German Large Combustion Plant Regulation signalled a strategic re-orientation of European clean-air policy. The quality-related concept was abandoned in favour of an emission-oriented approach guided mainly by technical capabilities. This 'technology-centred' element had already found clear expression in the BATNEEC principle of the framework directive. The strategic change was not only in the German interest but also in keeping with Commission objectives. It is to be assumed that the detailed supranational regulation of emissions served the interests of the Commission in permitting it to enhance its regulatory authority vis-à-vis member states (Eichener 1992, 51). Precisely in environmental policy, the vested institutional interests of the Commission were very clearly articulated. This policy field, for which the Treaties of Rome had originally provided no competence for the EU, was more or less opened up by the Commission. The latter, with its political scope restricted by the extensive tie-up of budgetary resources in agricultural, structural and regional fund budgets, is obviously interested in the opportunity offered to expand its regulatory scope (Majone 1989, 167). Whilst the approach previously taken had left it up to individual countries to decide on the means for attaining a specified quality level, the setting of emission limits for specified substances and industries substantially narrowed this room for manoeuvre to the advantage of the Commission.

The Reaction of Member States: the British Blockade. Member states reacted very differently to the proposed directive. Whereas Germany, the Netherlands, and Denmark supported the proposal, it met with far-reaching opposition from the United Kingdom, Italy, Greece, and Ireland. France and Belgium, by contrast, remained relatively neutral, due to their comprehensive nuclear energy programmes (Boehmer-Christiansen/Skea 1991, 238). France generates seventy-five per cent of its power in nuclear stations and only a meagre five per cent in coal-fired plants. This source structure meant that the abatement requirements and limits presented the French with no problems whatsoever, since the predominance of nuclear power had already largely eliminated SO_2, NO_x, and particulate emissions from large combustion plants. France accordingly had no interest in blocking the directive, because it hardly affected French clean-air policy. Nevertheless, the benefits were also minimal, since France's geographical position precluded the importation of large quantities of pollutants from foreign power stations. Most foreign pollutants came (and come) from Germany, where very stringent standards were in force anyway, regardless of the EU directive.

Britain, by contrast, rapidly developed into the principal opponent of the directive. A major reason for British opposition was the fundamental discrepancy between the British view of the issue and the Commission's 'German' perception, which continues to this day to hamper consensus formation at the European level (interview with EU Commission DG XI, March 1993). The divergent British viewpoint was largely determined by the particular constellation of national context variables we have described. The country's geographical position, the structures of settlement and industrial density, the lack of scientific evidence, and the general economic situation prevented development of awareness that action needed to be taken. In addition, the Conservative policy of cutting government spending blocked definition of more stringent standards for the nationalized power-supply industry, because investment in the appropriate control technology would have further increased the burden on public finances. Growing international pressure since the early eighties could do little to affect this attitude. Thus, despite their relative isolation (in 1984 the Eastern Bloc countries has also declared their willingness to join the '30% Club'), the British opposed all measures to reduce transboundary air-pollution flows. This was attributable primarily to the effects of another context variable of importance in relation to the Large Combustion Plant Directive: the structure of the energy sector.

In 1947 all power supplies in the United Kingdom had been nationalized. After some degree of more or less marginal reorganization, the structure of the British energy sector was established in 1957 in the form that was to persist until the early eighties. Electricity generation in England and Wales was placed in the hands of a single organization, the Central Electricity Generating Board (CEGB) (Rüdig 1991, 157). This body was thus responsible for a vast proportion of British pollutant emissions (70 per cent of SO_2 and over 30 per cent of NO_x emissions), especially with coal supplying over

80 per cent of the energy input in British power stations due to the low use of nuclear energy (Knoepfel/Weidner 1985, 19; Boehmer-Christiansen/Skea 1991, 142f.).

In a first reaction to the draft directive, the CEGB cast doubt on the scientific arguments advanced by the Commission and drew attention to the unusually high costs of implementation (Weidner 1987, 96). As far as the scientific angle was concerned, the lack of evidence for a direct causal connection between SO_2 and NO_x emissions and the acidification recorded was emphasized. The CEGB justified its scientific scepticism with reference to the results of a comprehensive research programme it had financed, which, running since the mid-seventies,[11] had failed to confirm Scandinavian and German findings (interview with DoE, January 1993). These scientific arguments were backed up with economic reservations. The CEGB was especially concerned that expensive retrofitting of existing plant would be unreasonably costly. A further consideration made heavy investment in avoidance technology appear dubious at the time. The CEGB was convinced that the future belonged to nuclear power, which would automatically bring down pollutant emissions (Rüdig 1991, 159).

However, the CEGB also had fundamental doubts about the quality of FGD technology. It has been running tests on such equipment since the thirties without achieving convincing results: 'You take it out of the air and put it into the water' (interview with former CEGB employee, September 1993). Later reports of success from the United States and Japan in the early eighties did nothing to mitigate CEGB scepticism (which can be attributed both to its own years of experience and to long-term planning in the British energy sector). Since it is much cheaper to equip new power stations with FGD than to retrofit old plants, the CEGB had no interest in the technology, since no new plants were on the drawing board at the beginning of the eighties. Technological research and development in this area was accordingly neglected. Changes on the international technology market hence found no echo in the United Kingdom. The focus there was on expanding nuclear energy (Boehmer-Christiansen/Skea 1991, 147; 209).

Although the CEGB can be considered a governmental actor, in the mid-eighties it pursued — as representatives have stressed — a policy independent of government, but which diverged less in aim than in motive from the Whitehall line. While the government was mainly concerned to minimize government spending and was thus interested in preventing such high outlays on the relevant technology, the CEGB wanted primarily to find an economically effective solution that was scientifically substantiated.

The CEGB was perceived to be the major opponent against the directive. But the government was equally against it for a whole set of different reasons. The CEGB was against it because it didn't believe it would necessarily resolve the problem in the most cost-effective way. What the British government were fighting for were a whole lot of internal priority problems related to public expenditure and economic problems (interview with former CEGB employee, September 1993; see also interview with British utility, September 1993).

The close relations between the CEGB and the Department of Energy (DEn) permitted the latter to adopt a strong position vis-à-vis the DoE in inter-departmental conflicts. Moreover, DEn arguments were supported by British industry, whose major lobbying address in Whitehall was the Department of Trade and Industry (DTI).[12] And there was no environmental protection equipment industry with a vested interest in more stringent emission standards. The National Coal Board (NCB), a public body as well, also upheld the position of its biggest customer (Boehmer-Christiansen/Skea 1991, 215). The Environment Directorate of the DoE held the weaker cards in the face of these strong economic and industrial interests. The Directorate's relatively meagre staffing and financial resources permitted only reactive behaviour in the political process.[13] Moreover, the AI, the body responsible for controlling large combustion plants, had had its scope for action curtailed by being placed under the Health and Safety Executive, which maintained close ties with industry.

Acid rain not being considered a serious problem in Britain as it is in Germany, there was no strong public pressure to back the case for more stringent emission reductions. In any case, the options open to environmental organizations to influence the political process were very limited, mainly due, as we have shown, to the institutional peculiarities of the British governmental system. 'On acid rain, the influence of environmental associations has been marginal. ... Because British political culture is not issue-based or policy-based to the extent that other political cultures are' (interview with DoE, January 1993). The concentration on a central political arena imposed by the unitary nature of the state reduced the opportunities for such organizations to undertake political activities at various levels (Müller-Rommel 1992, 198). Access to Parliament was restricted by the first-past-the-post electoral system, which gave small parties little hope of winning seats (Budge/McKay 1988, 56). Finally, government could not be forced into action by being taken it to court. The constitutional principle of the supremacy of Parliament vests sovereignty within the political system in the legislature, to which the courts must bow. This is the only possible explanation why the 1984/1985 environmental protection campaigns, which pilloried the United Kingdom as 'the dirty man of Europe', had so little initial impact on the political decision-making process (Boehmer-

Christiansen/Skea 1991, 210). It was only with a certain delay that the efforts of environmental groups and this campaign to influence the political and administrative decisions bore fruit.[14]

The closed circuit of responsibility in the Westminster model gives the government more clout in the political process, and results in many decisive policy innovations emanating from Whitehall. Parliamentary intervention often has little or no effect on the substance of later decisions. This is reflected especially in the strong concentration of the power of decision in a small number of people within the executive ('the bottleneck of Whitehall').

You go like a supplicant to a medieval court. You can do things very, very quickly, like the Prevention of Terrorism Act which passed Parliament within two days or you can have a law like the Environmental Protection Act which takes 12 years to realize the ideas which were presented in 1976 by the Royal Commission (interview with AMA, January 1993).

Against this institutional background it is hardly surprising that Parliamentary criticism initially had no impact on British clean-air policy. This criticism was made manifest especially in a report to the House of Commons Environment Committee, which had modified its fundamental view of the issue — to some extent in the course of a political learning process — and which now called for Britain to adopt a far more active environmental policy. The members of the Committee had visited various research centres in Scandinavia and Germany, and had been very impressed by the evidence of environmental damage they had seen. In their report they declared themselves to be 'deeply disturbed over the United Kingdom's current policy position on acid rain' (quoted in Boehmer-Christiansen/Skea 1991, 212). The House of Lords Committee on the European Communities and the RCEP also recommended retrofitting FGD to two power stations for testing purposes[15] (ibid., 211). The criticism found support in the parties, which demanded stronger emission controls. Sections of the Labour Party — despite reservations about possible implications for coal mining[16] — and both the Conservative 'wets' and the 'think tank' (Centre for Policy Studies) argued for greater consideration of green issues (ibid., 213f.). Greatest pressure for higher environmental standards came from the Social Democrats and Liberals (now merged to form the Liberal Democrats) (interview with DoE, January 1993). Because of the strong concentration of the power of decision within the executive, Parliamentary criticism led to no immediate modification of the British position.

The impact of specifically British and international influences on the workings of the British policy network strengthened British opposition to the proposed directive. Mainly to blame was the wide discrepancy between British and German issue perception. Whereas the British persisted in their science-centred approach ('scientific evidence about pollution as harmful effect'), the Germans sought to impose their precautionary principle at the European level. There was further divergence in the typical policy instruments deployed, as we have described for individual countries. The British approach in clean-air policy — the quality-oriented bpm principle — which had stood the test of time for over a century, was called in doubt by the proposed directive, which followed an emission-related strategy (interview with RCEP, September 1992). The institutional context in which the bpm principle was embedded influenced not only the potential economic cost of administrative restructuring but also the goals the British would envisage. Objective, abstract and general criteria for decisions as contained in statutory emission standards appeared scarcely compatible with the British tradition. 'It's a very British thing to think that actually legislating makes things worse' (interview with AMA, January 1993).

Commission Strategy: a Failed Attempt at Isolation. From the perspective of the Commission, British concurrence was crucial if the goal of reducing emissions from large combustion plants to the greatest possible extent was to be attained. As the biggest pollutant exporter in Europe, the United Kingdom contributed substantially to the transnational problem of acid rain. In 1984 and 1985, the Commission accordingly attempted to isolate Britain in negotiations at the EU level, with the aim of putting them under greater pressure. This was to be achieved mainly by making certain concessions to the other opponent countries. However, the Commission's strategy foundered with the accession of Spain and Portugal to the Union on 1 January 1986. Spain in particular considered it impossible to implement the proposed directive because of its rapidly growing energy needs, which were met largely by constructing coal-fired power stations. Spanish resistance freed the British from their isolation (Boehmer-Christiansen/Skea 1991, 238f.). In view of the Spanish-British alliance, agreement thus seemed relatively remote at this period. 'The unholy alliance between Spain and the United Kingdom cost us more than a year and made it possible for the British to get off too lightly' (interview with EU Commission, DG XI, March 1993).

This impression was strengthened by British tactics during supranational negotiations. The hard line taken by Whitehall left no room for any concessions whatsoever. 'Not without reason have insiders likened the British environment negotiators during the 1980s to a soccer team with eleven very good goalkeepers' (Boehmer-Christiansen/Skea 1991, 250). While the Federal Republic stuck stubbornly to its position, negotiations were just as consistently retarded or blocked by the British. Number games and stalling tactics spun out the negotiations: 'It was worse than haggling in the bazaar' (interview with EU Commission, DG XI, March 1993).

In the further course of negotiations, however, the British unexpectedly relaxed the deadlock, improving the chances of reaching a compromise after all. The reasons are to be found on two levels. First, changes in the British network ensured that the British government now came increasingly under domestic pressure to take action; and second, the supranational decision-making process took on a new dynamic from the sequencing of presidencies in the Council of Ministers. The coincidence of domestic and European pressure increased British willingness to make certain concessions to its negotiating partners, bringing a long battle to a conclusion with the adoption of the directive by the Council of Ministers in June 1988.

Developments in the British Network. International and domestic factors encompassing scientific evidence, international pressure, public environmental awareness and the environmental movement, the economic situation, and the structure of the energy sector changed the general setting for domestic political processes in the British network, hence favouring British acceptance of the Large Combustion Plant Directive.

In early 1986 the first unequivocal scientific findings on the impact of transboundary air pollution on soil and water in Scandinavia became available. They had been made mainly in the course of research undertaken in the context of the international RAIN project (Reversing Acidification in Norway), which had begun in 1983 to run for five years (Sheail 1991, 260; interview with former CEGB employee, September 1993). Although the project basically confirmed the earlier CEGB doubts about the direct causal links between British SO_2 emissions and the damage recorded, it nevertheless underlined the need to reduce emissions of this pollutant for the sake of long-run improvements. Scandinavian soil acidification had been shown to be not the result of current emissions alone, but the cumulative outcome of widespread air pollution stretching over decades since the beginnings of western European industrialization in the nineteenth century. This invited two conclusions. It was clear that soils would recover only if European SO_2 emissions were immediately reduced; and, in view of the slow accumulation process, regeneration could quite evidently not occur overnight even were immediate reductions forthcoming. The recovery process, too, would take decades (interview with former CEGB employee, September 1993).

Corroboratory conclusions were reached by the Royal Society Surface Water Acidification Programme (SWAP)[17] (Sheail 1991, 260), and by the CEGB chairman Lord Marshall and his Environmental Research Director Peter Chester after visiting Scandinavia in June 1986. They gathered on-the-spot information about the current state of research and found developments highly disconcerting: 'They were convinced that there was enough scientific evidence to justify a complete turn of policy' (interview with former CEGB employee, September 1993). These research findings brought about a lasting change in the CEGB position.

It was this evidence that finally convinced the CEGB that it would have to do something about stopping SO_2 emissions. Up to that point we had taken a very firm stand against control technology because we had a lot of experience with it. We were the only ones that had at that time because we had run sulphur removal plans on certain London power stations since the early 1930s (interview with former CEGB employee, September 1993).

Thereupon the CEGB proposed retrofitting FGD to two power stations that were to be commissioned in 1993 and 1997. At the same time it announced that all new coal-fired power stations would in future be fitted with appropriate desulphurization equipment[18] (Boehmer-Christiansen/Skea 1991, 218; interview with British utility, September 1993).

The emerging 'scientific consensus on acid rain' (interview with DoE, January 1993) not only occasioned a degree of rethinking within the CEGB but also intensified the international pressure on the British government which had already begun to increase appreciably some months earlier with the 1985 Helsinki conference. Twenty-one countries had signed a protocol to the Convention on Long Range Transboundary Air Pollution committing themselves to reduce SO_2 emissions by 30 per cent by 1993. The recent scientific evidence now brought greater pressure to bear on the United Kingdom, which had refused to join the 30% Club.

Domestic demand for the government to take action also increased with the growth of environmental awareness among the British public. A series of surveys conducted in 1986 and 1987 showed that the public was giving more and more salience to environmental protection. The majority of respondents regarded the environmental programme of a party as decisive for their electoral decision, which, with a general election coming up in 1987, put the political parties on the spot. Several studies also indicated widespread public dissatisfaction with British clean-air policy, especially with regard to acid rain[19] (ENDS-Report 1987/155, 3).

Another reason for growing public ecological awareness was the general improvement in the economic situation that made itself felt from 1983. The economic upswing was accompanied by increasing interest in better quality of life, which enhanced the priority given environmental matters (interview with DoE, September 1993; Greenpeace, January 1993). 'It was a time of boom and everybody was doing well commercially, so these issues became important' (interview with a British environmental consultant, September 1993).

This greater public awareness forced the political parties to adapt their policies accordingly. This was abundantly plain from the platforms of all parties for the 1987 general election. 'All the manifestos devote more attention to environmental issues than at the 1983 election' (ENDS-Report 1987/148, 3). Whereas more swingeing environmental rules had previously been quite beyond the pale of Conservative policy styles — stricter emission limits for nationalized power stations would have entailed higher government spending —environmental policy, power strategy and Conservative ideology now began to converge. The Conservative party manifesto contained proposals for reducing emissions from British power stations, but at the same time it announced the imminent privatization of the CEGB.[20] The government was thus able to kill two birds with one stone: it firstly took account of increased social concern about acid rain, which was certainly wise from the perspective of electoral tactics; and secondly the planned privatization of the CEGB was not only in line with Conservative thinking, but also obviated public spending on abatement technology.

Distinctive aspects of British environmental organization activities also contributed to ameliorating conditions for environmental action. Not least of all because of the peculiarities of the British governmental system, the success of these organizations lay less in the direct political influence they exercised than in the general awareness of environmental issues they stimulated in society and administration, thus gradually creating a new context for action.

The real success of the environmental associations has been in creating this tremendous body of support for the environment[21]. ... They have changed the climate for other forces to have more room for manoeuvring than they would otherwise have had. What they have been doing is successfully colonizing the [political and administrative] mainstream. ... You don't have the kind of aggressive action that you get in France or Germany (interview with DoE, January 1993).

In the early eighties, environmental groups had no appreciable influence on political decision-making processes, but since then their continual activities have ultimately changed the political conditions for action. Green ideas have become more and more institutionalized on the political agenda. They thus effected not only a change in Conservative policy but also made this change politically practicable. The weight of the DoE in Whitehall decision-making grew in parallel to the salience of environmental issues (interview with DoE, September 1993). 'By 1986 it was actively trying to formulate positive policy steps' (Boehmer-Christiansen/Skea 1991, 218).

The greening of the political and administrative system has been facilitated by a number of factors. Apart from greater competition between parties on environmental issues that had arisen with the new Liberal Democrats,[22] particularly important factors have been more effective lobbying by environmental organizations and their greater political professionalization. Big environmental groups regularly meet informally to coordinate strategy in their dealings with government (White-Grove 1992, 121f.). Pursuing a uniform strategy allows the organizations to exert greater pressure on government and to promote their interests more efficiently. At the same time it increases the willingness of the politicians and administrators to integrate environmental interests more strongly in decision-making processes. These networks, which are coordinated by full-time employees of the various environmental groups, provide the 'key links' (ibid., 122) between the green lobby and the political and administrative system.

Given the institutional characteristics of the British governmental system, such contacts usually operate in extra-parliamentary channels. Informal personal contacts ('backstairs channels') play just as important a role in feeding environmental goals into the government department bureaucracy as official hearings and consultations (ibid., 107). As British observers point out, using such channels to advance environmental interests has been facilitated by an optimal mix of radical organizations (Greenpeace, FoE) and established associations (Council for the Protection of Rural England, National Society for Clean Air). The radical groups with their spectacular activities improved the negotiating position of the established groups, whose moderate demands consequently found a positive response in government circles. As Greenpeace and FoE began to become established from the late eighties, the negotiating potential of British environmental organizations deteriorated (interview with British environmental consultant, September 1993).

Since the mid-eighties British environmental associations have used the opportunity for action offered them by British membership in the European Union. They seek increasingly to influence domestic policy developments via EU contacts. Highly important in this connection is the Environment Subcommittee of the House of Lords Select Committee on the European Communities, the work of which is strongly influenced by the environmental associations. This committee — a sort of clearing-house for the positions of British interest groups — examines all proposed EU legislation and prepares reports on it. It has a considerable reputation with the EU Commission, which in its turn can influence the position of the British government. 'The results of [the committee's] work have repeatedly contributed to influencing attitudes and option within and without the British government' (White-Grove 1992, 107).

In a period when environmental awareness was increasing, resulting in corresponding environmental policy proposals in the Conservative party programme, the government decided to privatize energy supplies. This step was determined in essence by two factors. First, it was in line with the neoliberal Conservative strategy to reduce the public sector share in GNP and to place greater emphasis on market-economy concepts.[23] Power political aspects also played a weighty part. Market economy incentives were intended to promote the expansion of nuclear energy for the long term purpose of curbing the conflictual potential of the National Union of Mineworkers. A twin aim was to deprive the Labour Party of a section of its traditional support by weakening the trade unions (Rüdig 1991, 158).[24]

In response, the CEGB launched a large-scale campaign to secure its organizational survival. The heart of the argument was the thesis that the planned expansion of nuclear power could be successfully carried out only by a large state enterprise like the CEGB. Although with hindsight this argument proved correct — nuclear energy remains in government ownership — the CEGB was no longer able to prevent privatization[25] (ibid., 167).

The privatization debate, which had been going on since 1987, changed the perspective of the British government on the Large Combustion Plant Directive. Current developments weakened the CEGB position in negotiations with Whitehall. Whereas in 1984 the CEGB had still been able to assert its interests at the highest level (interview with former CEGB employee, September 1993), it could now no longer count on government support, 'when the most fundamental assumptions about its own structure and identity had been overturned' (Boehmer-Christiansen/Skea 1991, 220). These changes in the national context ran parallel to the supranational network processes, the dynamics of which were determined essentially by the sequence of Council of Ministers presidencies.

The Presidency of the Council of Ministers: Political Responsibility versus National Interests. The official proposed Commission directive having been submitted to the Council of Ministers in late 1983, and the original strategy of isolating the United Kingdom having for the moment come to nought with the accession of Spain and Portugal, the role of the Commission in the negotiations has since 1986 been that of moderator rather than initiator. Decision-making priorities were set largely by countries holding the presidency of the Council of Ministers for successive six-month terms. Among the controversial points tackled in the negotiations was the extent to which SO_2 emissions from existing plant should be reduced (interview with EU Commission DG XI, March 1993).

The Dutch presidency (first half of 1986) suggested two important modifications to the original proposal. First, SO_2 should be reduced in two steps: by 45 per cent by 1995 (the Commission proposal had been 60 per cent); the remaining 15 per cent would be eliminated by 2005. The second change suggested was in the form of the 'bubbles'. 'National bubbles' were now to be superseded by a 'European bubble'. The reduction rates for individual member states were to be fixed in terms of objective criteria such as GDP, emission quantities produced, the 'trade balance' on pollutants, and the measures so far taken to reduce pollutant emissions. Under this system, some countries (Germany the Netherlands, France, Denmark, Belgium), which had already introduced far-reaching control measures, would not be required to take further steps. The Dutch concept would have imposed a 40 per cent reduction in SO_2 emissions on the United Kingdom by 1995. The British, although the latest scientific findings had increased pressure on them to act, escaped having to react at all because of the strong opposition articulated by other countries. Spain and Ireland in particular could not see their way to effecting the 10 per cent reduction required of them. Despite a dilution of the values proposed by the Commission, the Dutch suggestion was hence doomed to failure (Boehmer-Christiansen/Skea 1991, 239ff.).

The situation was to become far trickier for Britain from July 1986, when the United Kingdom took over the Presidency of the Council. Although it is perfectly legitimate for the country holding the Presidency to adjust the European agenda to take account of national interests, for a number of reasons there are relatively strict limits on how far this can go. In the first place, the framework set by the Commission with a proposed directive can generally be only marginally modified in the Council of Ministers (interview with DoE, September 1993). More comprehensive changes could provoke the Commission to withdraw the proposal, a political risk which the country holding the Presidency can scarcely afford to take — especially if it is the main opponent of the directive in question. Furthermore, the Presidency carries with it a certain obligation to safeguard not only national interests but also Community-wide interests (Kirchner 1992, 109). In the last resort, this means that the incumbent must — if for no other reason than to retain integrity in the international arena — seek to establish consensus in the Council.

This is illustrated by the particular problem confronting the British Presidency in negotiations on the Large Combustion Plant Directive. Particularly in view of the latest scientific findings, Britain was under strong pressure to submit a constructive compromise proposal to the Council of Ministers. On the other hand it was clear that any compromise would be contrary to the British way of thinking. The concern with emissions embodied in the proposed directive, which implied strict reliance on technological capabilities, would in any case involve institutional changes in the quality-oriented British regulatory system. Central to the directive was the fixing of emission limits, which could not then be altered at Council level without taking the objectives of the proposal to absurdity. The dynamics of the European decision-making process, increasing international insistence (enhanced by research), and rising domestic pressure (as issue awareness had grown) forced the British government into a balancing act. They had to present a more or less acceptable compromise proposal, which necessarily required them to make certain concessions. But they also had to give this proposal a form that would make immediate agreement unlikely. After all, the directive was contrary to British thinking; a further delay or even the subsequent failure of negotiations continued to be in the interests of the British government. However, this could not be explicitly aired while Britain was chairing the Council.

On the basis of the CEGB announcement that FGD was to be retrofitted to two power stations, the British proposed a 30 per cent EU-wide reduction in SO_2 emissions by 1995, and a 45 per cent drop by 2005. At the same time they wanted the basis for calculation to be extended beyond emissions from large combustion plants to include all other sources. This would have suited the British, since SO_2 emissions from such sources had fallen substantially in the United Kingdom as a result of the economic crisis and the more widespread use of gas. It was also suggested that all plants with a capacity under 100 MW should be exempted from the emission limits, an idea put forward primarily on the urging of British industry, which often operated low-capacity power stations to supply in-house power. With this concept the British had reached a first station in their balancing act. They had proposed a compromise that from their perspective was acceptable and did not endanger their prestige.

They achieved their basic intention of dragging out negotiations as long as possible by making a proposal that necessarily provoked opposition from other countries. The British threw out the objective criteria for fixing 'national bubbles' introduced by the Dutch, suggesting instead that the political acceptability and practicability of emission reductions in individual countries were more appropriate criteria. This would have meant that countries that had already been very active in the field would be required to continue making greater efforts than others. France with its nuclear alternative, for example, would have been required to reduce SO_2 emissions by 91 per cent by the 1995 deadline, whilst the United Kingdom would have had to meet a target of only 28 per cent (Boehmer-Christiansen/Skea 1991, 242). It is hardly surprising that, despite British concessions, no agreement was yet in sight in the Council of Ministers at the beginning of 1987.

Isolation and Bilateral Negotiations. In the course of its presidency (first half of 1987), Belgium tried to achieve agreement by making special concessions to the principal opponents Britain and Spain. The essence of their proposal was the instrument dubbed 'emission credits'. The idea was for emission reductions prior to the base year stipulated by the directive — 1980 — to be credited to individual countries' future 'SO_2 bubble'. This would have suited the United Kingdom in particular, where SO_2 had been substantially reduced in the course of the seventies. Furthermore, emissions from power stations built between 1980 and 1987 were to be deducted from 'national bubbles'. This was a special concession to recently industrializing countries, which lacked the economic strength to be able to afford FGD. Although the Belgian concept re-introduced the original uniform bubbles, the 'emission credits' it proposed permitted individual national conditions to be taken into account (Boehmer-Christiansen/Skea 1991, 243).

According to the Belgian concept, Britain by 1993 would have had to reduce SO_2 emissions from coal-fired power stations by 26 per cent. The figures for 1998 and 2005 were 46 per cent and 60 per cent respectively.[26] Despite these further concessions, the British were unhappy with the concept. This was because even these less onerous requirements could not be met with the measures put forward by the CEGB (ENDS Report 1987/147, 20). Since the Spaniards were also demanding further concessions (especially financial aid with the construction of flue-gas desulpherization plant), pressure on the British government was still within limits — 'only if the Spanish opposition is dealt with [is] the heat ... likely to turn on the UK' (ibid., 22). While Denmark failed to move in this direction during its presidency (late 1987), its successor, Germany, managed in early 1988 largely to isolate the United Kingdom.

From the very beginning, the German presidency was under great pressure to achieve agreement, since little was to be expected from the next two presidencies, Greece and Spain. Should the Germans fail to find an acceptable compromise after five years of negotiations, it seemed inevitable that the project would fold. At an informal meeting of the Council of Ministers in February 1988, the Germans managed to isolate the British delegation completely. They did so largely by offering new concessions to Spain, whose reduction targets were set substantially lower than in the Belgian and Danish proposals. 'British negotiators found themselves isolated in discussions on the LCP Directive. There appeared to be no way forward in Europe unless some further concessions were made by the UK' (Boehmer-Christiansen/Skea 1991, 220f.). Pressure was thus intensified on the United Kingdom, finally forcing the British to give some ground to avoid finding themselves completely offside in Europe (see Peters 1992).

At the same time, the German delegation tried in bilateral talks to get the British to concede further ground. While European pressure on Britain continued to increase, changes within the United Kingdom reduced British room for manoeuvre in the negotiations. Once the CEGB had decided to fit FGD to two power stations, the government could no longer play the scientific card. And growing public sensibility to ecological issues forestalled legitimation of such isolated oppositional behaviour. Finally, the DoE, which was urging more stringent environment controls, gained more and more influence in political decision-making — thanks not least of all to the changes that had been brought about in the political and social context for action by the activities of environmental organizations.

Quite another development, the privatization of energy supplies, made the British government more willing to compromise. To avoid frightening off potential investors and buyers, any speculation about the SO_2 reductions to be imposed on the future private-sector utilities had to be nipped in the bud.

This increased British interest in bringing negotiations to a rapid conclusion. 'At least then, regardless of which way it comes out, the industry will know where it is going. The thing that the stock market doesn't like is uncertainty. They don't mind high costs if they can budget for them. What they don't like are surprises' (interviews with British utility, September 1993; IEEP, December 1991; HMIP, December 1991). An important argument put forward in the privatization debate was that only private enterprise could raise the money needed to install control technology (interview with multinational pharmaceuticals manufacturer, January 1993; interview with EU Commission, DG XI, March 1993).

Despite this coincidence of national and supranational constraints, negotiations remained a tug-of-war to the very end (interview with EU Commission, DG XI, March 1993). During the German presidency, the Commission was very much the 'policy broker' (Sabatier 1988). It supported the British position, especially as regards the still controversial issues of measuring and monitoring procedures. By making concessions in these areas, it sought to win British agreement to the reduction targets that had been proposed by the Federal Republic (Boehmer-Christiansen/Skea 1991, 245). The meeting of the Council of Ministers on 16 June 1988 at last achieved a breakthrough. In a session that went on until four in the morning, a compromise was finally found, focusing not on scientific but on political requirements: 'At the end of the day it was a political decision. The number that came out was as much a surprise to the CEGB as to anybody' (interview with British utility, September 1993).

The Final Version of the Directive: a Late Compromise
The Large Combustion Plant Directive that was ultimately adopted in June 1988 after almost five years of negotiations diverged markedly from the original Commission proposal, especially with respect to SO_2 reduction rates. Both nominal values and time scheduling had been substantially modified. Limits for new plants, however, were largely adopted as originally proposed. The directive contained the following arrangements:

1. The British 'bubble' for reductions in SO_2 emissions from existing large combustion plants was to be 20 per cent by 1993, 40 per cent by 1998, and 60 per cent by 2003. British NO_x reductions by 1993 were to be 15 per cent, and by 1998 30 per cent. The corresponding figures for Germany and France were 40, 60, and 70 per cent for SO_2, and 20 and 40 per cent for NO_x (see tables 14 and 15).
2. The directive was to apply for all plants with a capacity in excess of 50 MW. However, emission limits for plants with a capacity between 50 and

100 MW were to be fixed only after a Commission report on the availability of low-sulphur fuels.

3. Emissions were to be measured on the basis of monthly averages.

4. By 1 July 1990 member states were required to draw up appropriate programmes for the gradual reduction of total annual emissions from existing plants. These programmes were to determine not only the timing but also implementation details.

5. By 1 July 1995, a general adjustment of emission limits was to be carried out taking account of the best available technologies and environmental requirements. The Council adopted this amendment unanimously on the proposal of the Commission (European Communities 1988).

In addition, the directive contained a number of special provisions in concession to the interests of individual member states. For example, new plants with a capacity of over 400 MW may exceed the emission limit for SO_2 by double the figure if they operate for less than 2200 hours per year. This provision favoured the French in particular, since their power stations have widely varying rates of capacity utilization. At peak times they far exceed emission limits while keeping below the limit values on average for the year. This was why France had initially urged the adoption of limit values in the form of annual averages. But Britain, too, profited from this arrangement, since the British have many high capacity plants to take advantage of economies of scale (interview with former CEGB employee, September 1993).

In addition, new plants fired with indigenous coal were permitted to exceed limits if the particular properties of the fuel used demand disproportionately expensive abatement technology to keep within them. This provision was a further concession to Britain, where domestic coal has a high sulphur content (ENDS 1988/161, 23). Finally, member states could apply to the Commission for changes in reduction requirements if unexpected complications were to arise regarding energy demand or supplies of certain fuels, provoking serious technical difficulties in implementing the directive. Milder emission limits for new plants were allowed in Spain until 2000 (European Communities 1988).

In summary it can be said that the decision-making process on the Large Combustion Plant Directive was decisively affected by the incompatibility of two diverging problem-solving strategies. Negotiations were dominated by the contradiction between the German perspective — characterized by a technology-centred understanding of emission reduction at source and by the precautionary or anticipatory principle — and the British viewpoint, insisting on scientific causality and oriented on ambient air quality and demonstrable harmful impacts. After the Federal Republic had succeeded in introducing its strategy at the European level, British resistance to the priority given emission control and technological capabilities was more or less inevitable.

Table 14: SO_2 Emission Ceilings and Reduction for Existing Plants

Member state	Base value (1,000t)	Emission ceilings (1,000t/year)			Reduction against 1980 emissions in %		
	1980	1993	1998	2003	1993	1998	2003
Belgium	530	318	212	159	-40	-60	-70
Denmark	323	213	141	106	-34	-56	-67
Germany	2,225	1,335	890	668	-40	-60	-70
Greece[1]	303	320	320	320	+6	+6	+6
Britain	3,883	3,106	2,330	1,553	-20	-40	-60
Spain	2,290	2,290	1,730	1,440	0	-24	-37
France	1,910	1,146	764	573	-40	-60	-70
Ireland[1]	99	124	124	124	+25	+25	+25
Italy	2,450	1,800	1,500	900	-27	-39	-63
Luxembourg	3	1.8	1.5	1.5	-40	-50	-60
Netherlands	299	180	120	90	-40	-60	-70
Portugal[1]	115	232	270	206	+102	+135	+179

1) The positive percentage figures arise from the concessions made during negotiations to the recently industrializing countries that emissions from large combustion plants authorized between 1980 and 1987 were not to be included in calculating reduction targets.

Source: European Communities (1988)

Although other countries (especially Spain) also opposed the directive, the British position was the be-all and end-all for the directive. Whilst other opponents were concerned 'only' about the financial aspect, which could be largely settled by appropriate derogation, the United Kingdom had cardinal objections. The very essence of hallowed British regulatory practice had been called into question by the European requirements. The introduction of statutory emission standards was in radical contradiction to the British approach, with a century-old tradition operating without legally binding standards. The institutional adjustments looming for the British control system were reason enough for the British government to seek to stall the supranational decision-making process.

Moreover, the domestic context encouraged such behaviour on the part of the United Kingdom. In contrast to Germany, Britain's favourable geographical position and the non-occurrence of external environmental shocks produced neither problem-generated pressure on government to act, nor public awareness of the issues involved. Furthermore, the institutional structures of the British governmental system gave environmental groups little scope for initiating effective political action. Britain, which exported most of the pollutants it emitted, had little to gain by such arrangements, especially since it had no environmental protection industry worthy of the name. What is more, the strong actor CEGB, which had fundamental doubts about the scientific and technical arguments advanced by Germany and Scandinavia, played a significant role in the political arena. Finally, the Conservative philosophy of public austerity stood in the way of introducing more stringent environmental standards for state industries.

Table 15: NO_x Emission Ceilings and Reductions for Existing Plants

Member state	Base values (1,000t)	Emission ceilings (1,000t/year)		Reduction against 1980 emissions in %	
	1980	1993	1998	1993	1998
Belgium	110	88	66	-20	-40
Denmark	124	121	81	-3	-35
Germany	870	696	522	-20	-40
Germany[1]	36	70	70	+94	+94
Britain	1,016	864	711	-15	-30
Spain[1]	366	368	277	+1	-24
France	400	320	240	-20	-40
Ireland[1]	28	50	50	+79	+79
Italy	580	470	428	-2	-26
Luxembourg	3	2.4	1.8	-20	-40
Netherlands	122	98	73	-20	-40
Portugal[1]	23	59	64	+157	+178

1) The positive percentage figures arise from the concessions made during negotiations to the recently industrializing countries that emissions from large combustion plants authorized between 1980 and 1987 were not to be included in calculating reduction targets.

Source: European Communities (1988)

The gridlocked negotiations began to move again only once pressure had been put on the British from two directions. While at home growing environmental awareness and the incipient institutionalization of green ideas, as well as the privatization of the CEGB increased British willingness to compromise, supranational decision-making received new impetus from the initiatives put forward by the succeeding national presidencies in the Council of Ministers. This situation was a particularly tricky one for the British presidency, which had to find a tolerable accommodation between national interests and international political prestige. Finally, it was left to the initiator of the directive, the Federal Republic, to break-up the 'unholy' British-

Spanish alliance with further concessions to Spain and more intensive bilateral negotiations, and to induce the British to yield by offering additional concessions to them as well.

3.1.3   France as Friendly Bystander and Coalition Partner

In contrast to Germany and Britain, which in the eighties had made determined efforts to keep European clean-air policy on their course either by setting the pace or putting on the brake, the French watched their neighbours' doings with relative calm and reticence. The attitude they took towards proposed measures was neither particularly recalcitrant, nor did they seek to take their turn at initiating or promoting specific arrangements. French indifference and neutrality was independent of the strategic position taken by the Commission. To this extent, the shift in European strategy from a quality-based approach to an emission-oriented one left the French — unlike the British — largely unmoved. From the French perspective, apparently, there was no special incentive to force through or block proposed measures either to minimize the legal and institutional costs of adjustment, or to safeguard economic competitiveness or to secure the negotiating position of state actors vis-à-vis industry. A number of factors in French clean-air policy can explain this at first glance surprising situation: the multiplicity of regulatory instruments, the structure of the energy sector, and the under-developed environmental protection equipment industry. Whereas the many instruments at their disposal determined French equanimity, especially towards quality-related directives and the framework directive on industrial emissions, the structure of the energy sector in France permitted them to take a back seat in negotiations on the LCP directive. And the low profile of the environmental protection equipment industry presented the French government with no urgent incentive to improve French sales opportunities by promoting stringent EU-wide standards. A further factor contributing to French neutrality was the circumstance that — where their existing regulatory instruments proved inadequate — the French took early steps to anticipate European measures so as to obviate later adjustments. As a result, 'European environmental policy does not fundamentally differ from French policy' (Larrue/Prud'homme 1990, 5).

Like Britain, France knew no nation-wide emission standards or pollution-concentration limits. However, whereas limits in the United Kingdom were fixed in individual terms, taking account of local environmental quality and the economic performance of the plant in question, in France such standards were negotiated at the regional level.

Neither the 1961 law on air pollution nor the 1963 and 1974 implementing regulations, nor the ministerial orders implementing the provisions specify *a precise air-quality objective* [emphasis added] to be attained; this omission is deliberate, because it is *the job of regional policy to define these objectives* [emphasis added] in terms of what are considered reasonable targets. ... French legislation also fixes no emission standards imposing general restrictions on combustion plant emissions' (Knoepfel/Larrue 1985, 44).[27]

This flexibility — combined with the special instruments of the protected and alarm zones, which find application in industrial agglomerations — permit the French to implement the quality-oriented directives on SO_2 and particulate matter, and on NO_x without completely reorganizing existing practice. In order to keep legal and institutional adjustment costs to a minimum, they initially waived translating the directive into national law, relying instead — like the British — on internal administrative orders.

The introduction of these limits was contrary to the intervention philosophy hitherto followed by the French administration. Until the eighties, the French authorities considered this type of standard would lead in effect to "rights to pollute" being granted in zones where air-quality levels were within the limits. The Minister of the Environment consequently refused to set any standards (Larrue/Prud'homme 1990, 5).[28]

Since administrative orders are binding only on the administration, it was not possible for environmental organizations, for instance, to take legal action to enforce implementation of the rules (Hoppe 1984, 160). This contradicted ECJ rulings requiring European directives to be implemented in the form of national legislation. The Commission therefore threatened to take France to court for contravention of the Treaties (*Ministère de l'Environnement* 1991, 291). European pressure led to the two EC directives being confirmed by implementing regulation[29] on 25 October 1991. Leaving aside this formal problem, substantive compliance with European standards represented no problem for the French (ibid., 292). Only the measurement and monitoring requirements contained in the directives made certain changes in the national control network necessary. A further effect of the directives was to strengthen the position of state control authorities, which could now plead European standards, vis-à-vis industry: '[The standards] have given more clout to the French authorities charged with administering this policy, ... , which have been able to intensify their activities against pollution emitters, especially in regions where the limits had been exceeded' (Larrue/Prud'homme 1990, 6).[30]

If implementing the 1984 framework directive — the basis for a new emission-oriented strategy — posed no legal and institutional adjustment problems worth mentioning, this was also due in no small measure to the structure of the French regulatory system. The prerequisites for authorization set out in the directive (BATNEEC; publication of the application for and notice of authorization; classification of plants requiring an authorization) had already been laid down in the 1976 Industrial and Nature Conservation Act and the regulations issued pursuant to the act. The *'meilleure technologie disponible sans surcoût excessif'* for reducing pollutant emissions is required as well as extensive public hearing procedures (*'enquête publique'*) (Mousel/Herz 1990, 170). Unlike Britain, France had no particular problem in accepting the arrangements under the directive for 'daughter directives' setting certain emission limits to be adopted by a qualified majority. There were two reasons. First, the French — despite the absence of nation-wide standards — did not share the profound British dislike of mandatory statutory emission standards. The quality-related procedure was part and parcel of British institutional traditions and normative conceptions. In France, emission standards could perfectly well be set at the regional level. Second, as we shall see, the French had nothing to fear from the only daughter directive under discussion at the time, the Large Combustion Plant Directive.

The Large Combustion Plant Directive adopted in 1988 fixed limit values for emissions of SO_2 and NO_x from new plants, and, in the form of 'national bubbles', provided for the gradual reduction of such emissions from existing plants. Although mandatory, plant-related emission limits were a novelty in France, the legal and institutional adjustments needed to accommodate them were within tolerable limits, since France had few new plants within the meaning of the directive. French energy policy was the main factor. In contrast to Germany and Britain, most primary energy in France was nuclear generated. Over a period of just under 10 years (1980-1989), the share of nuclear power in French energy generation rose from 25 per cent to 69.1 per cent (*Ministère de l'Industrie* 1991, 13). The nuclear energy share in electricity generation was no less than 75 per cent (*EDF* 1990, quoted by Berg 1991, 25). Fossil fuels like oil and coal now make a comparatively small and declining contribution. The limits for new plants laid down by the directive are thus practically irrelevant, since the construction of coal-fired power stations is not envisaged. There was also relatively little need to adjust because France had anticipated European arrangements. When it became clear at the start of negotiations that the planned directive would impose stricter requirements on the few new large combustion plants than had hitherto been the case in French practice, the French government took appropriate steps. As early as 1983 the Environment Minister issued a *'circulaire'* instructing the controlling authorities to take account of the planned EU standards: '... The French Ministry of the Environment attempted to *anticipate the decision of the Council* [emphasis added], taking up the main restrictions envisaged by the directive in a proposed circular order applicable to combustion plants subject to the legislation on registered works' (Larrue/Prud'homme 1990, 6).[31]

As far as existing plants were concerned, France was also not obliged to make great legal and institutional adjustments. Unlike in Britain, where a binding 'national plan' was elaborated to comply with reduction requirements, France was able to resort to a well-tried regulatory instrument deployed in both protected and alarm zones: fuel substitution (interview with EU Commission, DG XI, March 1993). In France, fuels are categorized by quality in terms of sulphur content. Normal fuels (*'fioul ordinaire'*) with a sulphur content of 4 per cent and fuels with a lower sulphur content (2 per cent — *'basse teneur en soufre'*) may in principle not be used in protected zones.[32] In such areas only fuels with a maximum of 1 per cent sulphur content may be used (*'très basse teneur en soufre'*) (Roqueplo 1988, 102).

Apart from the relatively low costs of legal and institutional adjustment, the French had no grounds of economic competitiveness to block negotiations. Since fossil-fuel-fired power stations[33] had so little political and economic salience, little attention was paid to the protest of the affected sectors, coal mining and the oil industry (Roqueplo 1988, 191). Besides, the reservations of these industries were mitigated by the fact that most plants had no need to install expensive FGD equipment, since their capacity was below the required level. It was possible to comply with the requirements of the directive by implementing a policy of fuel substitution. Moreover, the directive contained a special provision taking into account the relatively irregular capacity utilization rates common in French power stations. At peak periods, emissions frequently exceed limit values while on average for the year they are far below. It was thus decided that power stations with an annual running time below a certain level would be subject to less stringent emission standards (Boehmer-Christiansen/Skea 1991, 237).

Because of the wide-ranging regulatory instruments at the disposal of the French authorities, it was clear that the EU directives adopted in the eighties imposed no particularly comprehensive legal and institutional adjustments. France accordingly had no reason to stall the supranational decision-making process, especially where the proposed measures did not endanger the competitiveness of national industry. The extensive nuclear power programme was important in this connection. The relative insignificance of fossil fuels in energy generation meant that the provisions of the Large Combustion Plant Directive has had little impact on French industry.

Nor did the French have reason to take action themselves in the European arena to impose their own regulatory approach on other member states in the form of European directives. In this respect, too, the multitude of regulatory instruments was a crucial factor. Whilst it was in the interest of many countries to impose their domestic arrangements Europe-wide in order to minimize the legal and institutional adjustments that European directives entail, the French were able to remain relatively passive, since their multifaceted domestic regulatory regime allowed them to implement such measures without modifying their control system to any considerable degree.

The exploitation of European legislation by certain countries to expand their own sales of advanced technology (Héritier 1993, 6) brings us to another factor that goes a long way towards explaining France's indifference to European directives. The hypothesis that member states with a highly developed environmental technology industry will be especially prone to using European legislation to expand their own markets presupposes that government or industry sees environmental protection and the related technologies as an area subject to competition and open to development. Until the late seventies this was not the case in France. The French government did not recognize the need actively to promote environmental technology in the clean-air field until the early eighties, when they set up the *'Agence de la Qualité de l'Air'* *(AQA)*.[34] One of its tasks is to promote the development of environmental technologies. To this end it appropriates a large proportion of the revenue obtained from the *'taxe parafiscale'* ('parafiscal charge'), benefiting undertakings in the form of subsidies intended to encourage the development and use of environmental technology (Olier et al. 1989, 55; *Ministère de l'Environnement* 1991a, 7). Despite this government investment aid, the development of clean-air technology is still in its infancy. France accordingly has little interest in seeking to expand the market share of the national pollution protection equipment industry by proposing emission and technology-related directives.

3.2   Britain as Pace-setter: Procedural Measures

Once Britain, long opposed to the European emission-based directives, had accepted the Large Combustion Plant Directive, it exploited the resulting pressure for institutional adjustment to achieve far-reaching rationalization and modernization of the national regulatory system in the clean-air field. The dynamics for change arising from the coincidence of supranational and domestic imperatives induced a redefinition of British behaviour in the context of supranational policy-making. In renewing their control system, the British, who had always found themselves on the defensive in negotiating directives, wanted to go beyond existing European requirements to take position for a more active role in European legislation, and to secure as far as possible the 'advantage of the first move'. Hence, with its new legislation Britain not only satisfied European demands, but also anticipated future EU legislation by introducing elements before other member states, like integrated pollution control or extensive public access to information in authorization processes. 'We are now likelier to be taking quite an advanced and technically well informed position in discussions of draft air quality legislation in the Community' (interview with DoE, January 1993). What is more, anticipating EU measures lowers the risk of later pressure from proposed directives for institutional adjustment. It is the British who are now able to call the tune in important areas, and it is in their interest to impose their innovative arrangements on other countries via the EU.

3.2.1 Modifying the British Conception of the State

The revision of the British regulatory system against the background of both national and supranational developments has brought about profound changes in the British understanding of the state in the field of clean-air policy. They have affected two aspects: policy instruments, and interaction patterns between state and private actors, which find expression largely in modes of implementation.

The Reasons Behind the Changes

The domestic discussion on the institutional reorganization of the regulatory system was characterized — as in negotiations on the Large Combustion Plant Directive — by the overlap of European pressure and national contextual changes. Whereas EU measures made institutional adjustment of the British concept inevitable in the end, the process thus triggered generated momentum that was enhanced by developments in the national network. This resulted in some aspects of the new arrangements going far beyond the adjustments required by the EU.

Pressure for Legal and Institutional Adjustment: Implementation of EU Directives
EU directives, which are legally binding on member states, have to be implemented at the national level. Depending on national regulatory structures, this requires existing institutional arrangements to be modified to a greater or lesser degree. The assumption that, in negotiations on a directive, member states not only pursue competition policy goals but also seek to minimize the need for institutional adjustments, permits conclusions on the ultimate implementation of the outcome of such negotiations. It is thus to be expected that individual countries will try to manage as far as possible with the instruments they already possess. Only when it is clear that, even where the

full potential of these instruments is exhausted, the Directives cannot adequately be implemented will the necessary institutional changes be made.

The quality-related directives on SO_2 and NO_x thus provoked no initial adaptation in the British regulatory system because it, too, was based on a quality-oriented strategy. Despite the introduction of statutory quality standards — also a proceeding foreign to British pragmatism — these directives generated relatively little pressure to adjust existing institutional arrangements, since the choice of means to comply with and attain the given standards was left to the member states: 'There was flexibility built into the Directives which was very important. ... So, if you would close all the industry down, it's up to you' (interview with DoE, September 1993). The British accordingly attempted to implement the directives without making any institutional modifications, referring the Commission to existing statutory provisions and informal administrative rules ('circulars'). The Commission, however, complained from the outset that Britain, like France, had taken only administrative and not legislative steps towards implementing the Directive. A purely administrative procedure would make the government dependent on the cooperation of the competent regulatory authorities. The Commission felt that implementation of statutory quality standards was thus not adequately ensured. This view was confirmed by the European Court of Justice. Nevertheless, Britain adopted the pertinent Air Quality Standards Regulation only in 1989, which required the Secretary of the Environment to ensure compliance with the limits laid down (Haigh 1990, 189; Bennett 1991, 82). Hence, the Directives induced institutional changes in the British control system — which had hitherto operated without statutory standards — to meet European requirements, albeit against a degree of resistance and with considerable delay. 'One of the major effects of the EC was to ... introduce for the first time as a legal element air quality standards into our system' (interview with RCEP, September 1992).

Interestingly enough, the SO_2 quality directive in Britain had an impact not only on the law but especially on implementation practices. This was evident in local authority clean-air policy under the CAAs 1956/1968. There were still some areas (mostly mining regions) lagging behind in the designation of smokeless areas. The main reason was resistance from miners, who had no wish to stop using the cut-price coal to which they were entitled. The EU quality standards were thus frequently violated. But since the Directive required government to inform the Commission of such areas and of plans for progressively improving the environmental situation, the smoke control programme was somewhat speeded up (interviews with DoE, September 1991; IEEP, December 1991). Although under the CAA 1968 the government had the power to impose such zones on local authorities, the potential costs for local authorities and the extremely tense situation in the coal-mining industry meant that using this instrument to the full would have amounted to 'political suicide' by the government (interview with DoE, September 1993). The EU directive provided the government the basis for legitimizing their powers. 'For the first time, they [the government] threatened these local authorities saying: "Sorry, if you don't do it, we are going to use our reserve powers because we are now obliged internationally" (interview with RCEP, September 1993). This threat alone persuaded most reluctant local authorities to take corresponding steps.

Whilst the quality-based EU measures presented the British system with relatively little institutional challenge, the situation changed drastically when the Commission adopted an emission-oriented strategy. It was clear that implementation of both the 'Framework Directive' and of the Large Combustion Plant Directive would render lasting modifications to the British regulatory system inevitable.

This was first demonstrated by implementation of the Framework Directive, to be achieved by 1 July 1987 (ENDS 1987/148, 14). Although by adding NEEC to the European BAT, the British had succeeded in closing the gap between the European approach and their own bpm procedure with its strong economic bias, some discrepancies persisted between the two principles. They were particularly evident in the requirement under Article 9 of the Directive that both the licensing application and the decision by the competent authority be made accessible to public inspection.

Taking account of all relevant emission and quality standards in connection with BATNEEC established clear and unambiguous conditions for authorization. Although the AI had also operated in the context of bpm with so-called 'presumptive standards', they had not been legally binding: 'BATNEEC is writing — bpm was always a matter of debate and argument' (interview with Bexley local authority, September 1991). This implied a certain formalization of procedure. Furthermore, the EU provisions required the control system to be more 'transparent' by making both the licensing application and the permit open to public inspection. Although the Directive did not specify how this was to be organized, Britain, with its informal, secretive practice was now under pressure to make considerable changes, enhanced by domestic diffusion effects from other policy areas: 'The Framework Directive actually put on a public register what the conditions were. So they had to begin to change their concept' (interview with RCEP, September 1992).

A further difficulty in implementing the Directive concerned the range of processes subject to authorization. In Britain such a licensing requirement applied to only a portion of the installations specified under the Directive. These were the so-called 'scheduled works' which were under the supervision of the AI. All other processes were supervised by local authorities, which disposed of no preventive controlling powers. They could intervene only if demonstrable damage and environmental degradation were caused by operation of the plant ('public nuisance' procedure). The licensing requirements of these locally controlled processes thus failed to meet the conditions defined by the Directive (Haigh 1990, 226). The due implementation of the Directive was therefore possible only through appropriate institutional modifications to the British regulatory system. There were two basic options. Either the processes under local authority supervision could be placed under the central control of the AI, or local authority powers could be expanded to satisfy European requirements.

The DoE reacted to this situation with a consultative document prepared in 1986 (Air Pollution Control in Great Britain. Review and Proposals). Interestingly enough, this document was decisively influenced by the regulative substance of the Framework Directive, although it was really an official response to domestic proposals for change based essentially on the 1976 RCEP report[35] — 'The DoE's priority is to comply with the 1984 EEC Directive on industrial plants' (ENDS 1988/161, 20). The modifications envisaged by the paper were thus to permit the adaptation of the British regulatory approach to meet European requirements. It covered the following points:

1. The Environment Secretary was to be empowered to issue statutory emission and quality standards. This would ensure that European standards were taken into account in national licensing procedures (Weidner 1987, 88). This would permit not only the due implementation of the quality-related directives on SO_2 and suspended particles, as well as NO_2 but also the more formal definition of licensing conditions in accordance with the bpm principle, which would otherwise be retained: 'The former system of annual registration of scheduled processes is to be replaced by a system of prior authorization as required by the EC "framework" Directive' (Boehmer-Christiansen/Skea 1991, 266f.).
2. To satisfy the publicity requirements under the Directive, a 'system of written consents' was to be introduced in which both application and permit as well as the relevant emission measurements were to be published. However, no details on practical implementation were given.
3. Local authority powers were to be appropriately expanded. They were also to be granted preventive control powers for the processes under their control, based — as with the AI — on the bpm principle. For this purpose all industrial processes were to be divided into two categories: part A would cover all technically demanding processes under the supervision of the AI. Part B was to cover all other industrial processes requiring less technical expertise for their control. The fact that it was suggested to en-

large local authority powers instead of transferring local responsibilities to the AI is attributable to two factors. First, it was indicative of a degree of institutional path-dependence, since the British regulatory system has always been based on this dichotomy of central and local control. Furthermore, AI resources would not have permitted it to implement this reorganization, whereas only comparatively small changes were needed at the local level (DoE 1986a, 22).

Whilst implementing the quality-related directives involved relatively slight modifications to British practice, thus favouring more hesitant action relying on old instruments, the Framework Directive forced the British to change their regulatory system more drastically. Besides proposing to enlarge local powers, the consultative document was concerned primarily with greater formalization and openness in the British system. The adoption of the Large Combustion Plant Directive in 1988 intensified pressure on the British regulatory system to adjust.

The introduction of statutory emission limits constituted an absolute innovation in British practice, which had been operating on the bpm principle with its emphasis on the individual case and its exclusive concern with local environmental quality. Moreover, it was necessary to elaborate suitable programmes for the progressive reduction of total annual emissions from existing plants. Besides time requirements, they also had to specify appropriate reduction targets and details on how they were to be attained (European Communities 1988, 3). Existing instruments could not satisfy this requirement, since under the bpm principle the AI enabled emission abatement only if both appropriate advances in control technology were available and the use of the technology was economically feasible for the plant concerned. To this extent, bpm was not the wherewithal for imposing progressive annual emission reductions on industry as required by the Directive. Moreover, the individual orientation of this principle was scarcely suitable for coordinating total national emissions.

While the introduction of statutory emission standards had been anticipated in the 1986 consultative document, the DoE proposed corresponding measures for implementing the abatement programme in 1989. It was proposed that the Environment Secretary be empowered to issue a legally binding 'national plan' in subordinate legislation laying down the progressive annual reduction of emissions by industrial sector.[36] Reduction rates were defined individually for each plant (so-called 'company bubbles'), with firms being permitted a certain degree of trade-off between various operating sites.[37] These target values were to be taken into account by the supervisory authorities in the licensing and monitoring processes: 'A broad duty will be laid on inspectors to ensure that emissions from prescribed processes

are in conformity with any national plan' (ENDS 1989/178, 12). In contrast
to Germany, which set uniform emission standards for all plants regardless
of location and local environmental situation (interview with EU Commission, DG XI, September 1993), the British attempted to render implementation of the abatement plans more flexible by means of 'company bubbles'
and distribution among various industrial sectors.

Nevertheless, the Directive imposed changes in state arrangements in
Britain, since the regulative powers of supervisory authorities were
strengthened vis-à-vis industry. In contrast to the old regulatory approach, it
was now envisaged to control industrial pollutant emissions by means of legally binding time and limit-value requirements. At the same time, this implied a stronger position for the regulatory authorities vis-à-vis industry.
They were no longer necessarily dependent on consensual bargaining with
industry. They now disposed of concrete requirements embodied in law to
which they could refer in performing their functions. Implementation of the
emission-related EU directives thus triggered lasting institutional changes in
the British control system: 'There is a strong argument that this action did
provide a momentum and did actually get things moving in a way in which
our earlier rather cosy bpm-approach had not done' (interview with DoE,
September 1993). Domestic developments provided further impetus to the
restructuring process.

Developments in the National Network
The domestic debate on change was triggered by an extremely innovative
report issued by the RCEP in 1976, entitled 'Air Pollution Control. An Integrated Approach' (RCEP 1976). It advanced a number of demands that
would have involved fundamental changes in existing institutional arrangements. The core element was the idea of an integrated control system in environmental protection. For the first time, attention was drawn to the problem that the control of releases to the atmosphere was not to be considered
in isolation from pollution of the other environmental media soil and water:
'The reduction of emissions to the atmosphere can lead to an increase in
wastes to be disposed of on land or discharged to water, and vice versa. If
the optimum environmental solutions are to be found the controlling authority must be able to look comprehensively at all forms of pollution arising
from industrial processes' (RCEP 1976, 3). In order to give practical effect
to such an integrated approach, it was suggested that a single controlling
authority be formed, which in licensing procedure would define the solution
most compatible with all environmental media on the principle of the 'Best
Practicable Environmental Option (BPEO)' (ibid., 76). The new authority
was to be attached to the DoE and would be formed by merging the inspec-

torates responsible for water, soil, and wastes. For this purpose it would first be necessary to retransfer the AI, which had been attached to the HSE, to the DoE.[38]

The RCEP report also recommended retaining in principle the division between central and local control. The integrated approach was accordingly only feasible with a central authority disposing of the technical expertise needed to implement the concept. Local authorities were therefore generally to be responsible for technically simpler processes and were to control them separately for each medium. However, the RCEP proposal radically to expand local authority powers was a decisive innovation. They were also to operate in accordance with the bpm principle and only secondarily in obedience to the 'nuisance' rules, which permitted intervention only when damage or degradation was already in evidence.

Finally, the report proposed several measures to improve the openness of the regulatory system. The definition of bpm in specific cases was to be made comprehensible to the general public. For this purpose it was recommended in the first place to make licensing procedure more formal and to specify authorization conditions with greater precision. In addition, the public was to be very much better informed. The RCEP called for authorizations and data on the local environmental situation and polluting emissions from industrial plants to be recorded in public registers (ibid., 65).

To begin with, the RCEP proposals were acted upon by neither the then Labour government nor the Conservatives, in power from 1979. It was not until in 1982 that the government expressed an opinion, only largely to reject the proposals (DoE 1982). What moved them to do so was that at the time the government had not yet been subjected to either international or domestic pressure on the issue. The RCEP proposals for institutional innovation were apparently ahead of their time: 'The proposals of the Royal Commission have been too unusual, too bizarre, too far ahead of the time' (interview with RCEP, September 1992). However, the situation changed when the RCEP report gained new relevance in the context of implementing the Framework Directive, the provisions of which partly coincided with RCEP ideas: 'there has been a congruence of influence' (interviews with DoE, November 1992; January 1993). There was broad agreement both with regard to the formalization and openness of licensing procedures and to the enlargement of local authority powers. However, the EU requirements gave no incentive to include the more thoroughgoing RCEP recommendations in the institutional reorganization, especially the introduction of a public register containing comprehensive information and an integrated regulatory approach. This step was to become possible only once the development of various national factors had given new impetus to revising the existing approach.

From the mid-eighties, further developments to various aspects of the general national setting heightened the chances for innovation in environmental policy. The nascent economic upswing increased the government's room for manoeuvre vis-à-vis a flourishing industry to impose more stringent environmental controls. Also particularly important was the institutionalization of environmental protection in the policy-making process, which continued to progress until the early nineties. Finally, the stronger emphasis placed on the values of the 'free and responsible citizen' ('citizenship', 'consumerism') by neo-liberal Conservative philosophy fostered a more open information policy by state authorities and institutions as demanded by the RCEP in the environmental protection field.

Improvements in the economic situation began to show in Britain from the early eighties. In 1982 a cyclical upturn began, which intensified from the mid-eighties. GNP grew between 1985 and 1989 by an average 3.8 per cent per year (peaking at 4.8 per cent in 1987). Only in 1990, with a growth figure of 0.8 per cent, did the boom begin to weaken. Prices experienced a similarly positive development. Average inflation between 1985 and 1989 was 5.2 per cent (compared with 15.6 per cent between 1973 and 1979), a relatively low figure by British standards. Unemployment also dropped over the same period from 11,5 per cent (1985) to 5,5 per cent (1990) (OECD Historical Statistics 1960-1990). These improvements in key economic figures gave the government more scope for pushing through more stringent environmental controls vis-à-vis industry. At the same time, the boom raised the general standard of living, bringing in its wake greater sensitivity to environmental protection issues.

Already during negotiations on the Large Combustion Plant Directive, the effects on the political agenda of the increasing hold of green ideas had become apparent. This development, attributable primarily to the persistent activities of the British environmental organizations,[39] gradually modified the government's political context for action. After acceptance of the Large Combustion Plant Directive, this process accelerated still further, finding its climax in the famous speech by Margaret Thatcher before the Royal Society in September 1988, in which she addressed the issue of environmental pollution in detail for the first time in her period of office (Boehmer-Christiansen/Skea 1991, 264; interview with British environmental consultant, September 1993). 'Given the concentration of executive power in the British constitutional monarchy, such a change of direction — to be more exact a U-turn — by the Prime Minister transmits rapid and lasting signals to many areas of the British political system. The Thatcher speech ... gave a clear signal that these processes [the activities of environmental organizations] have finally managed to intrude far into the political system of the United Kingdom. One environmentalist compared the process with a split tennis ball subjected to ever growing pressure, and which initially offers considerable resistance, but at the critical point an irreversible, accelerating process begins' (White-Grove 1992, 112ff.).

This political pressure was intensified with the unexpected electoral success of the Green Party in the election to the European Parliament in 1989, in which they won 14.9 per cent of the vote. Although the peculiarities of the British electoral system prevented the party from obtaining any seats in Parliament despite their high share of the vote, the success of the Greens nevertheless sparked greater environmental policy efforts by the established parties (interview with Greenpeace, January 1993). This was especially true of the Conservatives, who had suffered substantial losses (over 6 per cent) compared with 1984.[40] At the same time, new global issues like the ozone hole and the greenhouse effect provided grist to the environmental policy mill. Protection of the environmental enjoyed high priority on the political agenda for a number of years, but the recession that began to develop from 1990 onwards has since displaced it more and more in favour of other concerns (interview with AMA, January 1993). Despite this negative development, environmental issues continue to play an important role due to their institutionalization within the political decision-making process. 'What has changed compared to, say, the late 70s is that now environmental issues are institutionalised, they are now on the agenda in a way they weren't ten years ago' (interview with British environmental consultant, September 1993).

The established status of environmental policy ideas is given very clear expression in the change in political leadership at the DoE that occurred in 1989 in reaction to the electoral success of the Greens. Nicolas Ridley, who had proved very undiplomatic in his dealings with the environmental movement, was replaced by Chris Patten, a politician with ambitions in environmental affairs (Boehmer-Christiansen/Skea 1991, 264; interview with DoE, September 1993). This general development continued under the new Prime Minister, John Major, who in a speech in July 1991 had affirmed the crucial importance of environmental protection in economic development. One of the most significant decisions made by the new Secretary of the Environment Michael Heseltine was to appoint an environmentalist[41] as full-time advisor on environmental affairs. This was the first appointment to such a high position of trust (White-Grove 1992, 113).

Whilst the RCEP proposals to make the regulatory system in the clean-air policy area more 'transparent' and open to the outside world had initially been regarded with reservation by the government, a greater trend towards openness began to develop in the late eighties. One reason may be diffusion effects from other policy areas, where relatively comprehensive public information rights had long existed. Aspects of the Conservative's neo-liberal philosophy under which individual rights of the citizen were redefined also played an important role.

The citizen was no longer regarded as the passive recipient of state services but as an active consumer choosing by quality criteria among the range of services offered by the authorities ('consumerism'). This point of view is in keeping with neo-liberal thinking, which sees societal well-being as dependent on the preferential decisions of individual utility maximizers. '[This] shift away from collectivist solutions in public policy' (Gamble 1988, 124) is abundantly clear in the notorious Thatcher aphorism 'There is no such thing as society. There are individual men and women, and there are families' (quoted in Offe 1990, 10). If citizens were to make a rational choice in individual cases, they had to have the pertinent possibilities to obtain information and participate. Government authorities were thus obliged to render public account of their activities ('accountability') (Rhodes 1991, 102). At the same time, this accountability was expected to induce greater effectiveness and efficiency in public authorities in the pursuit of their duties.

These goals also played an important role in the introduction in 1990 of the so-called 'poll tax'. It replaced the old local rates system entirely and was intended to establish a direct link between the tax to be paid by the individual citizen and the quality of local authority services. The poll tax was thus more of a charge for public services than a general public levy. For this reason it was termed 'community charge' or 'resident's charge' (Crick/van Klaveren 1991, 408).[42] However, the introduction of the poll tax proved a disaster for the government. The tax met with vehement criticism from the public (there were numerous demonstrations that sometimes ended in violent confrontation), so it was decided to abolish it in 1992. It was replaced the following year by the so-called 'council tax'[43] (Farrington/Lee 1992, 1; Bramley 1991, 284ff.).

The Citizen's Charter (1991) defined appropriate standards for public services and even gave consumers a right to compensation if services were not up to standard (Oliver 1991, 25).[44] The general process of greater openness in the United Kingdom continued in the nineties. Following the 1991 Citizen's Charter, the government adopted a white paper in June 1993 entitled 'Opening up Government', which among other things proposed to extend public access provisions in the environmental field to health and safety at work (interview with DoE, September 1993).

The impact of these factors on policy-making increased the chances for acceptance and realization of RCEP concepts not yet taken into consideration. While growing environmental awareness and the greening of political and administrative structures primarily favoured the establishment of an integrated approach, the general trend towards openness had a positive effect on the substantive scope of the public registers.

Already in 1986 the AI was transferred to the DoE with the simultaneous announcement that a joint controlling authority was to be set up on 1 July

1987. The new body — Her Majesty's Inspectorate of Pollution (HMIP) — was formed by merging the AI, the Radiochemical Inspectorate and the Hazardous Waste Inspectorate.[45] This change occurred primarily against the background of the elections to be held in 1987 and of enhanced public environmental awareness (interview with RCEP, September 1992). A year later the government finally reacted to the demands reiterated in 1984 and 1988 for an integrated approach, tabling the necessary legislative measures for Integrated Pollution Control (IPC) (DoE 1988a). The DoE proposals had in essence two objectives. Besides introducing the new IPC concept, complete compatibility with developments at the EU level was to be ensured (DoE 1988a, 8). The consultative document took account of all requirements under the EU directives (emission-orientation, standards, licensing conditions, BATNEEC), and provided for these rules to be applied in the context of an integrated concept covering all environmental media, which went beyond the provisions of the EU directives. Further consultations in December 1988 (DoE 1988b; 1988c) served to distribute the processes to be controlled between central and local authorities and to concretize development of IPC.

Against the background of a general increase in the openness of state activities, the RCEP recommendations reiterated in 1988 (RCEP 1988) on the introduction of public registers in the clean-air field received new impetus, which was further strengthened by activities of the Campaign for Freedom of Information.[46] In addition, some DoE officials took a very positive view of improved openness in regard to environmental regulation, 'they have been very keen on opening things up' (interview with British environmental consultant, September 1993; interview with DoE, September 1993). The government reacted in 1989 to growing domestic pressure with a consultative document proposing a system of public registers for the processes controlled by the HMIP and local authorities (DoE 1989). However, the proposal met with fierce opposition (especially from the RCEP, the Campaign for Freedom of Information and local authorities), because publication of average annual emissions was planned only in summary form (ENDS 1989/176, 24). The government quickly gave in to the considerable weight of public opinion. The new Environment Secretary played an important role: 'The decision to have a more liberal disclosure regime was taken by Environment Secretary Chris Patten' (ENDS 1990/180, 23). The public registers are thus fully in line with RCEP recommendations, containing both licensing application and permit together with all the results of emission readings, which are constantly updated (Gibson 1991, 27). Like implementation of the IPC approach, these rules, too, went beyond existing EU provisions, which do not require disclosure of monitoring findings.

Looked at as a whole, it is clear that the early RCEP ideas were able to gain momentum only under European compulsion and changes in the domestic policy context. Whereas the government initially rejected the proposals outright, in 1986 partial acceptance was signalled, although only to the extent of establishing congruence with European provisions (especially the Framework Directive). In sum, the incipient greening of the policy-making process also came to bear at this point in time. For electoral strategy reasons the HMIP was set up, thus taking account of at least some aspects of the integrated approach advocated by the RCEP. Interestingly enough, administrative integration thus preceded the legislative implementation of IPC (interview with DoE, November 1992). However, the actual turning point in reforming British legislation came only after further changes within the national network and acceptance of the Large Combustion Plant Directive. The amelioration of the general economic situation, the ongoing greening process, and the increasing opening up of the 'secretive state' in connection with Conservative efforts to restructure the state expedited a momentum in domestic policy-making that was ultimately to bring the RCEP ideas to fruition, permitting the institutional transformation of British clean-air policy in excess of the changes required by European directives.

The Outcome of Changes in the State: The Environmental Protection Act and Its Significance for the Type and Form of State Intervention

The pressure for adjustment generated by implementation of European directives and the innovative domestic policy ideas advanced by the RCEP, which managed to impose themselves only in the interaction of additional factors such as improved economic conditions, a general greening and opening up of the political and administrative system, caused a fundamental modification in state arrangements relating to clean-air policy. This modification brought changes in the British negotiating position in supranational decision-making. How, under changed domestic conditions, did the British attain their goals of minimizing legal adjustment costs and safeguarding economic competitiveness? To answer this question we need first to establish the essential nature of the changes occurring in the British understanding of the state. As in the initial treatment of British clean-air policy in chapter two, our interest focuses on developments in legal regulatory structures, policy instruments, and modes of implementation.

Legal Regulatory Structures and Policy Instruments
With the 1990 Environmental Protection Act (EPA), Britain thoroughly rationalized and modernized its environmental policy.[47] At the same time the British reacted to the provisions of European directives and the innovative ideas put forward by the RCEP. Particularly important are the provisions

under Part I of the Act (Integrated Pollution Control and Air Pollution Control by Local Authorities), which regulate IPC and local authority clean-air policy.[48] Two fundamental innovations in the British control system were introduced:

1. Instead of separate pollution control for each environmental medium, all technically complex processes are now supervised by the HMIP in the framework of an integrated approach (IPC intends to ensure optimum load allocation among the various environmental media.
2. Local authority controlling powers for the remaining processes in the clean-air field have been greatly expanded. For the first time they now have the possibility of intervening preventively and not only when nuisances and damage have already occurred. At the same time, however, more detailed behaviour-related rules are to be laid down by central government as 'guidance notes'.

The Environmental Protection (Prescribed Processes and Substances) Regulations list the substances and processes subject to IPC or local authority control. Whereas under the former regulatory system, authorizations were negotiated between the AI and industry, highly specific conditions are now laid down for implementation of IPC by HMIP. An authorization may be granted only if several conditions are met. The primary condition is that BATNEEC (Best Available Techniques Not Entailing Excessive Cost) be used to reduce and avoid emissions of pollutants. In contrast to the Framework Directive, which is concerned only with technology, the concept of 'technique' applied in Britain is broader. Besides pure control technology, it relates to the type and manner of application (design, staff qualifications, working methods, supervision, and maintenance) (Gibson 1991, 24; interview with HMIP, September 1991). If a plant discharges emissions to several environmental media, the operator must meet not only the BATNEEC condition, which is monitored separately for each medium, but must also ensure that pollution of the environment is minimized as a whole. This requirement finds expression in the Best Practicable Environmental Option (BPEO) (interview with IEHO, March 1992).

Independently of these two principles, it must in all cases be ensured that all the relevant international and national limits are respected, regardless of whether they are quality or emission standards (DoE 1991, 6). Compared with former practice, which knew only non-binding internal administrative standards,[49] recourse to statutory standards is a significant change in British clean-air policy attributable to EU influence (interviews with HMIP, September 1991; IEEP, December 1991).

The introduction of BATNEEC and the setting of statutory limits also marks the shift from a quality-oriented approach to an emission-related one in British clean-air policy (interview with HMIP, September 1991). This re-orientation is evidenced by the way in which BATNEEC is specified. Instead of certain control technologies, only 'release levels' for pollutant emissions are prescribed, which in the opinion of HMIP can only be kept by deploying the best available techniques. This is intended to give plant operators a choice between several equivalent technologies (DoE 1991, 18; interview with HMIP, September 1991). Moreover, limit values serve only as a basis for defining BATNEEC, which, regardless of such standards, provides for incremental lowering of release levels as technological advance permits.[50] In this manner the flexibility of the former bpm[51] is combined with statutory standards defining minimum requirements (interview with IEHO, March 1992). For the rest, BATNEEC makes no provision for orientation on the local environmental situation as had been the case with the bpm approach. Qualitative environmental aspects thus no longer play a role in defining control technology for the individual firm. 'IPC says that any emission must be bad. ... There is no sort of assessment whether it is bad or not' (interview with multinational pharmaceuticals group, January 1993). To this extent, elements of the precautionary principle are given greater weight. 'BATNEEC is a way of implementing a precautionary principle because it's saying: you must reduce that substance as far as possible. And the only thing that stops you is the technology that is available' (interview with CBI, September 1992).

Licensing conditions are concretized in 'guidance notes' issued for every industrial process subject to IPC by HMIP.[52] They provide not only a schedule of the pollutants to be controlled, but also all relevant national and international standards, recommendations for the control techniques to be used, and the release levels that in the opinion of HMIP can be attained with the proposed technology. It is planned to update the guidance notes to keep pace with technological advance, but at least once every four years (DoE 1991, 18).

At present, the complete implementation of IPC is, however, subject to certain restrictions, since the authority responsible for water quality control, the National Rivers Authority (NRA), has not yet been subsumed under HMIP. This causes jurisdictional overlap between the two authorities in controlling the discharge of pollutants to bodies of water. A coordination agreement specifies that a final authorization under IPC is the province of HMIP, but that the NRA may impose certain conditions that HMIP may sharpen but not tone down. Furthermore, a permit may not be granted by HMIP if the NRA comes to the conclusion that certain water quality standards cannot be met (interview with HMIP, September 1991; Gibson 1991,

23f.).[53] Waste regulation, which is the responsibility of local waste regulation agencies (WRAs), is still organizationally separate from the IPC regime.

In the course of the coming years, however, it is intended to bring together the NRA and HMIP in an Environmental Protection Agency to permit full implementation of the integrated concept. At the same time, local authorities are to lose their waste regulation powers, which will be transferred to the new agency (interview with AMA, November 1994). The new authority is to have the status of an 'independent public body' (interview with DoE, November 1992). In this way it will have greater independence vis-à-vis political priorities and ministerial influence. 'There is a feeling that a separate agency would give more focus and independence to the administration' (interview with DoE, September 1993). Although the new agency will still be organizationally under the supervision of the Environment Secretary, it is to report directly to Parliament, and is thus released from direct political answerability to the government. Its functions and regulatory powers will not be affected by this institutional change (interviews with DoE, September 1993; November 1992). The transfer of executive function to independent bodies is, moreover, part of a general strategy with which the Conservative government is attempting to make public sector operations more efficient, and in the long run to reduce the size of the civil service (interview with DoE, November 1992; Oliver 1991). Although Prime Minister Major announced the establishment of the authority as long ago as the summer of 1991, the reorganization process is likely to take some time because of conflicts of interests in the government. Particularly problematic are competence issues between the DoE and the Ministry of Agriculture, Forestry and Fisheries (MAFF), which is also interested in taking over part of the NRA functions (ENDS 1991/198, 14). Problems have also been caused by the differing organizational structures of WRAs, NRA and HMIP (interview with AMA, November 1994).[54]

The 1994 bill on organizational restructuring places greater emphasis on cost/benefit considerations, a traditional element in British regulation, which had been pushed somewhat into the background by the 1990 EPA. The agency will thus have to take account of possible costs and benefits accruing from its action or inaction. However, the extent to which this provision represents a partial return to the 'old' regulatory practices is unclear for the present, nor can it be judged whether it creates a new basis on which environmental organizations could take legal proceedings to oblige the authorities to undertake environmental activities (interview with AMA, November 1994). Against the background of the present debate on deregulation in Britain,[55] however, the former could well be the case. In 1994 a deregulation act was passed, permitting, primary legislation to be modified by secondary legislation in certain cases, where procedure for adoption is simplified. It is thus conceivable, 'that parts of environmental legislation that are thought to impose a burden upon business and which do not constitute a "necessary" protection can be removed through a piece of secondary legislation which has a very brief debate in Parliament' (interview with AMA, November 1994).

Local authority controlling activities in the clean-air field differ little from the policy instruments available to HMIP. Local authorities, too, control the processes under their supervision on the basis of authorization conditions specified by BATNEEC and relevant emission or quality standards. Only the BPEO principle does not apply in this connection, since no integrated approach is practised at the local level. Nevertheless, the EPA substantially expanded local authority powers. For the first time they are now able to control the release of pollutants to the atmosphere by preventive means, whereas previously they had been able to intervene only when harmful environmental impacts had already occurred. Licensing conditions, as in the IPC procedure, are specified in guidance notes for the given industrial process (interviews with IEHO, March 1992; Bexley local authority, March 1992).

Implementation: Structure and Mode
The previous consensual, informal negotiation of operating conditions behind closed doors between regulating authority and industry, 'the old chumminess between inspectors and industrialists' (interview with DoE, November 1992), was replaced by a more formal and open style giving the inspectors greater regulative powers than under the old system (Jordan 1993, 413). This 'arm's-length approach' (interview with multinational pharmaceuticals group, January 1993) is largely attributable to institutional changes.

The introduction of public registers was an important factor, considerably increasing the openness of the regulatory system. These registers, which have to be established for both locally controlled processes and for IPC,[56] contain all the data relevant to licensing and operation that is in the possession of the controlling authority. It includes not only application, authorization, any legal proceedings and public authority objections, but also the results of emission measurements made by the authorities or especially by the undertakings. The firms may withhold such information only if they can prove that it is a matter of commercial confidentiality. Both local authorities and HMIP have dealt very restrictively with this exemption provision.[57] 'It isn't acceptable to argue that information shouldn't be available in the register because if the public saw what was being released from the process, it would embarrass the company. Embarrassment doesn't count. The only test is that the information would prejudice to an unreasonable degree the commercial interests' (interview with HMIP, September 1991; interview with Bexley local authority, March 1992).

The combination of openness through comprehensive public registers and an emission-oriented strategy based on statutory standards gives the controlling authorities less scope for cooperative negotiation, and favours greater severity towards industry. 'If information is appearing in the register about monitoring, there may be evidence that the releases are in excess of the authorized limits. If the Inspectorate hasn't noticed that then Greenpeace or FoE will. ... So I think we will find more enforcement notices than in the past' (interview with HMIP, September 1991; ENDS 1990/181, 21). This attitude is confirmed by industry, which has had initial experience with the new rules: 'The inspectors are more like policemen now' (interview with CBI, September 1991).

This arm's-length approach is also manifest in the far more formal elaboration of the guidance notes. In contrast to the earlier definition of the 'notes on bpm', which was done at an informal level by AI experts and industry, HMIP now draws up the rules by itself and communicates them to industry only afterwards in the context of two consultation rounds. Moreover, the final draft is made available to the public before coming into force. Thus not only industry but also environmental groups have an opportunity to raise objections in the final round of consultations (interviews with IEHO, March 1992; IEEP, December 1991; DoE, December 1991).

Despite this stricter mode of regulation, there are still certain cooperative elements in relations between the controlling authorities and industry. This may be because especially older inspectors, still used to the earlier bargaining practices, have not adapted to the new regime without further ado. 'You have to teach the old dogs new tricks and that takes quite a long time. It is something in the tradition of the administration that tends to go for keeping the things steady, going for compromise' (interviews with DoE, September 1993; November 1992). Putting an arm's-length approach into practice also presents certain problems. The mass of written information that has to be exchanged between the controlling authority and industry necessarily puts great strain on the resources of controlling authorities, thus slowing down licensing proceedings, which can place industry at a competitive disadvantage. In this situation it is in the interests of both HMIP and industry to hold informal talks before an application is filed so as to limit the amount of information that has to be communicated: 'The idea that the regulator sits in his office and never comes to the plant is just not going to work. You do need a dialogue, particularly when you are developing something. ... Nobody can tell us what information we need. [Under] the previous system the regulator actually came in and we discussed our problems and came to an agreement' (interview with multinational pharmaceuticals group, January 1993; see also interview with CBI, September 1992).

The definition of BPEO in individual cases shows particularly clearly that licensing conditions cannot be determined unilaterally by HMIP. First experience with implementation indicates that, in order to obtain authorization under IPC, industry has to provide HMIP with the relevant information. In contrast to the previous system, it is now a matter not so much of the quantity of industrial emissions but the impact of these releases on the environment. Only on this basis can HMIP check whether control measures are compatible with the BPEO concept or whether alternatives more compatible with the environment have to be pursued.[58] The new regulatory concept thus stresses responsible thinking by industry; they have to become better aware of the concrete impact of their activities on the environment: 'One of the main achievements of the system has been an increase of the self-knowledge of industry. ... It is not enough to send the application in saying: "This is what we do, please authorize". [HMIP] wants to know what the impact of what they do is ' (interview with DoE, November 1994).

The emphasis placed on industry's own responsibility is also shown by both local authorities and HMIP entrusting measurement and monitoring largely to industry, restricting themselves to spot checks. However, the measurement techniques used are inspected within the context of authorization proceedings (interviews with HMIP, September 1991; Bexley local authority, March 1992). Many firms have meanwhile introduced so-called quality assurance systems elaborated by the British Standards Institution (BSI).[59] They set standards for a complete industrial environmental management programme. This includes in particular current standards for control systems and monitoring technologies, rules on structuring operations, and requirements with regard to personnel qualifications and further training. On applying for authorization, firms vouch for their quality assurance, which obviates further inspections by the controlling authorities with the exception of spot checks (interview with IEHO, March 1992). From the perspective of the controlling authorities, this self-regulation by industry is advisable in view of the vast mass of information to be processed and the limited human resources available to do it. It can, moreover, be assumed that industry is concerned to handle the standards responsibly to avoid damaging their public image. Problems tend to arise rather with smaller firms under local control that have not adopted these quality assurance systems: 'Local authorities are faced with many cowboy operators and back street premises' (Carden 1992, 3).

With regard to relations between central and local government, the new Act contains provisions going to a certain extent against the general trend towards centralization pushed by the Conservatives. For example, the right granted local authorities under the EPA to authorize industrial plants in their jurisdiction in prospect is a fundamental enlargement of local powers (interview with DoE, January 1993; interviews with local authorities: Corporation of London, Bexley, September 1991; interview with IEHO, March 1992). 'EPA did confirm the local authorities' role. ... And that is not insignificant in a climate where local authorities were not regarded as partners of central government.... It was half a hand of friendship back to local authorities' (interview with AMA, January 1993). The release levels and BATNEEC definitions laid down in the guidance notes substantially expanded the range of local authority regulatory instruments and strengthened their position vis-à-vis industry: 'there is more support to go for a high and technically appropriate standard' (interview with Bexley local authority, September 1993).[60]

The increase in local authority powers in the clean-air field is in keeping with the general trend to be observed in relations between central government and the local level. The government is again concerned to involve local authorities more fully in political decision-making. The hitherto conflictual practice is giving way more and more to a consensual stance: 'the relationship with central government has improved significantly' (interview with AMA, November 1994). From the local authorities' point of view, there are three factors responsible for this change. First, it is clear that Prime Minister Major — unlike his ferrous predecessor, tends to rely more on classical British proprieties like consultation and consensus. Second, the Conservative government is in a weaker position than in the eighties, and can no longer afford 'to have local authorities as an enemy' (interview with AMA, November 1994). The third factor is the general shift in policy priorities within the Conservative Party: 'The great area of conflict in the Conservative Party now is not what to do with local authorities, it's what to do with Europe. ... So, the ideological conflict has moved on and therefore there is less reason to fight between central and local government' (ibid.).

Despite this strengthening of decentralized elements, the guidance notes also involve elements of central control. Although local authorities are given a certain scope for action and interpretation, the notes nonetheless — in contrast to the old system — markedly restrict local flexibility: 'It's no true devolution' (interview with AMA, January 1993). Local authorities can thus no longer decide by themselves whether and how to intervene. At regular intervals they have to report to central government on the state of implementation (interview with Bexley local authority, March 1992). What is more, with their meagre financial and personnel resources, local authorities frequently have difficulty in duly exercising their functions. 'It's a larger job with less staff because central government is continuously cutting back local authorities' (interview with LGMB, January 1993). Although the local authorities receive revenues from industrial licensing fees, they are for the most part insufficient. Since they have no say in how much is charged, duties pursuant to the EPA are often performed to the detriment of other tasks of the environmental health officers (interview with IEHO, March 1992).

The British debate on the distribution of central and local powers is particularly sharp in connection with the European definition of subsidiarity. While central government sees subsidiarity as 'getting Brussels from its back' (interview with AMA, January 1993), local authorities prefer a definition that gives them more scope for action and more autonomy in relation to central government (Héritier 1993). In order to improve their position vis-à-vis central government and to win support for a more active environmental policy, local authorities are turning more frequently to Brussels or applying to the European Court of Justice. 'Local authorities become more European in the way they are thinking' (interview with Bexley local authority, September 1993): 'There is tremendous detailed knowledge [on environmental affairs] on the local level, but there is no mechanism to connect them into the policy-making process. Therefore, in order to influence European legislation, the professional organizations of the local authorities joined the European Environmental Bureau' (interview with DoE, January 1993; interview with AMA, January 1993). This conduct of the British local authorities is to some extent in line with the proposals of the Fifth EU Action Programme, which provided for increased application of 'pressure from below' to improve the implementation of European legislation (interview with LGMB, January 1993). However, the Commission has not directly confirmed that it finds itself confronted by this sort of activity on the part of British authorities, and stresses that they would hesitate to encourage local authorities explicitly to embrace such a strategy (interview with EU Commission, DG XI, March 1993). This cautious statement by the Commission may be influenced particularly by fundamental objections of British central government to such ventures by local government.

It is very difficult and very unfortunate if you are finding a situation where interventions from Community level are substantially disturbing allocations between central and local government. If at EC level you are dealing with different issues from what you are dealing with at national level, that's fine. You can justify it. But if basically they are both tackling the same issue this does cause confusion and undermines confidence (interview with DoE, September 1993).

While the previously informal and in camera relations between regulator and industry gave the public practically no right of participation or information, this situation was substantially better under the new arrangements. For both the public and especially environmental organizations, more comprehensive rights of access to information increased the possibilities of intervening. The public registers containing all information relevant to authorization including emission readings are open to general inspection. This gives environmental organizations the opportunity to inform the public of any irregularities in licensing proceedings and to launch appropriate campaigns, or in the event of standards being infringed, to pressurize the British authorities via the EU. Large organizations like FoE obtain register data directly from HMIP: 'That saves them coming along and asking every week' (interview with HMIP, September 1991).

Despite these new possibilities, the general public has so far made little use of their informational rights. A survey by the NSCA in May 1992 showed that comprehensive inspection of the public registers had taken place in only 6 of the 286 local authorities questioned. In 148 authorities there had been low demand for data, and in the remaining 126 none at all (NSCA 1992a). A study carried out in 1993 by the NSCA[61] confirmed this general trend: of 288 local authorities one only reported that the public had made extensive use of the register, while 167 reported low use and 120 none. The environmental organizations, too, exercised their access rights to a greater extent only where certain problems arose. 'Pressure groups are interested if they are particularly targeting your organization for something' (interview with multinational pharmaceuticals group, January 1993; interview with IEHO, March 1992). Another reason for this reserve may be that the information contained in the registers is extremely complex and not appropriately processed (interview with LGMB, January 1993). It is interesting to note that the public registers are, however, much consulted by competitor undertakings, since the published data permits conclusions to be drawn on the parameters of corporate production planning.[62] But the question of how competitors have handled the complex authorization process also encourages such firms to take an interest. 'They used the public register as a way of comparing their own application with someone else's. And what they try to do is: "Well, can we do a bit better than that?" ' (interview with CBI, September 1992).

If one compares the EPA 1990 arrangements as a whole with earlier licensing practices, considerable changes in the type and form of state intervention in the field of clean-air policy become apparent. The first changes are in underlying policy instruments. By adopting the BATNEEC principle used in the framework directive and by accepting European quality and emission standards, the hitherto informal British practice, which knew neither formal authorization conditions nor statutory limits, has taken on a degree of formalization and legalization. Local authority powers have been accordingly enlarged. With the introduction of an integrated concept for technically complex processes, which is intended to minimize emissions to the environment as a whole in accordance with the BPEO principle, the British regulatory system has gained an innovative component going beyond the arrangements adopted to date by the EU. The changes in policy instruments caused readjustments in implementation. The public registers recording all authorization data on an industrial process are the central element. They signal a fundamental departure from the informal, secretive licensing procedure under the Alkali Act, and give environmental organizations more effective means to exert influence. At the same time, they oblige the regulatory authorities to adopt a certain formality in their relations with industry, which means that the 'old chumminess' between regulator and regulated has at least to some extent given way to an arm's-length approach. Finally, the enlarged local government powers are a development that in some measure runs contrary to the general trend in relations between central and local government in favour of centralization. This element is strengthened by general efforts on the part of local authorities to safeguard and augment their autonomy by establishing greater contacts in the framework of EU institutions.

Britain as 'Pace-setter' in European Policy

Innovation in national legislation, largely determined, as we have shown, by the coincidence of domestic and European adjustment needs, was accompanied by a change in the British attitude towards European policy. The British thus took the opportunity offered by the rationalization and modernization of national practices to redefine their role in supranational policy-making. 'British policy came up somewhat and wanted to make more positive contributions and actually get an outcome that was more sympathetic to us' (interview with DoE, September 1993). With the EPA 1990 they had not only satisfied European requirements but gone beyond them to anticipate EU developments yet to be realized at the European level or in most other member states. Especially important in this respect was the concept of integrated pollution protection and the far-reaching public informational rights in relation to authorization processes. In adopting the EPA, Britain threw off the defensive 'braking' role it had assumed in European negotiations since the early eighties, primarily in reaction to the 'German-influenced' EU Commission procedure. The new national arrangements allowed the British to be far more active in influencing European legislation.

EPA enables government to absorb EC legislation more easily and/or get ahead with national legislation and set their own standards in a way which would be more impervious and set Britain ahead of EC legislation. The argument from Government ministers throughout the Act was that Integrated Pollution Control was setting the standard, that we are ahead, that everybody else was welcome to follow, and that, as a matter of fact, good old Britain had done it again (interview with AMA, January 1993).

This was the background to a fundamental change in the British position in supranational decision-making:

The British deliberately changed their attitude 180 percent. It is not that they gave up their — in principal sceptical — attitude towards quick innovations. They still want to know 'what will be the costs, what will be the benefits' (interview with EU Commission, DG XI, March 1993).

Although the cost-benefit ratio in environmental policy measures continues to be a crucial component in the British problem-solving philosophy, the EPA marked a new departure for British conduct in European clean-air policy.

The rationale behind the new British stance is the same as that which had motivated the Germans in 1982 to push their Large Combustion Plant Regulation at the European level. Every member state is interested in 'imposing' essential elements of its own institutional arrangements on other member states via the EU in order to keep the cost of adapting its national legislation to future European directives as low as possible. Furthermore, a member state with particularly stringent regulatory provisions has an interest in making its rules binding on other countries via European directives to avoid harming the competitive position of its own industry. The application of an integrated permitting procedure, which was now mandatory in Britain for large plants, put British industry at a competitive disadvantage because taking account of and calculating all possible environmental impacts required a great deal more time than was needed under the old licensing system. This gave foreign competitors not subject to the strict British requirements a time advantage in introducing new products on the world market (interview with multinational pharmaceuticals group, January 1993). It is therefore not surprising that Britain has supported the preliminary draft of a European directive on integrated pollution control.

The British also urged adoption of the Directive on free access to environmental information in 1990. The provisions of this directive were largely paralleled by those of the EPA, which, with the introduction of public access, provides for comprehensive public informational rights. This was to the competitive disadvantage of British industry, since information in the public registers was also available to interested competitors: 'Then it becomes unfair if we have Germans for example looking at our industry in the registers whereas our industry cannot look into German registers' (interview with DoE, September 1993). Britain also advocated adoption of a directive on environmental auditing, which was to define certain corporate environmental management standards. Although this measure was mooted on the initiative of the Commission, 'once it was out, Britain was very interested' (interview with EU Commission, DG XI, March 1993). The Commission for its part took the standards of the British Standards Institution as the basis for elaborating the Directive (interview with BSI, January 1993; interview with EU Commission DG XI, March 1993; interview with EIA assessment association, October 1993).

In general it can be said that British policy initiatives at the European level concentrate primarily on measures designed to change administrative and industrial decision-making processes. This emphasis on procedural provisions is a fundamental element in the English legal tradition, which focuses on adjective rather than substantive rules (interview with DoE, September 1993). The British continue to evince a certain 'mistrust against technology-based approaches' aimed at tightening or introducing emission limits: 'We don't like the prescription of means in any context' (interview with DoE, September 1993). Whereas the British pace-setter role finds expression predominantly in relation to procedural measures, the British tend to put on the brake when it comes to substantive rules. The dichotomy in British behaviour has been especially evident in the negotiations on the Integrated Pollution Control Directive, where British and German problem-solving philosophies have collided.

However, the strategic re-orientation undertaken by the Commission in the Fifth Action Programme is basically in tune with the British philosophy. The programme now replaces strict emission-orientation with stress on quality goals, to be attained by relying more strongly on 'pressure from below'. The idea is to generate such pressure mainly through greater public access to information and through participation rights. Hence, although it is up to member states to determine how standards are to be complied with, the public is to be informed of the outcome of implementation. This, the Commission believes, will ensure that countries will be scrupulous in implementing European measures (interview with EU Commission, DG XI, March 1993). Current Commission plans for a combination of public 'transparency' and voluntary private-law agreements on emission reductions with industry (interview with EU Commission, DG XI, September 1993), are more in line with the British procedural orientation than the German interventionist philosophy.

In order to promote their interests, the British now seek to influence European policy formation at a very early stage, in contrast to their previous strategy. In connection with the Directives on free public access to environmental information and integrated environmental protection, for example, the British took action to get the topics on the agenda even before the Commission had come up with a concrete proposal (interview with DoE, January 1993; interview with EU Commission, DG XI, March 1993). 'We are trying to get into the debate at an earlier stage, before the issues are crystallized' (interview with DoE, September 1993). British procedure is to decide its own national position before a directive is discussed at the European level and to settle any conflicts of interest among various national actors in the run-up phase. This permits the British to exert rapid and effective influence on the European debate (interview with DoE, January 1993). Another variation is to approach the Commission directly with a specific proposal for a directive. Due to the paucity of their personnel resources, the Commission is more than happy to receive assistance in the form of member state expertise. Examples for this course of action can be seen with the planned petrol regulation or the ground-level ozone directive: 'The structure and approach of the last directive is strongly influenced by what we have been starting to do in the UK. In turn, we had to change our procedure only in a few details' (interview with DoE, January 1993). The British are also making increasing use of the possibility to influence the Commission through the delegation of national experts (interview with DoE, September 1993). The Commission encourages this procedure because without adequate personnel they must rely on the specialized knowledge offered by member states. Thus a DoE expert who had himself worked on the conception of British IPC was delegated to the Commission to prepare the draft directive on integrated environmental protection (interview with EU Commission, DG XI, March 1993). Finally, more effort is made to influence the European agenda in bilateral talks with countries holding the presidency of the Council of Ministers: 'We are going for much more bilateral discussions with other members' (interview with DoE, September 1993).

Whilst in the eighties the British distinguished themselves by applying the brakes on European clean-air policy, for the past six years they have been playing a far more active role in European policy-making. Having thoroughly modernized and rationalized their own legislation, they are now attempting increasingly to bring EU legislation into line with their ideas. This course of action is consistent with the rationality criteria formulated at the outset, to the effect that the behaviour of member states in European negotiations aims to minimize institutional adjustment costs and to ensure national economic competitiveness. The new British 'pace-setter' role is, however, not unlimited. There is dichotomy in British behaviour with respect to adjective and substantive arrangements. All measures designed to determine industrial and administrative procedures are keenly supported, whereas mistrust persists towards a technology-related approach based on emission limits. However, the new strategy of the Commission to combine quality-orientation and 'pressure from below' is much closer to British thinking.

3.2.2 Priority for Air-Quality Control and Industrial Self-Regulation

In the Commission's increasing preference for 'soft' control and in its greater efforts to give member states the scope to develop their own initiative in realizing Community policy, it has taken up other instruments. They are designed primarily to control air-quality, which is to be done through operator self-regulation accompanied by the broadest possible public access to information.

Access to Environmental Information: 'Prescribed Transparency'

Although both the European Parliament and various environmental organizations had been calling on the Commission since the mid-eighties to introduce legislation on better access to environmental information, policy-making had made little progress. It was only when Britain — following the thorough reform of its domestic environmental legislation — had abandoned its sceptical attitude to the Directive that the picture changed. Through the far-reaching public access rights provided under the EPA 1990, Britain had become 'first mover' at the European level in regulative transparency in the environmental field. This new legislative leadership by the British dynamized decision-making, leading in 1990 to adoption of the Directive on the freedom of access to information on the environment. This directive is in harmony with the Commission's increasing efforts to avoid the top-down imposition of rules and to rely rather on 'pressure from below' to attain environmental policy goals.

Objective and Content of the Directive
The objective of the Information Directive is to ensure free access to the information on the environment held by the authorities and the dissemination of this information, and to lay down the basic conditions under which such information is to be made accessible (Art. 1). Information on the environment within the meaning of this directive includes all data stored in written, visual, sound or electronic form on the state of bodies of water, air, soil, animal and plant life, and natural habitats. Activities or measures that influence or degrade this state, as well as activities to protect these environmental areas are also subsumed under the Directive's information concept. Public administrative authorities that perform functions in the environmental protection field, whether at the national, regional, or local level, and which dispose of such information, are required to make it accessible. Exempted from this requirement are institutions acting by virtue of their judicial or legislative competence (Art. 2). Individual member states are to ensure that

all natural or juristic persons are given free access on request to the environmental information held by the authorities without proof of a special interest. The files or data to be withheld from the public are also specified. They include data relating to national security interests, to matters which are *sub judice*, commercial confidentiality and personal privacy, as well as records communicated by a third party not legally required to do so. Information can also be withheld if making it public would increase the probability of damage to the environment. If the applicant is of the opinion that his request for information is wrongly rejected, he may challenge the decision in accordance with national legal procedure by taking either legal or by administrative action (Art. 4). Member states are permitted to charge a reasonable fee for transmission or inspection of data (Art. 5). Because of the subsidiarity principle, which leaves implementation to the member states, the Directive says nothing about how access to environmental data is to be provided in practical terms.

Since member states have an interest in reducing the legal and institutional costs of adjustment to European legislation, the positions and negotiating strategies in the discussion on the Information Directive can be understood only in the context of the respective national legislation on the access to records.

The Law Relating to the Inspection of Records in Britain, France, and Germany

As we have seen above, British environmental policy has changed radically over the past five years. Under the EPA 1990 public registers were established as recommended since 1976 by the RCEP. These public registers, open to inspection by all, contain all relevant permitting and operational data as well as the results of emission monitoring (see Gibson 1991). Firms could withhold data only if they could prove that disclosure would infringe commercial confidentiality (interview with Bexley local authority, September 1993). The introduction of the public registers signalled a fundamental departure from the secretive regulatory practices that had hitherto been characteristic of British clean-air policy.[63] With the provision of public access, British arrangements went far beyond the provisions of the Information Directive. The directive provided only a passive right of information on request, whereas the British rule granted an active right of access to information. However, the European directive applies to all environmental data, while the public registers cover only certain data pertinent to authorization procedure. Experience has so far shown that the British public has made limited use of the new information rights. At present firms are most interested in the data to gain insight into competitors' production processes.

France is the only one of the countries under review to have had legislation on the inspection of records providing for public access to environmental information since as long ago as 1978.[64] The French authorities are required to provide access to all administrative documents in all areas of activity. Hence everyone has a fundamental right to demand to inspect environmental data. However, this is tied to the existence of *'documents administratifs'*; there is no right to compile data, for example, data on emissions. And documents transmitted to the authorities only for their information (*'à titre d'information'*) are not open to public inspection. However, documents sent to the authority in connection with the exercise of their functions (*'par destination'*) or which has been produced at the instigation of the authority may be inspected by the public (Winter 1990, 186). Like the European directive, the French law on access to information provides for certain exemptions, protection of commercial confidentiality, public and national security, or 'incomplete, preparatory' documents.[65] The practice being long established in France, more use of the right to inspect records is made there than in Britain. In order to improve the exchange of information between administrative authorities and the public, multipartite institutions, *secrétariats permanents pour la prévention des pollutions industrielles* (*SPPPI*) and *associations pour la surveillance de l'air,* in which central and local government and associations are represented, have been created on the initiative of the *DRIRE* (interview with *Les Verts*, June 1993). Besides the general right to inspect records, there is also the institution of the *enquête publique* in France, which provides for the active information of the public in formal permitting procedures.

In Germany, the principle of 'restricted access to records' applies for the administration in general and thus also for the environmental authorities. Information in the possession of an authority is in principle confidential. The public has a possibility to obtain information only in special cases, for example in the context of licensing proceedings where public participation is required (*DNR* s.a., 2). The German rules on access to data distinguish three types of inspection: first, the 'active' duty of the administration to furnish information, which requires notice to be given and the information to be laid open to public inspection (*'Auslegung'*); second the 'passive' duty to inform, where only the disclosure (*'Offenlegung'*) of information is required, and finally the duty of registration, where processing and laying open to inspection is required but no notice of its availability (*'Bekanntmachung'*) (Burmeister/Winter 1990, 93). Although under formal licensing procedure the administration has an active duty to furnish information, interested parties have considerable obstacles to overcome even when the documents are open to public inspection during permitting proceedings (Haussmann-Grassel 1985, 50). It is often difficult to find the relevant office within an authority where the records can be consulted. Perusal of the files is subject to further problems. There are no special facilities set aside for the purpose. Inspection takes place only during normal office hours. Comments from participating authorities or expert opinions often appear belatedly. Data on expected emissions or technical data are often missing as well (Führ 1989, 70). Thus, under present conditions, exercising the legal right to inspect data is subject to practical restraints.

The Decision-Making Process

In a nutshell, what is striking about the decision-making process on the Information Directive is that — due to the procedural nature of the measure, which addresses no detailed technical issues — drafting in the Commission was politicized at an early stage. The subject-matter being dealt with required no expert consultations behind closed doors to settle technical or chemical issues of merit. What the planned measure meant for member states were directly evident. This circumstance, together with the fact that disclosure of environmental data was a highly sensitive political issue in many countries, explains why no official Commission proposal was so long in coming. The Commission initially found little political support for its project, although corresponding demands had been advanced by the Fourth Action Programme and by the EP and various environmental organizations. Against the opposition of many member states, the Commission was unable to launch a legislative initiative capable of finding consensus. This situation changed only after Britain, as one of the large member states and opinion leaders, had in the second phase abandoned its sceptical stance on the Directive and given its active support and cooperation to the Commission initiative.

The starting point for the development of the Information Directive is to be seen in particular in the 1987 Fourth Action Programme, where the improvement of public access to environmental data was an explicit goal. Demands by the European Parliament also contributed to advancing the decision-making process. The Environment Committee of the EP justified the need for a right of access to environmental information on grounds that 'environmental protection and the development of parliamentary democracy are well served by maximum openness on the part of government and industry with regard to information relevant to the assessment of activities which can affect the environment' (ENDS 1987/148, 23). The EP was seconded in this by a resolution of the European Environment Bureau (EEB), which represents more than seventy national environmental organizations. The Commission was open-minded on the question, but pointed out that, because of very different legal arrangements in member states, 'great political sensitivity attends this question' (EU Environment Commissioner Clinton

Davis, quoted in ENDS 1987/148, 23). But since they had committed themselves under the Fourth Action Programme to drafting an environmental information directive and were under increasing pressure from the EP, the Commission finally reacted by producing a proposal for legislation. However, it met with strong opposition from many member states. Only Denmark, France, Luxembourg, and the Netherlands were in favour of European regulation in this area, all countries that already had such arrangements at home.

The fundamental antagonism of the Council of Ministers towards the Commission proposal only began to ease when Britain, until then a declared opponent of the Directive, had had a change of mind, suddenly urging the Council of Ministers to accept the measure. The background to this at first glance surprising change of mind on the part of the British was the reorganization of the national regulatory concept under the EPA 1990. Until that time, environmentally relevant data had been handled by British authorities as highly confidential. This explains why the British initially reacted so sceptically to the Commission draft: if a European environmental directive were to have been implemented, the British would have faced substantial institutional and legal adjustment costs.

It was only with the prospect of a fundamental re-orientation of existing regulatory practice granting far-reaching public access to information that Britain's interest increased under the new, ecology-minded Environment Minister Chris Patten to play an authoritative part in shaping the proposed European Environmental Information Directive. 'We had a strong hand in the Directive' (interview with DoE, September 1993). With the reform of national legislation, which had been undertaken in reaction to growing international and domestic pressure, the British had had the opportunity to assume the role of first mover on the European level and shape the Directive on their own last. The EPA now being on the statute-books, the British government was also keen to avoid any competitive disadvantage British industry might suffer because of the public registers. For the disclosure provisions under the EPA were so broad that competing foreign firms had access to the British registers, permitting them to draw conclusions on production processes. This duty to disclose information would hardly have been defensible vis-à-vis British industry if the government had not sought Community-wide harmonization of public access facilities at the European level.

During their presidency in the Council of Ministers, the British pushed for the rapid adoption of the Information Directive. They called on the Commission to submit a draft. But opposition from a number of countries put paid to the new Commission proposal. Germany, in particular, showed tenacious resistance to such a directive. Far-reaching rights of access to information have always been anathema to the German authorities responsible for pollution control and to industry, because they impugn the principle of confidentiality (UBA 1991, 35). France, on the other hand, remained strictly neutral on the Directive, since it had already legislated on the matter and therefore anticipated no legal adjustment costs.

Having failed to obtain the consent of the Council of Ministers during their presidency, the British tried to bring about a decision through close cooperation with the Irish presidency. It proved strategically advantageous for Britain that the proposed directive had to be redrafted at the last moment to take account of a number of objections on procedural details. 'The final Directive was written half an hour before the Council meeting broke up' (interview with DoE, September 1993). Germany, finding itself isolated, finally gave up its opposition to vote for the Directive, not least of all to avoid losing its reputation as an front-runner in environmental policy (interview with EU Commission, DG XI, July 1994). Germany found it easier to give its consent because practical implementation was largely left to the member states. In implementing the measure, the Germans accordingly tried to minimize the impact of the Directive on their existing regulatory concept.

Implementation of the Directive at the National Level
Given the divergent policy positions that came to bear in policy formulation and the differences in national institutional conditions, implementation of the Directive at the national level varied greatly. This variance was further encouraged by the relatively large scope for interpretation allowed by the Directive. With their new environment act, the British had no implementation difficulties, since the provisions were compatible with their approach to regulation. Although the EPA rules on commercial confidentiality differ on a few points from the Directive provisions, no complaints have yet been heard about implementation problems (interview with DoE, September 1993). In transposing the Directive into national law, the Department of the Environment laid down detailed administrative rules. The authorities were advised to handle public access to data as liberally as possible. Rejection of requests for information were therefore to be carefully scrutinized when on grounds of commercial confidentiality, for which firms must present cogent reasons (interview with DoE, September 1993). Moreover, unlike in Germany, no charge is made in Britain for the provision of information (DoE 1992, 18).

The French, too, had no problem in translating the European Information Directive into national law, having already introduced appropriate legislation. Where documents are despatched, a charge is levied to cover copying costs, but not to cover administrative costs (Winter 1990, 200). Disclosure is facilitated by the activities of environmental protection associations in the multipartite bodies (*SPPPI, associations pour la surveillance de l'air*). They are concerned to raise the general ease of access to information (interview with *Les Verts*, June 1993). Nevertheless, in France as in Germany, there is a discrepancy between model and reality. Environmental associations thus complain that, despite a fifteen-year-old tradition, 'defensive' administrative practices still occur. It often proves extremely difficult to get at the relevant files. The information sought is frequently made available only after massive pressure has been brought to bear or after a complaint has been made to the Environment Ministry (interview with *Nord Nature*, January 1994). 'There's formal democracy, ... , there are procedures that exist on paper' (interview with *Les Verts*, June 1993). There is also a lack of staff to handle information requests, and a lack of systematic processing of documents to facilitate the smooth handling of requests (interview with *Ministère de l'Equipement*, June 1993). By contrast, implementation is furthered by the *'Commission d'Access aux Documents Administratifs'* (*CADA*), which as a sort of complaints authority watches over compliance with the right of access to files (interview with environmental protection association, January 1994).

Having resisted the introduction of the Information Directive, Germany then resisted its implementation. This was evident from its belated realization and from its obvious failure to comply fully with the provisions of the Directive. Although implementation of the Directive in national law was to have been achieved by the end of December 1992, the Federal government had only a ministerial bill to show for its efforts by the deadline (Wegener 1993, 17). It was only in mid-1994 that the *Bundestag* passed the Environment Information Act by which the Environment Information Directive was transposed into German law (Scherzberg 1994, 733).

It is, however, questionable whether the belatedly adopted Act meets all the requirements of the European legislation. Lawyers and environmental organizations have complained in particular that the circle of authorities required to furnish information has been restricted to a few special bodies, and that the exemptive circumstances justifying a refusal to disclose data are so vaguely defined that they far exceed in extent the residual access rights. The concept of pending proceedings, for example, which may justify the withholding of information, is interpreted very broadly by the German courts.[66] It is also complained that the threat of fees of up to DM 2000 calculated to defray costs frighten off potential applicants.[67] 'The Environmental Information Act is thus an inadequate implementation of the Directive. It is not designed to optimize the conditions for the public monitoring of implementation in the environmental field, but to maintain the status quo of restricted administrative openness to the greatest possible degree' (Scherzberg 1994, 745). However, not only the public is affected by the reticence of the authorities. The Federal Environment Office is frequently refused information by factory inspectorates (interview with *UBA*, January 1993). Local authorities are refused emission data on firms in their region which they need to set local quality standards. 'We [the local environmental authorities] often receive requests for information about emissions from a plant. ... We ask the factory inspectorate to furnish us with the relevant data, because they are the only people who have them. We invite them to the policy committee. But they never come and plead data protection' (interview with local authority environment department, August 1993).

There are many reasons for the restrictive attitude prevailing in Germany. First and foremost is the deeply rooted ideological concept of the confidentiality of official data, which the European directive has fundamentally challenged. It is foreign to the German administrative culture to disclose environmental data to the public. 'It's almost indecent to publish emission data. The authorities keep it under lock and key' (interview with *FDP*, March 1993; interview with Greenpeace, October 1994). This basic attitude is exacerbated by fears that, were broad use to be made of access rights, 'the administration would break down' (interview with Institute for Urban Studies, November 1992). The administration assumes that it would be overloaded if it had to deal with requests for data in addition to routine duties. 'I think there'll be a "joyful" awakening for some people if someone actually goes to court to get what he is entitled to under the Directive' (interview with environmental consultancy firm, September 1993). The tendency to be economical with data disclosure is also apparent in industry, which fears that a more extensive interpretation of access rights would lead to 'not inconsiderable delays in the planning of projects' (*BDI* 1992, 2).

The generally negative attitude has been further intensified by the economic and financial difficulties accompanying German unification. The Information Directive is thus completely at odds with the intention of the 'Acceleration Acts' designed to promote investment in the East by restricting public participation (interview with *DGB*, February 1993). The directive is implemented with corresponding enthusiasm: 'We go as far as we absolutely have to but not an inch further' (interview with *UBA*, November 1992).

Tardy implementation is also caused by coordination difficulties between Federal and state governments. A federal/state working group recommended adopting a federal act requiring the concurrence of all states. 'And this is of course far more complicated and much more difficult than if the Federation takes sole responsibility' (interview with *MURL*, February 1993).

Conclusion

The decision-making process on the Environmental Information Directive clearly shows how difficult it is for the Commission to push through a directive without the active support and initiative of at least one opinion leader among the member states. Although it was able to count on the support of the EP and could plead the provisions of the Fourth Action Programme, a breakthrough in the policy-making process was achieved only when the British abandoned their sceptical stance and, have performed a U-turn, attempted in the capacity of 'first mover' to propel the planned measure in their direction. Another peculiarity of the decision-making process was that, due to the procedural nature of the Directive, expert consultations were not necessary in problem-definition at the Commission level. The process shifted very rapidly to the Council of Ministers, where the only country seeking initially to block the proposed directive was the Federal Republic. Certain concessions having been made on implementation, the Germans finally gave their consent, fearing that failure to do so would endanger their reputation as pace-setters.

The divergent positions taken by Britain, France, and Germany in decision-making on the European Environmental Information Directive can be explained in terms of their respective national legislative and institutional cultures. Having reformed its own legislation, Britain attempted to make the new procedural rules binding at the Community level and thus to assume the role of front-runner vis-à-vis other member states. The imposition of the British mode of implementation at the European level saved Britain institutional and legal adjustment costs. The British were naturally also keen to safeguard their industry's competitiveness by installing a comparable right of public access to information at the EU level. Vis-à-vis industry, the British government was able to justify the granting of greater access rights only by offering the prospect that the rules would be Europeanized.

In this context the discrepancies between British and German regulatory philosophies also became evident. Whereas the British were concerned primarily with procedural reforms, the Germans were interested in substantive and technology-based measures. In keeping with the German view on regulation, regulative law principles were championed without considering any substantial regulatory function for the public.

The Commission, in contrast to Germany, was interested in going beyond the European Information Directive, which provides for an only passive data access right, to establish an active information policy. This was to be coupled with air-quality requirements to generate pressure 'from below' in the event of limits being exceeded (interview with EU Commission, DG XI, March 1993).

Integrated Pollution Control: 'Subsidiarity' versus 'Harmonization'

Controversy on the EU Directive on 'Integrated Pollution Prevention and Control' (IPC) strikingly reflects competition between emission and quality-based approaches to problem solving. As we will be showing, vehement controversy arose among representatives of governments and industry in EU member states on which regulatory philosophy is best suited and necessary to bring about harmonized, integrated environmental protection throughout the Community.

Analysis of the IPC Directive is, however, interesting not only with regard to the regulatory contest between countries, which was a decisive factor in negotiating and formulating the measure. A further important aspect is the change in Commission strategy. Confirming the trend that became apparent with the Environmental Information Directive, the Commission is progressively moving away from dirigistic forms of control. It is resorting instead to framework control, seeking to ensure conformity with European requirements via 'pressure from below'.

The Draft Proposals of the Commission
Prompted by the British 'Integrated Pollution Control' system, which had given the British a 'first move' advantage over other member states, the EU had since 1991 been pushing a directive to introduce a cross-medium, integrated approach to authorizing industrial plant and fixing emission limits (interview with BDI, March 1994). Britain, in favour of such a directive because it already disposed of such legislation, delegated a DoE official on the request of the Commission to the DG XI to work on drafting of the Directive. The other member states reacted to the proposal (parametric adjustment).

The still very controversial measure is intended to supersede the 1984 Framework Directive on air pollution from industrial plants. In contrast to the Framework Directive, the IPC Directive pursues an integrated concept. This concept, an essential element in the 1992 Fifth Environmental Action Programme of the Commission, is to be realized in three dimensions.

First, the time dimension: environmental protection measures are to be taken not only 'at the end of the pipe' but as early as possible, for example at the product level. Second; the spatial dimension: efforts to curb environmental pollution are not to be restricted to a single environmental medium but to embrace all media. Air, water and soil pollution are to be taken into account as a whole and with regard to their interaction. Third, a social dimension: in dealing with environmental issues, all the affected groups and individuals are to be involved as far as is feasible and early as possible.

In the course of the problem-solving phase in 1991, a first meeting of national experts took place on the subject of 'integrated permitting'. A first discussion paper was debated, which presented the underlying concept of the new directive: 'The ideal concept behind an integrated permitting Directive would be a single authorisation issued by a single authority regulating all the releases to land, air and water' (ENDS 1991/196, 35). In April 1992 there was a second meeting of experts, where a number of elements in the original draft were dropped on the instigation of industry. They included in particular the obligation of the operator to take out insurance against any degradations that might be caused by operation of the plant, and plans to introduce an environmental register. This schedule of the largest pollutant emitters was to be addressed by a separate directive (ENDS 1992/207, 29).[68]

According to the current draft of September 1993,[69] the object of the IPC Directive is 'to provide for measures and procedures to prevent, wherever practicable, or to minimize emissions from industrial installations within the Community, so as to achieve a high level of protection for the environment as a whole' (Art. 1). For this purpose, member states are to ensure 'that no installation shall be operated without a permit issued in accordance with this Directive' (Art. 3).

Details on load distribution among different environmental media are to be left to the individual member states, since the Commission feels that no satisfactory optimization models exist in this area, and that subsidiarity has to be respected (interview with EU Commission, DG XI, March 1993). In keeping with the subsidiarity principle, member states are themselves to define limit values for the substances to be controlled on the basis of the best available techniques: 'We must always ask ourselves what it does for the environment, if one cannot better solve the problem at the national level, why the problem has to be settled at the Community level'. From this point of view, the Commission feels it is under the urgent obligation 'to give local authorities the liberty to decide themselves within a certain framework in accordance with local possibilities and necessities' (interview with EU Commission, DG XI, March 1993). From this perspective it would be inappropriate and 'presumptuous of Brussels to tell every member state in every situation what the standard is' (interview with EU Commission, DG XI, September 1993).

The EU is rather to set quality standards with which member states comply by enforcing suitable emission limits. Where such EU-wide requirements do not exist, WHO standards are to be taken, which are much more stringent, 'since the WHO can be pretty unconcerned in announcing its recommendations, and has the precautionary principle ... built in ... This means that we ... become stricter' (interview with EU Commission, DG XI, September 1993).[70]

If the prescribed quality goals cannot be attained with national emission limits, member states are to be obliged to go beyond what is possible using the best available techniques and to prohibit certain production processes in strongly polluted areas. Vice versa, in regions where environmental quality is good, certain concessions may be made on the technology, if the resulting increases in emission levels are negligible (Art. 9 para. 3).

To avoid sacrificing the effectiveness of the Directive on the altar of subsidiarity, the Commission envisages the broadest possible public participation and access to information. Member states are to be required to publish applications for the authorization or alteration of plants and all emission figures, so that the affected public has the facts and can raise objections if need arise. Via access to these data, which in some countries — including Germany — have hitherto been subject to data protection, the public would have an opportunity to examine, weigh up and compare releases from different plants. Where a given plant is demonstrated to be responsible for high emission levels, it is hoped that public pressure will ensure that the firm reduces pollutant releases to safeguard its image and competitiveness.[71]

The concrete arrangements for implementing the integrated approach are primarily concerned with the authorization of industrial plants on the basis of 'best available techniques' (BAT),[72] specifying what information is to be contained in the permitting application (Art. 5), what organizational and institutional conditions are needed to ensure an integrated approach to authorization (Art. 6), and under what conditions a permit may be granted or refused (Art. 8). In order to achieve the approximation of technical standards throughout the Community, and thus to attain a similar level for limit values in the medium term, member states are required to report on the state of technical knowledge. This information, in the exchange of which the Commission 'wants to act as motor', is then to be formalized in 'technical notes'.[73] Whilst member states would thus be required to set emission limits themselves on the basis of the 'technical notes', the Commission or the Council of Ministers would 'fix common Europe-wide standards only in certain areas and for certain substances (interview with *BDI*, March 1992).

Member States' Negotiating Positions: Best Available Techniques Versus Local Environmental Situation
During the ongoing drafting phase at the Commission level, the costs and benefits of such a Directive are being discussed in detail and potential compromises sounded out. The criteria for defining and implementing best available techniques is the main bone of contention (interview with EU Commission, DG XI, June 1994). Four points in particular are at issue. First, there is no consensus on the extent to which local environmental quality ought to be taken into account. Closely related to this issue is the debate on 'European versus national emission standards'. A third controversial aspect is whether economic aspects should play a role and, if so, which ones.[74] And, fourth, it is not clear whether or how the public should be involved in the permitting process. The general constellation of interests is

that Germany, with its tradition of technological environmental protection, together with Denmark and the Netherlands, insists on a technology-based approach, while southern European member states and the United Kingdom favour a quality-oriented *modus operandi*. The French position, as so often, lies between these two poles (interviews with EU Commission DG XI, June 1994; *BMU*, October 1994; *Ministère de l'Environnement*, January 1995). Member states have hitherto agreed on only one point: the recourse to WHO standards envisaged by the Commission is categorically rejected. Member states do not wish to have limit values set outside their sphere of influence. They prefer to 'keep their finger on the trigger if need arise' (interview with EU Commission, DG XI, June 1994).

German criticism of the British-inspired Commission proposal is based on four considerations in particular: first, the Federal government has environmental policy reservations; second, German industry is seen at a competitive disadvantage; third, Germany sees legal discrimination problems; and finally there are fears that existing German regulatory practice may have to adapt to accommodate the public participation rights envisaged. The German negotiating position gives unambiguous expression to the interest in minimizing the costs of legal and institution adjustment accruing from a European measure. Germany thus points out that national air and water standards are so conceived that they take account of interaction with other environmental media as it is. They therefore claim that no revision of national practice is needed (interviews with *BDI*, October 1994; *BMU*, October 1994).

The Federal Republic complains primarily about the lack of EU-wide emission limits defined on a technological basis and the 'escape from BAT clause', which under certain circumstances allows less stringent limits in unpolluted regions (*BMU* 1994, 7ff.). The government, industry and environmental associations 'are united on this as on no other issue' (interview with *BDI*, October 1994).

Environmental policy aspects that bring the Federal Republic to urge 'a definition of the best available technique independent of prior pollution' play an important role in German argumentation. 'From the point of view of an environmentalist, the best available filter installation is just as appropriate in Ireland as in the Ruhr. We must do what we consider to be technically feasible and financially affordable and to make as little use as possible of natural resources.' For it is only 'a question of time until the environment can no longer cope' (interview with steel producer, March 1993). An approach based on local environmental quality would, in the opinion of the *BMU*, 'deal a heavy blow' to EU environmental policy. 'If the EU chooses this path — we'd never go along with it, but we can be outvoted — we foresee great problems in advancing environmental protection at all. And that's why

we're fighting tooth and nail against [the escape clause in] Article 9'
(interview with *BMU*, October 1994).

Another reason for German opposition lies in the competitive difficulties
that Germany would experience under the 'more stringent rules and high
environmental costs' (interview with *BMU*, July 1993; interview with *BDI*,
March 1994). Germany as an industrial country subject to strict environmental requirements would be at a disadvantage vis-à-vis 'cheaper' locations
in the low-regulative, southern countries. The gap between competitive
conditions in Germany and other European countries due to the differential
stringency of environmental requirements 'would be further cemented, if
under the IPC Directive air-quality standards were to be the yardstick'
(ibid.).

The economic advantages of a quality-based approach could of course be
exploited in Germany as well as in other European countries. However, an
EU-wide, quality-oriented approach to problem solving would bring no
change or easing of emission-related regulatory practices in Germany, since
the Federal Republic is 'a highly industrialized country with hardly any va-
cant land left for industrial development' (interview with *BDI*, March 1993).
Because of the high population density and the nation-wide distribution of
industry, Germany has far less margin for action than, say, Ireland or
Spain, where there are still large regions more or less uninhabited and with
no or low pollution from industrial sources. Compliance with the applicable
EU environmental quality standard in such areas, claims a *BDI* spokesman,
could allow emissions 'ten or twenty times higher than would be attainable
with the best available techniques in Denmark or Germany, for example'
(interview with *BDI*, March 1992). Although the limit values derived from
the environmental quality standards may deviate only within certain limits
from the values applicable under BAT, 'this is only a qualitative formulation
that means practically nothing at all. The "local man" in Spain can do what
he wants' (ibid.). Air-quality standards would permit 'two different levels of
industrial pollution rights. Undeveloped areas like Portugal and Spain would
have greater pollution rights and thus lower environmental costs' than developed areas like Germany (interview with *BMU*, July 1993). From the
German point of view, the EU ought therefore to set uniform limits at least
where competitive problems are to be expected (interview with *BDI*, October 1994).

Another reason why Germany is unlikely to depart from its regulative law, technology-
based interventionist philosophy 'has to do with the mentality of German public servants'.
Even if the authorities in Germany were to be served a directive giving them the option of
fixing less stringent limit values where the local situation permits, the authorities here
wouldn't do it. Even in Upper Bavaria, where there is little industrialization, our
authorities wouldn't abandon this philosophy of setting emission limits in terms of the
'best available technology' (ibid.).

A representative of the Federation of German Chambers of Industry and Commerce (*DIHT*) regards it as 'illusory to think that the purely quality-based approach can prevail in Germany. The emission aspect will be retained in any case' (interview with *DIHT*, March 1993). Moreover, from the German perspective, a quality-based strategy contradicts aspects of economic development and equal treatment before the law.

There is also a concept of economic development that points to using the best technology ... I gain leeway for subsequent investment by using the best available techniques. And it's a question of equal treatment. If I apply a prior-pollution philosophy, then the first to arrive is the best treated. ... That seems to us to be illogical (interview with *BMU*, October 1994).

The Germans are also wary of the right of public access to data provided for under the IPC draft proposal (disclosure by the competent authority of all data relevant to authorization). They point to the provisions of the Environmental Information Directive, and see 'no need to regulate this again explicitly in the IPC Directive, especially if the different arrangements are incompatible' (interview with *BDI*, October 1994).

What the Germans judge to be the 'costs' of the proposed solution are booked by the British on the benefits side. Orientation on environmental quality and the balancing of costs and benefits before adopting measures are the central British arguments apart from the extensive access to data provided for under the EPA 1990. The British, who in the eighties revised important aspects of their legislation under pressure from the 'German-influenced' Commission strategy and adopted emission-related elements, are unconditionally in favour of the IPC proposal. This is because they can to a certain extent enjoy the advantage of the first move: by reforming their national legislation and introducing an integrated environmental protection concept, Britain has became a pace-setter in European policy-making. On the other hand, it cannot be denied that, despite the modified concept, Britain continues to prefer an environmental policy that takes account less of technological abatement potential than of local environmental quality. The Commission's IPC proposal in so far represents a partial return to their former ways for the British.

The differences between the German and British positions have shown in the discussion on how account is to be taken of different environmental media in an integrated procedure. The British want to define a 'site-specific best practicable environmental option' (HMIP 1994, 10), whereas the Germans prefer integration at 'a metalevel and not at the local, case-specific level' (interview with *BDI*, October 1994). In other words, they would like to see EU-wide emission limits taking interaction between the various media into consideration. 'It's a matter of the fundamental approach, and from our

perspective can take only this form for reasons of justice and precaution'
(ibid.). By contrast, the issue of equal treatment ought in the British view
not be subject to general standards but relate to the impacts of the individual
plant on the environment.

So our system works on the best protection of the environment that is possible, and then,
the equal treatment for all is among the best possible that is affordable. ... The question is
how much environmental protection can you make them achieve in terms of what they can
afford. So there's an equality in a sense, but it's not the same as imposing exactly the
same requirements on every plant irrespective of the impact on the environment
(interview with DoE, November 1994).

In keeping with this view, the British are 'very strongly opposed' to EU-
wide emission standards:

If you try to assess what is best for the environment from a particular installation and
what the best package of releases into air, water and land is, you maybe have to trade off
between one and the other and make a balance. You cannot have an outside fixed constraint to start with from the European level because you are meant to be looking at the
particular site (ibid.).

In defending its position, Britain points to the subsidiarity principle and the
difficult and protracted negotiations that would result if 15 EU member
states had to agree on common limit values. This aspect is also considered
problematic by the Commission (interview with EU Commission, DG XI,
September 1993).

The British tradition of lending greater weight to adjective rules than to
substantive output requirements is apparent not only in their negative stance
on European emission standards. It is also evidenced by the keen support
given to rules on public participation and access to data. Britain, which has
amended its legislation in many areas to include such public participatory
rights, plays a very active role in introducing such arrangements at the
European level, in striking contrast to Germany.

The French have once again found themselves occupying the ground between the contrary German and British positions. Unlike their stance on the
Large Combustion Plant Directive, however, the French posture has in this
case been neither neutral nor disinterested. On certain points they have
leaned towards the Germans and on others towards the British. They feel
that in defining limit values the local environmental situation should be
given primary consideration, which need not contradict conformity with the
best available techniques. The French, too, aim at exploiting the technical
standards to the full, but like the British also wish to reserve the option of
fixing emission limits in terms of the local situation (interview with

Ministère de l'Environnement, January 1995).[75] The French feel that EU-wide emission limit should be set wherever distortion of competitive positions is to be feared.

> Our general position is the following: where there are competing sectors or there is at least competition at the international level, European limits have to be fixed. ... And it's our opinion that where it isn't competitive, where international competition is not involved, we feel that the European Community could possibly recommend limit values where such values could be necessary to meet a global objective to reduce certain releases (interview with *Ministère de l'Environnement*, January 1995).

Because of the broad range of instruments available to French environmental protection, the French assume that the Commission's IPC proposal is largely congruent with their existing arrangements, and that the potential costs of legal and institutional adjustment — as in the case of other directives — will prove relatively low. 'For us this directive will in point of fact reflect our national regulations' (ibid.). The following glance at the state of negotiations in the Commission will show how member states attempt to promote their respective positions.

The Decision-Making Process in the Council of Ministers

Discussion in the Council of Ministers, marked by negative coordination, negotiations, and compensation payments, has so far demonstrated how decision patterns are generally governed by institutional conditions. In addition to the general increase in bilateral coordination and alliance formation between member states and the growing importance of the European Parliament, the role of the successive presidencies of the Council of Ministers must be mentioned. Developments under the SEA and the Maastricht Treaty have enhanced the significance of the first two aspects.

The possibility of reaching decisions by a qualified majority demands more intensive coordination between member states, whether to organize majorities or vetoing minorities. Bilateral negotiations are also important because in the official COREPER meetings the points of view of national governments are often simply reiterated without compromises being sought in an effort to solve the problem in hand. Negative coordination occurs in the sense of assessing the compatibility of one's own stand with that of other countries This frequently amounts to little more than a diplomatic affirmation of a government's position without that much attention being paid to analogous statements from other member states (interview with Council of Ministers, Environment Committee, January 1995). Bilateral talks between member states are consequently all the more intensive. The Germans thus sought to bring about agreement on their terms in parallel to negotiations in the Council of Ministers: 'As far as tactics are concerned in the EU, what you have to do is first talk round the party relatively close to you. And when you've got him on your side, you try with the next one. And you tackle the party furthest away from you last of all' (interview with *BMU*, October 1994). The Germans contacted the British particularly frequently in an effort to persuade them to accept deletion of the 'escape from BAT-clause' they so feared. Although such a demand runs counter to their classical quality-oriented thinking, the British obliged the Germans on this point. This concession can be explained by a certain willingness to compromise on the part of Germany in including cost/benefit aspects in BAT, but may also be attributable to this concession have lower salience for Britain. The Commission draft otherwise corresponded largely to their own national arrangements.

At the same time member states and the countries holding the presidency have sought to sound out the views of the EP, which has gained in influence through the codecision and cooperation procedures, and to take them into account in their proposals (interviews with EU Commission DG XI, July 1994; DoE, November 1994; *BMU*, October 1994). One consequence is that the positions of national representatives in the EP and the national participants in expert groups at the Council of Ministers level are often identical. Parliament's enhanced influence is also apparent in the growing attention paid to it by national associations. Germany was thus keen to lend greater weight to its technology-based ideas in the IPC negotiations by seeking cooperation with the EP. Talks were held on more than one occasion between members of the EP Environment Committee and both the German government and the *BDI* to explain German ideas on the issue. That these efforts were crowned with a certain degree of success is shown by the official opinion adopted by the EP in December 1994.[76] The changes sought by the EP are largely in keeping with the German position. Parliament demanded complete deletion of the 'escape from BAT-clause', and EU-wide emission limits in accordance with the best available technique principle for all substances.[77] One particular reason for the 'German-influenced' EP opinion was that 'German parliamentarians take a lively interest in this issue'. 'It is simply the case that Members from other countries did not give such high priority to the matter. And that's why the demands so clearly bespeak a German outlook' (interview with *BDI*, October 1994).[78] Moreover, the EP has a reputation for advocating stringent environmental regulation. In the case of IPC, for example, the EP affirmed requirements going beyond German proposals and even adverse to them where it considered them inadequate. Particular demands were for the IPC Directive to be linked with an environmental register (a negative ranking of major polluters)[79] and the new Environmental Impact Assessment Directive.

The presidency of the Council of Ministers can be used as an institutional resource in negotiations. The Federal Republic thus made an enormous effort to influence the redrafting of the IPC Directive. While the Greek presidency (first half of 1994), diverted discussion to comparatively noncontroversial matters, Germany gave high priority to the Directive during its six months in office (second half of 1994):[80] 'The Germans obviously are very, very keen to try and get as much agreement on it before the French get their hands on' (interview with DoE, November 1994).

Shortly before the COREPER meeting, the German delegation unexpectedly tabled a new draft formulated only in German. An English version was forthcoming only in the course of the meeting: 'The English version of the text arrived right as we were discussing it in the German version. ... And some delegations, particularly the Spanish were very rude about that. [They] said this is not a compromise text, it's a German piece of paper' (ibid.).[81] Another problem was that the Germans had simply deleted numerous articles from the Commission draft. 'What the proposal did was to delete a whole number of articles which are actually vital to IPC' (interview with DoE, November 1994). The Federal Republic wanted not only to push through their technology-based philosophy (i.e., EU-wide emission standards, deletion of the 'escape from BAT-clause'), but also to bring procedural provisions into line with their own national rules.[82] It was hence hardly surprising that the other member states reacted with such aversion: 'The German text was very heavily criticized by all other member states and by the Commission. ... Everyone said "completely unacceptable. ... We just have to wait until the French take over and we will go back to the Commission text and just forget it" ' (ibid.).

Faced with concerted opposition from the other negotiating partners to this attempt to use the presidency to reverse the compromise so arduously achieved by the Commission, the Germans backed down and submitted a new proposal closer to the Commission draft. Nevertheless, Germany, with the support of the European Parliament, insisted on a number of aspects they considered crucial, such as EU-wide emission standards. They also insisted that, in keeping with the German polluter-pays principle, the responsibility for the environmental impact of industrial activity should be placed squarely on the shoulders of the operator ('fundamental responsibility of the operator'), while the Commission proposal made the member states accountable.[83]

After the partial backdown by the Germans, the chances for agreement in the Council of Ministers increased. The French presidency returned largely to the original Commission proposal, with the exception of the controversial 'escape from BAT-clause'. To compensate deletion of this provision, the southern member states were offered the prospect of local environmental aspects playing a bigger role in the licensing of industrial plants (interview with Council of Ministers, Environment Committee, January 1995). On the question whether national or European emission standards should apply, the French position also seemed to be to seek a compromise. Thus European standards could be fixed wherever necessary for reasons of competitiveness — a standpoint shared by German industry (interview with BDI, October 1994). A further authorization criterion in addition to the BAT principle has been advocated especially by the Nordic countries and Austria, namely taking possible trade-offs into account between different pollution sources, such as stationary and mobile sources. This proposal has been generally accepted by all member states (interview with Council of Ministers, Environment Committee, January 1995).

Conclusion

This is not the place to decide which is the 'better' or 'more appropriate' of the problem-solving approaches confronting one another in the debate on the IPC Directive. Given the prevailing legal tradition, nation-wide distribution of industry and high population density in Germany, it is 'rational' from the German point of view to advocate an emission-related, best-available-technology approach. For Britain and other, particularly southern European countries with no comparable comprehensive and well-developed regulative law system, including those which are much less industrialized or escape environmental problems because of geographic or demographic circumstances, a quality-based approach concerned with the local environmental situation appears appropriate. Given the frequent criticism of the 'centralism' and 'rage to regulate' of the Brussels bureaucracy, it is understandable that, pleading the subsidiarity principle, the Commission should want to leave the setting of limit values to local authorities. This decentralized problem-solving variant is doubtless closer to the course favoured by the Germans. Its price, however, is that the goal of harmonizing environmental protection at a relatively high level throughout the Community has receded into a remote future. There is a very real risk of 'the subsidiarity principle serving as a vehicle to throw out environmental norms' (interview with *FDP*, March 1993), and 'being used in a misunderstood way to attach non-binding targets to projects and to hand them over to the local level, which then does nothing' (interview with *CDU*, March 1993.

The discrepancies in problem-solving approaches — best available techniques versus air-quality standards — and the underlying multifarious motives and goals — subsidiarity versus harmonization to protect the environment or to equalize competitive opportunities — explain the vehemence with which the IPC Directive has been discussed among representatives of European governments and industry.

As far as policy making is concerned, a clear pattern can be discerned in the generation of this important directive on European environmental protection: in a first phase, Britain and the Commission, whose perception of the issue and problem-solving philosophies converge, interacted in defining the problem. In the next phase of consultation among experts, 'problem-solving' was dominated, to Germany's chagrin, by the philosophy of the British and the Commission: elements of an environmental quality orientation, of integrated permitting, and of greatest possible openness were to the fore. However, when it came to preparing a concrete decision in a third phase, and the positive and negative concerns of member states were clearly presented and brought into negotiated balance, Germany threw all its resources, especially institutional ones (Council presidency, relations with the EP) into the scales to extract regulatory concessions closer to the German approach. Much to the regret of the British, German efforts in this final round of negotiations have not been unsuccessful. A good start does not mean the race is won.

The Environmental Audit: 'Industry Will Handle It'

The Eco-Audit Regulation[84] adopted by the Council of Environment Ministers in March 1993 provides a further illustration of the divergence of issue perception particularly between British and German environmental policy. While the British put their money on the self-regulatory capacity of the economy, German environmental policy is characterized by regulative law, the most important control instruments of which are the impositions and prohibitions. Against this background, the positions taken by the individual countries in drafting the Regulation and the specific opposition that has arisen in implementation are understandable.

Although the initiative for an environmental audit regulation came originally from the Commission,[85] Britain played an extremely active role in the policy-making process. The European environmental audit concept thus evidences a high degree of British influence, the British having set the pace in standardizing environmental management systems[86] and thus succeeded in putting their mark on European standardization. In drafting the Directive, the Commission had recourse to the provisions of the British standard, especially with regard to the concrete formulation of individual procedural rules (interview with EIA association, October 1993). Britain as 'first mover' was thus in a good position to promote its interests in the EU decision-making process.

The Objective and Substance of the Environmental Audit
The EU Eco-Audit Regulation was designed to create a voluntary Community system for assessing and improving the environmental performance of

industrial activities and furnishing the public with adequate information. It is the explicit aim of the audit system, which is to come into force in 1995, to improve environmental performance in firms. The measures taken to achieve this include the elaboration of an in-house environmental protection system, validation by external verifiers of the effectiveness of company environmental protection measures, and annual publication of information on the environmental performance of the firm in the form of 'environmental statements'.[87]

For a firm to participate in the environmental audit system, the sites must first be registered that it wishes to subject to environmental impact assessment. After registration, the firm concerned is obliged to review its environmental management system and to install corporate environmental protection instruments that satisfy the requirements of the EU regulation. Assessment of this environmental protection system is the chief function of the environmental audit. Environmental performance assessment is carried out by internal verifiers with adequate training, and who must be informed about the industrial activities to be audited.[88] The outcome of the audit is made available to the public in the form of an environmental statement. Besides a description of company activities and environmental impacts, it contains an account of corporate environmental protection instruments, the deadline for submission of the next environmental statement, and the name of the certified verifier.

All environmental protection instruments at the disposal of the firm, including the draft environmental statement are then validated by a certified verifier. The verifier examines the extent to which the provisions of the Eco-Audit Regulation have been complied with, and whether the information given in the environmental statement is complete and correct. The verifiers, who have to be technically qualified as well as independent and impartial, are selected under the provisions of the Regulation in an accreditation procedure to be elaborated by individual member states. Moreover, member states are required to establish a competent authority to register corporate environmental statements and to forward them to the European Organization for Testing and Certification. This supranational authority undertakes to coordinate national accreditation systems. Lists of the recognized verifiers as well as the participating sites are published in the Official Journal of the EU. If a site has satisfied all the conditions of the Regulation and the firm has submitted a validated environmental statement to the competent authority, the firm is entitled to use a statement of participation for publicity purposes.[89]

The Eco-Audit Regulation thus stresses self-regulation by industry with regard to the installation and monitoring of efficient environmental management subject to external, non-public assessment. In contrast to the Information Directive, which contains only a passive duty to disclose information, the Eco-Audit Regulation prescribes an active duty of publication (interview with BSI, January 1993). The provisions of the Regulation thus conform with the objective of the Fifth Action Programme of the EU to improve public access to information and ensure greater participation by affected actors.

Because drafting in the Commission was dominated largely by the need to coordinate questions of technical procedure and detail, the debate was not politicized at an early date as had been the case with the Information Directive. The costs and benefits of the Regulation were not immediately calculable for member states. Moreover, consensus was more easily reached at the Commission level because, as subsequent negotiations in the Council of Ministers were to show, the measure was accepted by most countries without much demur. During expert consultations, criticism was directed only at the mandatory participation in the audit scheme initially sought by the Commission. The Commission thereupon revised its concept, now permitting each firm to decide whether or not to participate. In compensation for this concession, however, the Commission insisted on planning the measure no longer as a directive but as a — directly applicable — regulation with very detailed requirements on implementation of the audit (ENDS 1991/ 194, 33).

The Decision-making Process in the Council of Ministers
In the Council of Ministers, too, where bargaining, negative coordination and compensation characterized decision-making, the representatives of the member states were unusually like-minded. Britain, which was a warm advocate of the Regulation, urged its rapid adoption (interview with EU Commission, DG XI, March 1993). The other countries also generally supported the Commission concept. Only Germany was opposed, thus blocking the decision-making process, since the measure was based on Art. 130s, which until the coming into force of the Maastricht Treaty prescribed unanimity in the Council of Ministers. What was the background to the divergent reactions among member states?

Britain's active role on the issue was attributable to the low costs of legal and institutional adjustment accruing to it from implementation of the Regulation. The self-regulation of industry and the procedural character of the arrangements conformed essentially to British regulatory tradition in environmental policy. Publication of corporate environmental data was also compatible with British ideas on regulatory openness as emphasized by the EPA 1990. Furthermore, active British participation was encouraged by the fact that provisions of the 1992 national British standard BS 7750 provided the basis for the EU Regulation. Unlike the Germans, the French, and a number of other European neighbours, the British had begun elaborating a standard as long ago as 1990, immediately after the Commission had announced its ideas about an environmental audit, thus securing a certain regulative lead for Britain. A second explanation for the British negotiating position is to be found in competition policy. The British, whose industry had been subjected to far-reaching data publication obligations under the EPA, had an interest in imposing these requirements on its European competitors. For this reason Britain advocated a relatively extensive interpretation of corporate information obligations (ENDS 1991/197, 36). Also important were the market interests of British consulting firms, which scented lucrative sales opportunities for environmental consultancy and appraisal services (interview with EIA association, October 1993).[90]

Although France did not take as active a part in negotiations as Britain, it joined most other member states in approving the Regulation: 'We were in any case very much in favour' (interview with *Ministère de l'Environnement*, January 1995). This stance is explained especially by the fact that the French 'had not stood idly by with their hands in their pockets' (interview with *AFNOR*, March 1993), but had been working on a standard of their own for an environmental management system, which was directed less towards making its own mark as conforming to the British standard: 'It's very close to the British standard. In the spirit of the ISO 9000 standards it's also to do with quality because there are a tremendous number of common points between environmental management and quality management. They're very close indeed' (ibid.). Because of their early anticipation of European legislation, the potential costs for France of legal and institutional adjustment have been reduced.

In contrast to the French and the British, the Germans faced substantial adjustment under the Regulation. This was a direct result of the diametrically opposed regulatory philosophies underlying German environmental law and the Eco-Audit Regulation. Whereas German environmental law is characterized by a command and control philosophy, the Eco-Audit Regulation is based on the self-regulation of industry and regulatory openness. It is thus hardly surprising that the German government, 'who had always thought that regulative law was the clou, and which had meddled around in firms from above *'par ordre de mufti'* (interview with EIA association, October 1993), initially blocked the Eco-Audit Regulation. Although both the Environment and Economics Ministries rejected the Regulation, they did so for different reasons. Whilst the *BMU* considered that the Regulation did not go far enough, the Economics Ministry criticized the additional burden on German industry, which was already subject to stringent standards (interview with *BMU*, October 1994).

The Ministry of Economic Affairs espoused the reservations expressed by industry, which feared 'that with the environmental audit we'll have a system imposed on us on top of the existing one that will make no additional contribution to environmental performance but only cause more bureaucracy' (interview with chemical company, February 1993; interview with *VCI*, March 1993; *BDI* 1993, 1). The provisions of the publication of environmental statements are also anathema to the Germans. Industry argues that the environmental audit can fulfil 'the task of ruthlessly analysing weaknesses' only 'if it is used as a purely internal management instrument without its findings being bruited about the market place and pubs the next day' (Meurin 1992, 302). If one 'tackles such a spring-clean internally, one doesn't want others looking on' (interview with steel producer, March 1993).

One point of criticism advanced especially by the *BMU* was concerned with the purely procedural nature of the measure. This was in contradiction to the traditional predominance of technology in German environmental policy.[91] With the carrying out of an environmental audit, firms are being required to reduce the predominance of technology:

They now have to pursue a progressive type of environmental policy. They have to formulate their own environmental guidelines, establish their own environmental policy. They have to draw up a schedule of measures. They have to structure their management is such a way that organizational responsibility can be shown if anything happens. This means that new management structures have to be created (interview with EIA association, October 1993).

The Germans therefore demanded that greater stress be placed on the best available technology in the Regulation, 'especially by setting substantive standards for assessing environmental performance by plants' (Führ 1993, 2; interview with *BMU*, October 1994). Industry was nervous that 'the standards required of individual plants in different member states would diverge strongly ... and that the demands made of an eco-management system' would differ 'from country to country and plant to plant', since there were as yet no Europe-wide common and concrete standards, which would be 'necessary to equalize environmental protections requirements at a high level' (*DIHT* s.a., 1; interview with chemical company, February 1993; *BDI* 1992a, 10).[92] This explains the circumstance that 'the Germans are fighting tooth and nail against having to accept the British standard' (interview with EIA association, October 1993) and 'want to develop a German standard of their own' (interview with steel producer, March 1993).[93] Since, however, 'German standardization missed the boat' in the development of corporate environmental protection instruments (Spindler 1993a, 87), *NAGUS* standardization committee on the Foundations for Environmental Protection, which has existed only since February 1993, is unlikely to catch up with the British by the time the Eco-Audit Regulation is to be implemented in April 1995.[94]

One could object that the environmental audit is carried out on a voluntary basis, and that hence no firm is obliged to subject itself to any unpleasantness or superfluous activities. This is a fallacy. No firm 'enjoying a certain public reputation in environmental protection matters' can afford 'not to appear in the annual Official Journal list' (Meurin 1992, 301). And, for reasons of competitiveness, the possibility of being able to advertise with participation in the environmental audit system 'takes the voluntariness of participation ... to absurdity' (*BDI* 1992a, 7). Undertaking an environmental audit, 'which is sensible from many points of view, ... ought not to be under constraint ... , but everyone should handle it himself, because it's in his own interest' (interview with ICC, July 1993). The consequence is pressure to adjust and a positive bandwagon effect.

A further reason why unconstrained voluntariness becomes 'voluntary constraint' (interview with EIA association, October 1993) is the pressure exerted by insurance companies and banks on firms to carry out an environmental audit. Especially in view of the tightening of the law relating to environmental liability, insurance companies very careful note measures taken to reduce ecological risks in firms, adapting their premiums accordingly (interview with insurance company, August 1993; interview with Third-Party Liability Association of German Industry, August 1993; interview with ICC, July 1993). In the insurance business, the convincing environmental policy motto 'precaution rather than clean-up' takes the form 'prevent damage to avoid damages' (Spindler 1993c, 12). The banks in their turn are increasingly granting loans only to firms that act not only rationally from an economic point of view but also with a sense of ecological responsibility (interview with EIA association, October 1993).

In discussions in the Council of Ministers, characterized by compensation and negative coordination and bargaining, the Federal Republic abandoned its earlier opposition despite its numerous objections, and voted for the Regulation in March 1993. Two factors are primarily responsible for this surprising change of mind, the anticipation of a changed general institutional setting and the willingness of other parties to compromise. Thus it was remembered that, once the Maastricht Treaty came into force, it would be possible to adopt the Regulation by qualified majority. This thought made the Germans more willing to compromise. If they wished to influence the drafting of the Regulation, they knew they would no longer be able to brandish their veto (interview with *BMU*, October 1994). Another factor making it easier for the Germans to accept the measure was that other countries were willing to make certain concessions in relation to Germany's technology orientation. On the urging of the Federal Republic, a further article was included in the Regulation providing for the continuous improvement of corporate environmental protection, 'which has to be based on the best available technology economically feasible. And through this provision, which Germany popped into the Regulation in the final phase of negotiations, the environmental audit moved away from being a purely management approach to take more of a performance-related direction' (interview with *BMU*, October 1994). German industry, too, thereupon modified its position, no longer asking 'whether they ought to approach the Regulation but how they could react most efficiently to it' (Annighöfer 1993; quoted in Henn 1993, 89).[95]

Although the compromise solution in the Council of Ministers had succeeded in persuading the Germans to give their approval, it became evident in the course of implementing the Regulation at the national level that conflicts inadequately overcome in the policy-making context can reemerge more strongly in the implementation phase. This conflict was to find expression particularly in the divergent views held by the German Environment and Economics Ministries on the environmental audit concept.

Implementation of the Regulation at the National Level
Although regulations generally restrict member states' autonomy more than do directives, the environmental audit provisions leave a comparatively broad margin for action in national implementation, especially in regard to concrete arrangements for accreditation procedures. The Eco-Audit Regulation merely requires that the accreditation of external environmental verifiers is to be effected in an independent and impartial manner.

The question of the accreditation procedure at the national level proved relatively unproblematic in both Britain and France. In Britain external auditors are accredited by the National Accreditation Council for Certification Bodies (NACCB), which was set up to implement the national quality management system under BS 7750. The NACCB is a multipartite body, composed of experts from industry, science, politics and administration (interview with DoE, November 1994). Whilst the British were able to take recourse to existing institutions in their national quality management system, the French could use the outcome of two pilot studies coordinated between the Ministries of the Environment and Industry which had been carried out in relation to implementation of the Regulation. 'The aim of this experimental phase was really to run in the whole process of this regulation ... at the industrialist level ... But it was also sought to simulate an accreditation body (interview with *Ministère de l'Environnement*, January 1995). As in Britain, the French accreditation procedure was multipartite to ensure the greatest possible objectivity. The *Comité Français d'Accréditation* (*COFRAC*), whose job it is to accredit environmental verifiers, 'is an institution with an administrative council that is to have several sections, each composed of interested parties'.

In Germany, by contrast, where the procedural arrangements for the environmental audit are in strong contradiction to classical German interventionist philosophy, two implementation models were hotly disputed: the state-oriented concept of the Environment Ministry and the industry-oriented concept advanced by industry. The Ministry tried to deal with the legal and institution adjustments required by this concept new to German regulatory tradition by falling back on classical forms of state control. The preferred model was to entrust ascertainment of the reliability and impartiality of verifiers to the Federal Environment Office. Only the technical assessment of verifiers was to be assigned to a body 'close to industry', the Association for Accreditation (*Trägergemeinschaft für Akkreditierung*). Industry, having withdrawn their opposition to the environmental audit, wanted to see less classical control and more self-regulation. Their counterproposal, which also found the support of the Economics Ministry, aimed to establish a self-administrative, decentralized accreditation system by chambers of industry and commerce and chambers of handicrafts[96] (interview with *BMU*, October 1994).

After more than two years of discussion, the Federal government finally tabled a bill in April 1995 implementing the Eco-Audit Regulation, which took account of both models. Although the desire of industry to have an accreditation authority 'close to industry' was given its due, the *Deutsche Akkreditierungs- und Zulassungsgesellschaft für Umweltgutachter mbH* (German Accreditation Company for Environmental Verifiers Limited) founded by the *BDI*, *DIHT*, the Central Association of German Handicrafts[97] and the Federal Association of the Professions,[98] was placed under the legal tutelage of an Environmental Verifier Committee.[99] This committee prepares general guidelines for elaborating admission procedure, which are binding on the accreditation body. It also manages the pool of environmental verifiers accredited by the company from which firms can select their external verifier. Represented on the committee are the Federal and state governments, industry, verifiers, trade unions, and environmental associations (*Bundesrat — Drucksache* 210/95). The adjustment constraints facing the traditional German regulatory philosophy in implementing the Eco-Audit Regulation have thus been dealt with in a compensation solution containing elements of both state control and industrial self-regulation.

Another factor accompanying the implementation debate in all member states is the extent to which firms participating in the environmental audit ought to be granted permitting facilitation. German industry in particular hopes to gain certain concessions in the field of emission monitoring and the authorization of smaller plants (interview with *BDI*, October 1994). In Britain, easier conditions in emission monitoring have now been available for some time for firms applying quality management under BS 7750. The French stance is more reserved. According to the Environment Ministry, the results of implementing the Eco-Audit Regulation are to be awaited before any move is made (interview with *Ministère de l'Environnement*, January 1995).

Conclusion

Negotiations on the Eco-Audit Regulation confirm the hypotheses guiding our study on the behaviour patterns of member states in regulatory competition at the European level. Britain, with quality management systems and corresponding procedural standards already in place, was able to make its mark in the policy-making process as 'first mover'. The emphasis put on the self-regulatory force of industry and the procedural character of the Eco-Audit Regulation are in line with the British legal system, which focuses not on the setting of firm technical standards as targets but on regulating procedures. The British were accordingly keen to advance the environmental audit system as a self-regulatory instrument that could be integrated relatively easily into their regulatory system. However, British strategy was grounded not only in legal arguments but also in economic interests. The battle is now on to win the biggest possible share of the hotly contested consultancy market. While German consultants are comparatively wary in their approach to the European market, their British counterparts have actively forced the pace in gaining accreditation as environmental verifiers and thus in opening up new, lucrative sales opportunities; a further incentive for the British government to play pace-setter. Germany, typically relying on the use of state control, the substantive aspects of which are based on technical advance, initially blocked negotiations in the Council of Ministers. Certain concessions by negotiation partners and the change in the voting modalities introduced under the Maastricht Treaty, which deprived Germany of its veto option, finally induced the German government to abandon its resistance to the measure.

Whilst the German and British positions were clearly at opposite poles, France's posture was one of hazy, unruffled and circumspect approval. In the case of the Eco-Audit Regulation, too, France persisted in its accustomed position of neither particularly urging nor impeding events. This is understandable given the fact that France had anticipated the possible effects of European regulation at a relatively early date, and had sought to coordinate its own standardization efforts with activities taking place at the European level.

The low costs of legal and institutional adjustment made implementation of the Eco-Audit Regulation easier for both Britain and France. It was much more difficult for Germany, where the prevailing regulatory tradition was confronted with a completely new concept. While the Environment Ministry attempted to persist in traditional forms of state control, industry urged adoption of implementation 'close to industry' including a greater self-regulative element.

3.2.3   France as Friendly Onlooker and Coalitionist

During the eighties, the French remained broadly neutral in the supranational regulatory process. Regardless of whether the Commission was pursuing a quality-based strategy or an emission-related one, the French did not distinguish themselves by a particular determination either to support or to impede legislation. As we have shown, the neutral French attitude can be attributed to two essential factors: first, the multiplicity of domestic regulative instruments permitted implementation of EU legislation without excessive legal and institutional adjustment costs; and second, there was no economic reason to promote or ward off any measures.

Towards the end of the eighties, the Commission once again changed strategy with the Fifth Action Programme. It returned from its emission-based, best-available-technique approach to a quality-related policy, albeit complemented by an important component: broad public access to information as a lever for more effective implementation of quality standards. The new Commission approach found expression especially in measures adopted in the early nineties on the freedom of access to information on the environment, on integrated environmental pollution (IPC), and on environmental auditing.

However, the change in strategy on the part of the Commission did not affect the general negotiating position of the French, who were the well-disposed, neutral bystanders in the regulative contest fought out principally between the Germans and the British. The reasons for the French stance were the same as in the eighties. They still had no special incentive either to minimize legal and institutional adjustment costs or to safeguard the competitive position of French industry by taking a more active part, be it as pace-setter or damper.

Many of the provisions of the Environmental Information Directive are thus in accord with French law on the inspection of records that has existed since 1978. Both the Directive and the French measure provide for a passive duty of disclosure, that is to say the competent public authority must disclose information only on specific request. The question of when information may be withheld (for example where matters are *sub judice* or where commercial confidentiality is affected) are regulated in an identical manner. The French thus remained neutral on the preparation of the Directive, because the right of access to records was already regulated in France, alone of all member states at the time, and because no legal and institutional adjustment costs would thus be incurred. However, although national practice suffered no formal amendment, the Directive proved a useful tool, especially for environmental associations, in facilitating access to data. This was rendered difficult in France by the frequently uncooperative attitude taken by the authorities (interview with *Les Verts*, June 1993; interview with environmental protection association, January 1994). Nor was there any economic incentive to push through or block a European directive. The right to inspect files was a relatively minor concern for French industry, since with the *'enquête publique'*, firms are subject to much more sweeping (active) obligations to disclose information in plant licensing procedures. What is more, environmental organizations tend to obtain data via informal channels (Spanou 1988, 132). Particularly important in this respect are the multipartite bodies at the regional level (like *SPPPI* or the air-quality monitoring associations), in which both industry, administration, and environmental interests are represented.

The Directive on Integrated Pollution Control (IPC) also imposed no comprehensive changes on French regulatory practice. In the 1976 Act on the classification of industrial plants, a trans-medial approach had already been prescribed, covering air, water and soil: 'The Act on classified installations and the integrated approach tally in general principle on almost all points. ... France thus observed elaboration of [this directive] with a certain satisfaction'[100] (interview with EU Commission, DG XI, March 1994; interview with *CITEPA*, March 1993). France disposes not only of the legal preconditions for the practical implementation of an integrated approach, but also the necessary institutional potential. This is true for both the provisions on public participation, afforded in France by the public enquiry system: 'There are already administrative arrangements providing for coordination between authorities, and the public already has access to proceedings'[101] (interview with EU Commission, DG XI, March 1994). France thus faces only low legal and institutional adjustment costs through the IPC Directive: 'There'll be some minor new rules that won't be a great blow to France'[102]. France's favourable attitude has been encouraged by the fact that the quality-based *modus operandi* permits it to retain its existing practice of negotiating control requirements for industry at the regional level in terms of local environmental quality. The introduction of Community-wide emission limits would, by contrast, have rendered such negotiatory processes superfluous. The subsidiarity principle taken account of under the IPC Directive thus favours regional weighting in France.

Moreover, the IPC Directive accommodates the economic interests of French industry. Because of the strong regional discrepancies in settlement and industrial density structures, industry profits from a quality-oriented concept involving differential control stringency from site to site. In contrast to Germany, where such benefits can scarcely be exploited because of the high demographic density and nation-wide distribution of industrial activity, France is in a position to take full advantage of a broad margin for action. France also has a very limited environmental protection equipment industry to profit from the enhanced sales opportunities Community-wide emission standards would bring. These legal, institutional, and economic conditions explain France's favourable attitude towards the Directive: 'France is very favourably inclined to the Directive as a whole' (interview with EU Commission, DG XI, March 1994).

The French evidenced a similar pattern of behaviour on the Eco-Audit Regulation as on the Large Combustion Plant Directive, 'neither pushing the Regulation nor opposing it'[103] (interview with *Ministère de l'Environnement*, March 1994 but seeking to anticipate it. The French have thus prepared their own eco-management standard designed to be compatible with the British Standard 7750, the probable model for the EU Regulation (interview with *AFNOR*, March 1993). The proximity of the French standard to the British one reduces the potential institutional and legal adjustment costs of the European legislation. In addition, the stress placed on self-regulation by industry corresponds in a certain sense with French regulatory tradition. It finds expression, for example, in the frequently deployed instrument of the so-called 'industrial agreement' between the government and individual industrial sectors under which industry voluntarily undertakes to reduce pollutant emissions by a given amount.

Overall, it is apparent that the multiplicity of regulative instruments available to French environmental policy has ensured that the new strategic orientation of the Commission in the early nineties has imposed relatively little need for legal and institutional adjustment to satisfy European requirements. Moreover, the French seek to reduce adjustment costs as much as possible in sectors where they do not yet dispose of corresponding national arrangements by anticipating European legislation. This behaviour, noted already in the eighties in connection with the Large Combustion Plant Directive, was particularly evident with the Eco-Audit Regulation. French goodwill was, however, motivated not only by the relative lack of need for legal and institutional modifications, but also by the relative lack of negative impact on national competitiveness. French industry has had no particular interest to ward off arrangements at the supranational level that it faces at home anyway. Vice versa, these national arrangements demand no 'special sacrifices' of industry such that it has incentive to impose French rules on Europe to remain competitive. And there is little in the way of a French environmental protection equipment industry, the sector that could have an interest for sales reasons in stringent EU-wide standards.

3.3 France as Hesitant Pace-setter

As we have seen, France has neither set the pace nor put the brake on clean-air policy like Britain and Germany. The French have been able to afford a neutral negotiating position, neither enthusiastically embracing or categorically rejecting proposed measures. The French attitude was not shaken by the Commission's change in direction from an emission-based strategy to a quality-oriented one. Whereas the British and the Germans took this occasion to swap their respective roles of pace-setter and hinderer, the French stayed where they were. They apparently find themselves in a decidedly favourable situation that allows them to stay aloof from the supranational regulative contest while keeping a benevolent eye on the hubbub in the regulative marketplace.

They owe their position primarily to the wide range of regulatory tools at their disposal, which allows them to implement widely differing measures without difficulty. Their regional approach to controlling industry also imposes no uniform control requirements as is the case in Germany. With these regionally differentiated regulatory requirements, the French have no need to make their rules binding on the entire Community to secure economic competitiveness. And with a weakly developed environmental equipment industry in the clean-air field there is no incentive to open up new markets by pushing stringent limit values. Finally, the predominance of nuclear energy gives France a relatively flexible position on the abatement of pollutant emissions from fossil-fuel combustion.

That French neutrality is no immutable disposition is evidences by the measures to be examined in this chapter in the fields of air quality and SO_2 product standards for liquid fuels. In both cases France sought to influence the supranational regulatory contest in its favour — albeit with greater reticence than Germany and Britain usually show. The rather hesitant pace-setter role assumed by the French in this context is, however, differentially motivated. French interest in legislation on SO_2 products is due primarily to France being particularly affected by the issue, certain coastal regions being polluted by SO_2 emissions from ships. But economic interests deriving from the structure of the energy sector also encourage the French to take an active role. It is interesting to note that the French pacesetting role in the field of ambient air quality stems from a regional initiative, the effects of which are regarded with a favourable eye by central government because they would reduce still further the already low costs of legal and institutional adjustment.

Ambient Air Quality Assessment and Management: 'Mobilization From Below'

Two air-quality measures are at present on the Commission agenda: a framework directive containing general monitoring and procedural criteria and a 'daughter' decision on 'Exchange and Information about Air Quality'. While problem-definition and agenda-setting on the 'daughter' have been decisively shaped by French initiative, the Framework Directive is being promoted mainly by the Commission.

An important factor explaining the active Commission role is the special significance of such framework arrangements. They give the Commission an important enabling basis on which to establish future activities. They thus further the interest of the Commission in expanding its regulatory competences. Their 'autonomy-friendly' (Scharpf 1993b) nature and the absence of substantive requirements mean that member states tend not to oppose them too fiercely. This was made clear with the emission-oriented directives in the early eighties. Opposition from individual countries concentrated on the daughter Directive on large combustion plants, while negotiations on the Framework Directive were surprisingly rapid and unproblematic. Commission initiative to adopt framework directives is thus to be explained in terms of two factors: first, they offer it a good basis for enlarging their powers, and second they tend to find more favour with member states than substantive regulatory proposals.

In contrast to many other directives, in the development of which the Commission mainly responded to action taken by individual member states, in this instance of the framework directive, the Commission is attempting to elaborate a new and comprehensive concept for its hitherto fragmentary policy on air quality.[104] Although some member states and especially Britain have indicated their support, the Commission policy proposals are not based on concepts already realized by a member state. The roots of the initiative are in the Fifth Action Programme, which proposes the continuous extension of the list of polluting substances for which quality standards are set (European Communities 1994, 1). The Commission is also guided by conceptions that have gained in salience particularly through the Treaty of Maastricht: the demand for greater transparency in European policy and the subsidiarity principle. At the 1992 Edinburgh summit, EU heads of state and government reaffirmed their commitment to these principles in clean-air policy. This triggered the Commission initiative, which is concurrent with the French proposal (interview with EU Commission, DG XI, November 1994).

Another motive of the Commission was the experience gathered in the early eighties with a quality-based strategy. It had shown that setting quality standards was not enough to ensure due implementation. Further requirements were necessary, especially in two areas: comparable monitoring procedures throughout the EU, and — fully in line with the new Commission strategy — public information on compliance with or infringement of quality targets. 'If you're going to have limit values, they have to be checked and not simply exist on paper, because that serves only to reassure the public and flies in the face of transparency and the principle of regaining the confidence of the public' (interview with EU Commission, DG XI, March 1993). These goals are now central to a framework directive on air quality at present being discussed at the EU level and to a daughter directive dealing with the exchange of information between member states, the drafting of which is under strong French influence.

The Commission Concept
In order to give the proposed air-quality measures a uniform strategic orientation, the Commission submitted a draft framework directive in 1994[105] on 'Ambient Air Quality Assessment and Management'. The proposal contains general criteria on defining limit values, on uniformly assessing air quality in member states, and on informing the public, as well as on the steps to be taken by member states in the event of quality standards being exceeded. It also lists the polluting substances for which admissible atmospheric concentrations are to be regulated by daughter directives. Quality standards, and concrete monitoring details, are then to be laid down by daughter directives (interview with EU Commission, DG XI, November 1994). In keeping with the subsidiarity principle, Community-wide quality standards are then to be set where this serves the regulatory goals better than could national or regional measures. The Commission believes this is the case especially for substances for which there are already EU-wide quality or emission standards or which have transboundary impacts. Substances classified by the WHO as hazardous to health are not to be left to national regulation. The Commission argues that 'there is no evidence to suggest that populations in different parts of the Community vary in their tolerance of air pollutants' (European Communities 1994, 3).[106]

Since, in addition to environmental considerations, the protection of human health is a principal aim of the Directive, only substances scientifically proved to have a detrimental impact on health are to be dealt with.[107] Such proof, which is unproblematic for most atmospheric pollutants, is still controversial for CO_2. For this reason the Commission does not intend to introduce a Community-wide quality standard for this substance. Another reason is, however, that negotiations would have become far more difficult if the complex and controversial issue of CO_2 abatement were to be included (interview with EU Commission, DG XI, November 1994).

The draft proposal distinguishes three types of quality standard in terms of member state obligations: long-term quality goals, 'permitted margins of exceedance', and 'alert thresholds'. A basis is to be provided by a long-term quality goal for each pollutant, to be attained within 10 to 15 years in all regions. Permitted margins of exceedance define the leeway for exceeding long-term quality limits. It is intended progressively to restrict this margin. Member states are to be required to ensure that air quality remains within the permitted margins. In the event of exceedance, member states are to publish plans of action containing the measures necessary to improve air quality. Where emission values exceed the alert threshold, member states are to inform the public on radio, television and in the press. Table 16 shows the different measures member states are to take in terms of air-quality.

In connection with the far-reaching publication obligations for member states arising in the event of quality targets being exceeded, the Commission plans to publish an annual map showing air quality in the individual monitoring area. The categories 'poor, improving, good' are to be used. The Commission believes that such a rough categorization is easier for the public to grasp than concrete measurements (interview with EU Commission, DG XI, November 1994).

In addition to the subsidiarity principle, the comprehensive information of the public is the second element in the Commission's strategy on the Directive. This concept is evident in many current European environmental policy measures like the Environmental Information Directive and the Eco-Audit Regulation. A combination of quality-orientation and 'pressure from below' is intended to ensure compliance with and implementation of quality targets. 'What's the use of a limit value without implementation? It just reassures the public. ... We want to ensure that monitoring is taken just as seriously as the limit value itself. There's no sense in it merely pretending' (interview with EU Commission, DG XI, March 1993). In keeping with the subsidiarity principle embodied in the Treaty of Maastricht, procedural arrangements on informing the public are the task of member states.

Table 16: Requirements under the Framework Directive on Air Quality

Air-quality value	Classification of monitoring area	Duty of member states
Below long-term quality target	area of good air quality	– Notification of the Commission
Between quality goal and permitted margin	area of improving air quality	– Notification of the Commission – Appropriate measures to reduce air pollution
Between permitted margin and alert threshold	area of poor air quality	– Notification of the Commission – Appropriate measures to reduce air pollution – Publication of a plan of action
Above alert threshold		– Notification of the Commission – Appropriate measures to reduce air pollution – Publication of a plan of action – Information of the public through radio. television, press

On monitoring methods the proposed Framework Directive restricts itself to general criteria to be filled in by daughter directives. The latter are to deal with the location of measurement instruments, measurement methods and the minimum number of monitoring stations (European Communities 1994, 24).

Besides providing for public information to mobilize 'pressure from below', the Commission plans to establish a comprehensive monitoring network to improve the exchange of information among member states. Processing of the data collected and coordination of the network are to be entrusted to the European Environment Agency in Copenhagen. The aim is to record the precise distribution of air pollution in Europe over time, which has hitherto been impossible. It is hoped that this knowledge will permit EU clean-air measures to become more efficient and purposive (interview with EU Commission, DG XI, November 1994). Detailed procedure is to be laid down in a Daughter Decision on 'Exchange and Information about Air Quality', at present being negotiated in parallel to the proposed Framework Directive at Commission level (interview with EU Commission DG XI, September 1993).

The Commission work on this daughter measure have, as we have mentioned, been under strong French influence. *AIRPARIF*, a regional air-quality monitoring association from Greater Paris had begun to develop a new monitoring network at the same time. In order to assure the compatibility of its concept with Commission plans, and to reduce the costs of legal and institutional adjustment, the *AIRPARIF* had addressed an enquiry to the Commission on the matter:

> We'd been trying to work things out on our own, and after a while we decided to see what was going on elsewhere. We went to see the Commission and told them if people have been thinking about these things along the same lines in Europe — then we could perhaps go and see whoever has been doing the thinking and talk it over with them[108] (interview with *AIRPARIF*, March 1993).

Quite independently of central government intervention, regional initiative was taken to influence the Commission. For a unitary state like France, this was absolutely untypical behaviour: *'Mais alors, ce n'est absolument pas typique au point de vue comportement'* (interview with *AIRPARIF*, March 1993). The French Environment Ministry — the state actor generally responsible for environmental policy activity at the supranational level — is in a certain measure indulgent of the *AIRPARIF* initiative, since it has no financial or technical resources of its own in this field.

The Commission — always open to new ideas and initiatives because of its sparse personnel resources — was very interested, and asked an international group of experts to investigate whether the new Parisian air-quality monitoring network could serve as a model for corresponding European legislation (interview with EU Commission, DG XI, September 1993). 'Basically, we [*AIRPARIF*] were the Commission's guinea-pigs' (interview with *AIRPARIF*, March 1993). As a consequence, the provisions of the Commission draft correspond in many points with the concept elaborated by *AIRPARIF*. However, the Commission also made use of the findings of other pilot studies carried out in Brussels and Madrid.

The Positions of Member States in the Decision-Making Process
The 'autonomy-friendly' framework nature of the proposed Directive, the absence of substantive limit values, and the exclusion of politically sensitive issues like the CO_2 problem favoured a very constructive policy-making process at the Commission level (interview with *Ministère de l'Environnement*, January 1995). 'The Directive was welcomed by all member states' (interview with EU Commission, DG XI, November 1994).

Nevertheless, the problem-solving phase brought divergent member-state interests to light. However, because of the broad leeway for implementation at the national level, differences manifested themselves in unassertive form, and, the Commission believes, are unlikely to produce any fundamental obstacles to agreement in negotiations in the Council of Ministers (interview with EU Commission, DG XI, November 1994).

Since the Commission's quality-based strategy is largely in line with British problem-solving philosophy on clean-air policy, Britain takes a very positive attitude towards the measures (interview with EU Commission, DG XI, September 1993). British costs for legal and institutional adjustment are relatively low, since under the subsidiarity principle only quality targets are to be set, the ways and means of achieving them being left to the individual member state (interview with DoE, September 1993). Nor do the British see any problems with public information, since they have introduced extensive data disclosure obligations under the EPA. To this extent, the EU provisions represent the further development of the trend towards openness already established in Britain. The proposed Commission Decision on the exchange of information is also compatible with British traditions. It prescribes only procedural rules on recording and processing information, and not concrete technologies.

Germany, by contrast, is somewhat uneasy about the planned measures. There are two main reasons. The cost of legal and institutional adjustment to an alien regulatory tradition accruing under the Directive, and the economic disadvantages for German industry from a quality-based approach. One particular problem is the extensive public information rights prescribed by the Directive in the event of limits being exceeded. This is quite alien to German practice, such data being kept confidential in Germany.[109] The Directive would therefore cause high adjustment costs for the legal and institution organization of the German regulatory system (interview with EU Commission, DG XI, September 1993). What is more, an exclusively quality-based approach is fundamentally contrary to the German 'state-of-the-art' technological approach. A quality-oriented approach, setting varying emission standards in terms of local environmental conditions, would furthermore disadvantage German industry vis-à-vis foreign competition, because it is subject to relatively stringent emission standards. The highly developed German environmental equipment industry, the sales opportunities of which grow with strict international technological requirements, suffers economic disadvantage from a quality-based approach. The 'very obvious German policy' (interview with EU Commission, DG XI, March 1993), which is also regarded sceptically by the Commission, is being increasingly criticized by other member states.

The Germans demand that a so-called 'standstill principle' be included in the Directive. This would not allow any deterioration of air quality, even if compliance with quality targets were not to be at risk (interview with EU

Commission, DG XI, November 1994). The German demand, based on the precautionary principle, contradicts the right of economic development laid down in the Fifth Action Programme, which is stressed by southern member states in particular. Since, however, the Commission intends to prescribe very stringent quality goals, it expects the tension between the two principles to diminish (interview with EU Commission, DG XI, November 1994).

France — to some degree co-author of the Directive — has no legal, institutional, or economic reservations about the planned measures. This is partly attributable to the pacesetting initiative the French took via the regional air-quality monitoring association *AIRPARIF*. French arrangements are consequently largely compatible with the European measures. Once again, the multiplicity of French regulatory tools in clean-air policy proved important. This wealth of instruments and the broad public information rights provided under the 1976 Industry and Nature Protection Act obviate major reorganization in applying this quality-based concept. With its still embryonic environmental protection equipment industry, France is also not dependent on opening up markets in this area.

However, the French are at the same time very critical of German attempts to improve sales opportunities for German environmental protection technology via more stringent EU-wide standards:

Could there be a commercial angle to the proposals for more stringent rules? My answer to your question would be 'Yes, yes, yes'. ... Because Germany has concentrations of industry strict thresholds have to be used and machines to comply with these thresholds. Once you've got the machines, it's very tempting to say that the rest of Europe ought to use them, too. ... That's why certain Directives have a strong German flavour to them[110] (interview with *CNPF*, June 1993).

Conclusion

The present discussion in the Commission on the planned air-quality measures shows that the neutral French position is far from immutable. They play the role — albeit on regional initiative — of a 'hesitant initiator', which the legal basis and policy instruments at their disposal do not render imperative. On the other hand, France is understandably not unwilling where the Commission decides to orient its regulatory proposals on French arrangements.

Initiative in the decision-making process has been double. France has been a tentative 'first mover' on the Information Exchange Decision, while on the Framework Directive the Commission has been promoting a European regulation independently of national initiative. Adoption of a framework directive is particularly important for the Commission because it can thus provide itself with an enablement basis for expanding its regulatory powers in the future. Member states, by contrast, are not particularly interested in such measures, because the absence of substantive prescription presents them with less risk to their national autonomy.

Britain and Germany have been consistent in their behaviour in supranational decision-making processes. Whereas the British find a quality-based approach to their taste and have thus given their broad support to the measures, the Germans, with a radically opposed regulatory concept, have tended to put on the brakes. Only the French have exhibited a certain deviation from their otherwise neutral and disinterested negotiatory stance.

Product Standards for Liquid Fuels: 'Making a Virtue of Necessity'

France has taken action in a further area, applying the strategy of the first move. It instigated current consultations at Commission level on regulating the sulphur content of liquid fuels. Whereas since 1988 the use of solid fuels has been subject to plant-specific reduction requirements under the Large Combustion Plant Directive, Community-wide regulation of liquid fuel use has been fragmentary. So far there have only been product standards limiting the sulphur content of so-called 'light' fuels like fuel oil and diesel fuel. There have been no standards on 'heavy fuels' (so-called 'bunker oil'), which is used primarily in industrial production or to power ships.[111]

In 1990 the French delegation submitted a memorandum in the Council of Ministers calling on the Commission to fill the 'gap in regulation'. The French wanted to see product-related emission limits introduced for all liquid fuels containing sulphur. Such limits were to apply not only to the final product but to the entire chain of production.

There are two factors behind the French wish to see the Community 'think about a complete control of sulphur emissions along the entire liquid fuel chain, starting with the refinery with its releases up to the sulphur content of all fuels and residual substances that leave the refinery' (interview with EU Commission, DG XI, March 1993). France is particularly affected by the problem, and has certain economic interests relating to the structure of the energy sector. Because of the major shipping routes in the Channel along the French coast and the prevailing west winds, France is particularly exposed to SO_2 emissions from passing vessels, which use fuels with a very high sulphur content. And France would welcome an increase in the prices for fossil fuels from more stringent regulations, because it would permit the competitive disadvantages French industry faces from its high degree of reliance on relatively expensive nuclear power to be balanced out (interview with EU Commission, DG XI, September 1994).

The Commission reacted to the French proposal by commissioning various studies,[112] with the task of evaluating the costs and benefits that such a strategy would involve. For, to ensure greater efficiency in European policy, the Treaty of Maastricht requires the Commission to carry out a cost-benefit analysis for all measures planned (interview with EU Commission, DG XI, July 1994). The issues under investigation were both the economic burdens on industry and consumers and the qualitative progress that would result for the environment (interview with EU Commission, DG XI, September 1993). The thrust of these cost-benefit analyses 'was very much in favour of regulation' (interview with EU Commission, DG XI, September 1994), since the potential ecological benefits far outweighed economic costs.

On the basis of these expert opinions, the Commission intends to adopt a directive limiting the sulphur content of all fossil fuels through product standards. How drastic reductions are to be for each type of fuel is to be decided by three criteria: the share in total emissions of a category of fuel, the sulphur content of the fuel type, and the costs of attaining the respective product standard (interview with EU Commission, DG XI, September 1994). In contrast to the French concept of combating emissions along the entire production chain, the Commission is concentrating on the final product. Processing in refineries is thus not to be covered by the proposed legislation, but dealt with under the plant-specific provisions of the IPC Directive (ibid.).

Despite the unanimous opinion expressed by the experts, a viable compromise has yet to be found on the subject at Commission level that could serve as a basis for negotiations in the Council of Ministers. The main reason is opposition from the multinational oil companies and diverging national interests.[113] However, the findings of the scientific studies, which are a fundamental in the expert consultations at the Commission level, have considerably restricted the spectrum of policy options, thus 'framing' the process. As a consequence, the necessity of regulation is to a large measure accepted by the actors involved (interview with EU Commission, DG XI, September 1994). Conflict concentrates rather on the details of the substantive provisions.

The oil industry has played a very big role in shaping the Directive, and has attempted to push its ideas in constant dialogue with the Commission. This conduct is all the more striking since the oil industry has hitherto been notably absent in the debate on environmental policy issues at the European level (interview with EU Commission, DG XI, September 1994).[114] The oil industry has three objections to the planned measure. It fears distortions in competition from the regionally differing qualities of crude oil. Refineries primarily processing North-Sea oil would have an advantage over those using Arabian oil, since the North-Sea product contains less sulphur and thus requires less control. It is also feared that price increases caused by high investments in control will lead to a loss of market share to gas and coal suppliers. Since consumption of heavy fuel oil has been falling for some time anyway, the oil industry asks itself whether this trend makes high investment economically worthwhile at all. It is therefore not surprising that the oil industry is keen to obtain the weakest possible limit values (interview with EU Commission, DG XI, September 1994).

Four basic positions have been taken by member states on the issue. The French, particularly affected by the problem and with vested competitiveness interests, are in favour of relatively stringent limits. They are supported by a group of countries favouring very strict standards for purely environmental reasons. They include Germany, the Netherlands, and Denmark. Purely economic considerations predominate in a third group, constituted largely by the southern member states of the Community. They fear market losses for their oil industries, which mainly use high-sulphur oil from the Middle East, and therefore advocate less stringent limit values. A fourth and final group headed by Britain sees no need for Community-wide regulation of the matter. The British argue that the reduction of SO_2 emissions can be adequately achieved under the planned framework Directive on air quality and the provisions of the Geneva Convention on Long Range Transboundary Air Pollution (interview with EU Commission, DG XI, September 1994). This view gives clear expression to the British preference for a quality-based approach. They continue to regard concrete limits with a degree of unease.

The current discussion on the Directive shows that interacting economic and ecological interests can cause countries that generally take a passive line on clean-air issues at the European level to assume an active role in policy making. Where a country — in this case France — is particularly affected by deleterious environmental impacts, there can be strong motivation to take the initiative in the supranational regulatory process. This is especially the case where national regulation is unsuitable for handling the problem. The French must rely on supranational or international regulation, since national rules could do nothing to curb SO_2 emissions from passing ships. However, economic considerations also encouraged the French to take action. Strict controls on liquid fuels offer an opportunity to balance the competitive disadvantages French industry suffers because of its dependence on comparatively expensive nuclear power.

The course taken by the decision-making process to date also shows that cost-benefit analyses can facilitate consensus formation at the problem-solving stage by excluding certain policy options from the outset. Because of the unequivocal vote of all scientific studies on the issue in favour of Community regulation of the sulphur content of liquid fuels, member states are largely agreed on the need for corresponding measures. Policy-making will therefore not be impeded by individual countries. Discussion is focused on concrete details of the envisaged product standards.

It is worth noting, however, that it is the otherwise reserved multinational oil companies rather than member-state governments that are attempting within the framework of this minimal consensus to promote their interests in the policy-making process by exerting massive influence on the Commission.

3.4 German and British Interventionist Philosophies: An Abiding Conflict

Although the British, having revised their environmental legislation, are now making greater efforts to push through their quality-oriented, procedural approach, a glance at the more recent EU legislative proposals shows that they are by no means successful in all fields. Despite the high measure of compatibility between the British approach and EU Commission strategy, which is particularly directed towards combining quality-orientation and public participation, current developments in European legislation are a mixed 'strategic' bag. We find a mixture of intervention philosophies with both typically British and typically German traits. The 'mix', which varies from directive to directive, has a decisive influence on the positions taken by member states in the supranational decision-making process. If, for example, the Germans get the upper hand with their technology-based approach, the British express certain misgivings about the planned measure. Because of the revision of British regulation, which now also takes account of emission-related elements, Britain is under less pressure to make legal and institutional adjustments. The Germans fiercely combat proposals betraying a quality-orientation or serving to improve public rights of access to information, regulatory concepts fundamentally opposed to their own philosophy. Among current European measures exhibiting such a mix of German and British approaches are the directives dealing with the incineration of hazardous waste and the reduction of volatile organic compounds in the atmosphere.

The Incineration of Hazardous Waste: The German Interventionist Philosophy *Redivivus*

The background to elaboration of the Directive on the incineration of hazardous waste adopted in December 1994 was a Council Resolution of May

1990 in which member states urged the Commission to make proposals on the regulation of pollutant emissions from incineration plants for industrial wastes.[115] Some member states with relatively strict standards on waste incineration had called on the Council to adopt such a resolution. Germany was to the fore as 'first mover' in promoting and initiating this project.

Because suitable landfill capacity has almost been exhausted, waste incineration is an important alternative solution. Since controls on incineration differ from country to country, national industries subject to more stringent standards and the consequent higher costs are at a competitive disadvantage. The problem has worsened as opportunities to evade such disadvantages by exporting waste have been reduced.

The purpose of incineration is to reduce the volume of waste by a high factor, and to immobilize pollutants in the incineration residue, which then has to be disposed of in landfills anyway. Although you can't get rid of the harmful substances in wastes in a controlled landfill, you can concentrate them in the residues. But there are fewer and fewer suitable landfill sites in all member states. We have the best example in Germany, where individual towns have been facing a waste disposal emergency. For instance they all exported refuse to France, and then the French closed the borders. And now this waste has to be distributed somehow in Germany (interview with EU Commission, DG XI, March 1993).

The strong German interest in regulation was one of the main reasons why the Directive on the incineration of hazardous waste bears the features of the technology-oriented approach espoused by the Federal Republic.

The Content of the Directive
It is the purpose of the Directive to introduce measures and procedures to prevent, or where this is not possible, to reduce as much as possible impairment of the environment, especially pollution of the air, soil, surface and ground water, as well as dangers to human health from the incineration of hazardous waste. To attain these goals, the Directive lays down appropriate operating conditions and emission limit values for incineration plants for dangerous substances in the Community (Art. 1). Member states may authorize such plants only if they are equipped and operated in such a manner as to take account of all measures available through the deployment of the best available techniques to prevent pollution of the environment, especially measures to avoid or to minimize emissions.

The Directive sets limit values for different substances as minimum standards. They apply to particulate matter, organic compounds, SO_2, and various heavy metals (lead, cadmium, thallium, mercury, etc.). To ensure that the latest control techniques are taken into account in authorization, the Directive lays down that, regardless of compliance with emission limits, an operating permit is to be granted by the authorities only if the best technical methods available at the time of application are used. 'The Directive is not only based on limit values. ... The focus is on the dynamic development of techniques' (interview with EU Commission, DG XI, March 1993). The Commission justified this built-in adjustment dynamic with reference to the extremely rapid technological advances made in recent years.

Emissions are after all not prevented or reduced by setting limit values but by installing waste gas purification equipment. In recent years these techniques have made tremendous progress. ... So we want to capture this dynamic element in the Directive and have said that an authorization may only be granted if the application shows that suitable measures to reduce emissions as available at the time have been made. If I just stick to limit values I almost hinder technical progress (interview with EU Commission, DG XI, March 1993).

To permit national authorities to implement this dynamic adjustment to 'state-of-the-art' technology in practical terms, they must have access to information on the latest technical developments. To this end, the Directive requires member states to provide the Commission annually with comprehensive data on the measures taken by them (including the control technologies used). On the basis of this information, the Commission is to prepare a report to inform member states on current technical advances. The member states are required to forward these data to the competent permitting authorities.

The Directive focuses not only on the emission of polluting substances to the atmosphere but also stipulates that clean-air measures may not lead to increased water pollution. Although soil pollution control, to be dealt with in an appropriate landfill directive, has largely been excluded, the Directive nevertheless contains at least some aspects of an integrated control approach (interview with EU Commission, DG XI, March 1993).

On emission monitoring, the Directive lays down very detailed quality standards on measurement technology and procedure. The introduction of a so-called confidence interval, i.e. a limitation on measurement imprecision, is to ensure that measurements within the EU are comparable. 'We thus take account of measurement imprecision, but at the same time we restrict this confidence interval' (interview with EU Commission, DG XI, March 1993). This is intended to avoid competitive disadvantages due to divergent measurement accuracy from country to country. A particular problem in this context is the qualification of measurement facilities for monitoring very low limit values, as for dioxins or furans. However, the Commission assumes that the European standardization institute CEN will be able to elaborate an appropriate measurement method by the time the Directive comes into force (in late 1996). This will then automatically become part of the Directive, so that uniform methods will be ensured throughout the Community (interview with EU Commission, DG XI, September 1993).

As far as public information is concerned, the Directive is guided by existing national and Community law. In contrast to the IPC Directive with its active disclosure obligation, only a passive Community obligation is provided, deriving from the Directive on the freedom of access to information of the environment. The Directive refers to the publication procedures laid down by member states. This applies with regard to both public inspection of permit applications and permits issued by the competent authorities and to public access to emission data.

The Decision-Making Process

The decision-making process on the Directive on the incineration of hazardous waste, unlike other directives, was characterized by conflict between the EP and the Council of Ministers on the choice of legal basis. Despite this discussion, which considerable delayed the process, consultations in the Commission and negotiations in the Council were rapid.

Problem solving in the Commission was strongly influenced by the Germans. By taking an active role, the Federal Republic succeeded in imposing its emission and technology-based approach as the basis for consultations at the European level.[116] Two factors favoured German influence on the Commission draft. First, the Federal Republic as 'first mover' had corresponding domestic arrangements that could serve as a model for European legislation.[117] Second, the need for an emission-based procedure because of the special health risks arising from the burning of dangerous wastes was not called in question by the other countries. Even Britain, which despite certain changes in the EPA 1990, still relied more heavily on quality-related measures, advocated such an approach. 'In this case an approach from the impact side was never under discussion. ... Everyone agreed that the problem had to be tackled at source' (interview with EU Commission, DG XI, March 1993).

Apart from its interest in minimizing the potential cost of legal and institutional adjustment, Germany, in taking an active role, hoped to secure the economic competitiveness of German industry. Competitive disadvantages were on the horizon, since the low landfill capacities in Germany had made waste incineration measures inevitable. The relatively strict limit values defined under the 1990 17th Regulation Implementing the Federal Pollution Control Act imposed heavier costs on German industry than those its foreign competitors had to bear with their less stringent national standards, who frequently still disposed of sufficient landfill capacities. Another German motive was to promote the marketing interests of the national environmental protection equipment industry, which is a global market leader in this area (interview with EU Commission, DG XI, March 1993).[118]

However, Germany's assumption of the first mover role did not by any means produce identity between the German rules and the European draft. The technological requirements contained in the Commission proposal went far beyond the German provisions. These standards had been quite deliberately made so demanding by the Commission. It was the Commission's assumption that further advances in control technologies would be forthcoming in the nineties, which would permit greater reductions than were possible at the time. The limits were set in such a way that they could be complied with only by using future techniques and techniques still in the development stage. 'We're making a Directive that only has to be transposed into national law in mid-1995. So we have to set emission limits for the nineties based on technologies of the nineties' (interview with EU Commission, DG XI, March 1993).

In response to an EP opinion, the Commission also planned stricter controls on dioxin and furan emissions. Because the measurement methods were not yet determined, it was initially not intended to set limits for these substances. Instead, the Commission had proposed a guide value of $0,1$ ng/m^3, to be complied with using the most advanced technologies (interview with EU Commission, DG XI, March 1993). In its opinion on the draft, however, the EP had urged adoption of a limit value, because the location of waste incineration plants and the related risks to human health in the member states were extremely sensitive political issues (EC-Commission 1993, 3). The Commission thereupon modified its proposal and determined that the guide value was to become binding as a limit value as soon as CEN was able to come up with a suitable measurement concept (interview with EU Commission, DG XI, March 1993).[119]

Consultations in the Commission led to rapid agreement on an emission-based procedure, but discussions in the Council were marked by compensation considerations and bargaining on the concrete definition of individual limit values. Initially, member states agreed only that the emission limits proposed by the Commission were too stringent (interview with EU Commission, DG XI, March 1993). However, there were conspicuous differences of opinion only on the extent to which the standards should be eased.

Germany and the Netherlands wanted the limit values to be based not on future but on current technical possibilities. This would have brought them closer to the German and Dutch standards. For cost reasons, German industry was particularly opposed to technological standards stricter than those currently in place in Germany (interview with EU Commission, DG XI, September 1993).

France also advocated less stringent limits to be defined in terms of currently available technology. The main reason was the restructuring of waste incineration being undertaken in France at the time.[120] In the course of this process, many smaller, technically obsolete plants were to be shut down. At the same time France was planning the comprehensive technological retrofitting and expansion of a few remaining plants. (interview with *conseil régional*, March 1994; interview with *Ministère de l'Environnement*, March 1994). Under the impression of the risk that technologically stricter requirements would present of having to repeat the retrofitting of plants that were in the process of being extended, the French pushed for limit values in keeping with existing technology. They accordingly supported recourse to German standards, compliance with which they regarded as unproblematic. 'These limit values ... seem to us to be a good compromise, and our industrialists were willing to keep to them' (interview with *Ministère de l'Environnement*, March 1994).

Such moderate demands were too little for the British, who wanted to see the limit values eased to a far greater degree. The central elements in British problem-solving attitudes found clear expression. Despite certain changes under the EPA 1990, criteria like established scientific evidence, economic proportionality, and local environmental quality continue to play an important part in British environmental policy. The British had corresponding scientific and economic doubts about the 0,1 ng/m^3 standard for dioxins and furans. They argued that according to current scientific knowledge about the protection of human health, a limit value of 1 ng/m^3 would suffice, and that complying with a stricter standard would entail disproportionately high costs.[121] Britain regarded the stringent limit values as placing excessive emphasis on technological development, which would benefit the German environmental technology industry most of all: 'There was a strong drive for German technology which is something very difficult for the British ... It seems very odd to set a standard that cannot be achieved in order to drive technology to achieve it' (interview with DoE, September 1993).

In spite of these reservations, British opposition was far more subdued than it had been some years previously in discussion of the Large Combustion Plant Directive. One important reason was that the British faced little pressure for legal and institution adjustment. With the EPA 1990, Britain has created all the preconditions for 'absorbing' future European measures at the national level without difficulty. As we have seen, the British also recognized that, despite their partiality for quality-related measures, an emission-based approach was rendered necessary in this case by the particularly problematic nature of the issue[122] (interview with EU Commission, DG XI, March 1993).

Thanks not least of all to Britain's greater willingness to compromise, agreement was reached in the Council in July 1993 (a year after the official Commission proposal). Member states agreed to take current technological potential as the yardstick for setting limit values rather than future technology. In effect, this meant taking over the limit values set under the German implementing Regulation, which were more or less half way between the Commission proposal and the British demands. In view of the stringent Commission standards, less severe limits would scarcely have been politically practicable without risking the Commission completely withdrawing its proposal (interview with Tübingen *Regierungspräsidium*, March 1994; EU Commission DG XI, March 1993). Although a solution had thus been found in the interests of Germany and France, rapid technological advances in the field are also reducing implementation costs for Britain. When the Directive is due for implementation, current technological possibilities will, environmental technologists believe, be largely obsolete.[123] The conventional technologies that have to be used in Britain to comply with European standards are correspondingly cheaper (interview with Tübingen *Regierungspräsidium*, June 1995).

Despite the relative rapidity with which agreement was reached in the Council of Ministers, formal adoption of the Directive took some time. The delay was caused by a change in the legal basis for the Directive decided by the Council and the conflict this triggered with the EP. Two different legal bases are possible for European environmental measures, namely Article 100a, the main purpose of which is economic harmonization in relation to the common market, or Article 130s, which focuses on the realization of environmental policy objectives. Following a ECJ ruling, the provision is to be chosen that corresponds to the objective of the measure. The choice of legal basis has important implications for the decision-making process. Article 100a provides for participation by the EP under the so-called cooperation procedure. The Council of Ministers acts by a qualified majority. Under Article 130s, which prescribes unanimity, the EP is only to be consulted.

The Commission originally based its proposal on Article 100a. Since the Council of Ministers felt that the main objective of the Directive was not realization of the internal market but improvement of environmental protection, it decided in July 1993 that Article 130s was the more suitable legal basis. This would have deprived Parliament of any opportunity to influence the substance of the Directive, since it had already been consulted on the official Commission proposal. The EP needed only to be consulted now on the change in legal basis. The European Parliament, which saw itself losing the possibility of a second reading under the cooperation procedure, used the institutional changes provided by the Treaty of Maastricht to force a second reading. In the March 1994 hearing, intended to deal only with the legal basis issue, the EP once again expressed its opinion on the substantive aspects of the Directive. It referred to the Maastricht Treaty, which had by then come into force, which in relation to Article 130s now provided for the European Parliament to participate under the cooperation procedure (interviews with EU Commission, DG XI, July 1994; Tübingen *Regierungspräsidium*, June 1995). In the second reading thus imposed, the EP rejected the limit values under the German implementing Regulation as inadequate. It demanded that the more stringent standards the Commission had originally proposed be reconsidered (interview with *Ministère de l'Environnement*, March 1994; interview with Tübingen *Regierungspräsidium*, March 1994). However, the procedural delays did not affect the content of the Directive, since the Council of Ministers rejected the EP demands. The Directive was thus adopted with more than a year's delay in December 1994, although the Council of Ministers had reached agreement as long ago as July 1993.

Conclusion

The Directive on the incineration of hazardous waste shows particularly clearly that elements of German-influenced regulatory strategies continue to make their mark in European clean-air policy despite the shift in Commission strategy. The exchange of roles in setting the pace and braking developments at the European level that occurred between Britain and Germany in the early nineties and that is to be noted in many fields, has not been unconditionally valid in all areas of supranational legislation. One important factor — which the British acknowledge — is that toxic substances like dioxins can be effectively controlled only if tackled at source. It is practically impossible to control dioxin with a quality-based approach, since the environment and human health suffer considerable injury from the smallest dose. This argument and the relatively low costs of legal and institutional adjustment for national practice were instrumental in persuading the British to accept the measure. But it also showed that they continue to have certain misgivings when confronted by technology-based regulatory strategies that take little account of economic proportionality and scientific causality.

The decisive influence on the Directive was Germany, which by taking the initiative — for mainly economic reasons — succeeded in imposing many aspects of its national control approach at the European level. Stringent limit values are in the market interests of the German environmental protection equipment industry and balance out locational disadvantages caused in other industries by the strict German rules under the 17th Regulation Implementing the Federal Pollution Control Act, or which threatened to occur because of the lack of landfill capacity. France's position, determined by national restructuring of the waste incineration system, was less neutral than usual. The French also advocated adoption of the German standards, which reflected the current state of the art in technology. Since the French are busy retrofitting many incineration plants, an orientation towards future technological developments could possibly have given rise to substantial follow-up costs if new technical adjustments were to become necessary within a brief space of time to comply with European standards.

The decision-making process showed, moreover, the increasing influence of the European Parliament on European legislation. With its opinion, the EP delayed the coming into force of the Directive and pushed through a second reading to which it would have been entitled under Article 100a of the Maastricht Treaty, the original legal basis of the Directive. However, this action on the part of the EP was possible only in exceptional circumstances: the coming into effect of institutional changes under the Treaty of Maastricht in the course of the decision-making procedure.

Volatile Organic Compounds: New Markets for German Environmental Technology

Since 1991, the Commission has been working with the active participation of the Federal Republic on a strategy to reduce volatile organic compounds (VOCs) in the atmosphere. VOCs are hydrocarbons that contribute substantially by photochemical reaction to the formation of ozone, with its deleterious effects on climate and human health. Traffic and transport is responsible for 55 per cent of VOC emissions. About one third of total emissions of these hydrocarbons is caused by the use of solvents in certain industries and plants. A further 14 per cent of releases are from oil refineries (European Communities 1993, 1; Olier et al. 1989, 123).

A directive intended to reduce VOC emissions from stationary sources has been under discussion in the Commission since 1993.[124] Despite intensive negotiations, consultations among experts have hitherto failed to produce a viable compromise. The complexity of the subject matter, which involves numerous regulated industrial sectors and strongly differentiated geographical conditions, makes consistent negotiation and agenda-setting by the Commission difficult, and the sheer number of industries concerned, differing in level of development and importance from country to country and region to region precipitates conflicts between interests that are very hard to resolve. The general outcome of this constellation is that distributional aspects, which tend to be handled at Council level, have exerted an early influence on problem solving at the expert-consultation level. Tackling substantive and distributional issues simultaneously has impaired elaboration of a Commission draft capable of finding consensus, and has produced 'cafeteria' policies on the principle of 'a bit of everything'.

First Thoughts from the Commission
In 1993 the Commission produced a first draft Directive on reducing VOC emissions from industrial plants and processes. It was based on objectives formulated in the Fifth Action Programme on protection of the climate and the Earth's atmosphere (European Communities 1993, 1), and was a response to EU commitments under international treaties. The Directive was designed to put the Community in a position to fulfil the protocol on the reduction of VOC signed in the context of the Geneva Convention on Long Range Transboundary Air Pollution. This provided for a 30 per cent reduction in VOC releases between 1988 and 1999 (ENDS 1991/193, 39; Vernier 1992, 37).[125] Besides combating ozone formation, the Directive aims to protect human health against harmful volatile organic compounds (interview with EU Commission, DG XI, July 1994).

The list of industrial plants and processes covers over 20 different industrial sectors. Particularly affected are printing, varnishing, surface cleaning, impregnation, and the processing and treatment of rubber, leather and textiles. The Commission proposal, strongly influenced by the German regulatory philosophy, provides for industrial plants in the listed sectors to operate only with the authorization of the competent national authority. An operating permit is to be granted under the draft only if the prescribed EU-wide emission limits are not exceeded. The various volatile organic compounds are grouped in terms of dangerousness, with uniform standards for each group.

The limit values are based on the BAT principle[126] to be defined as in the IPC Directive (European Communities 1993, 4; interview with EU Commission, DG XI, July 1994). The Commission is convinced that such a technology-oriented approach is compatible with the subsidiarity principle, since the global nature of the issue gives a local quality strategy little prospect of success. The Commission also points out that in many cases sufficient margin for action will remain in choosing between several equivalent technologies in terms of costs. 'In many cases the Directive provides several options for meeting the emission reduction target. The practical implementation, including the identification of the most cost-effective measures, is therefore in the hands of member states and operators of the installations concerned' (European Communities 1993, 6). In order to give member states the possibility of taking costs and benefits into account, the draft lays down limit values in some areas that are below the current state of the art in technological development. As a result, more stringent standards could apply in certain countries than those envisaged by the Commission (interview with EU Commission, DG XI, July 1994). The Commission justified their course of action on the grounds that compliance with the reduction targets agreed in Geneva was not adversely affected, and that Article 130s, the legal

basis for the Directive, also allowed every member state to take more stringent measures.

As indicated by the analogous definition of BAT, the Commission intends
the VOC and IPC measures to be complementary. Whilst the IPC Directive
regulates emissions from large industrial plants, the VOC proposal targets
smaller and medium-sized enterprises the emissions of which are too low to
fall under IPC rules (interview with EU Commission, DG XI, July 1994).
The VOC draft thus contains some elements of an integrated approach. In
choosing the best available techniques, account is to be taken not only of the
release of harmful substances to the atmosphere but also of the pollution of
water and soil. The choice of suitable techniques is to include consideration
of both medium-specific interactions and economic proportionality. Indicative of an integrated approach is also the obligation of industry to prepare a
so-called 'solvent management plan (SMP)'[127] for each new plant. The SMP
identifies the individual substance and records the respective emission volumes. On this basis a programme is to be prepared for the progressive reduction of VOC emissions, taking account of technical possibilities and economic proportionality (*CITEPA* 1993/107, I.76). The SMP requirements
stress the salience of the integrated approach. 'The solvent management plan
looks at all media to minimize cross-media pollution and is in line with the
so-called "integrated pollution philosophy" ' (European Commission 1993,
28).

The integrated philosophy underlying both measures means that the VOC
Directive 'is tied to the fate of the IPC Directive in all matters still under
discussion' (interview with EU Commission, DG XI, July 1994). A central
point at issue still to be settled for both measures is whether EU-wide harmonization of control requirements should in principle be sought or whether
local environmental quality should not be preferred as the definitive regulatory criterion (interview with *BDI*, October 1994).

Member states are to be given wide margin for action in implementing
the Directive. It contains relatively broad specifications of measurement
methods that can be used by member states in monitoring VOC
emissions.[128] However, the SMPs to be submitted to the Commission by
industry give the Commission greater facility in controlling implementation
of the Directive (interview with EU Commission, DG XI, July 1994).

*The Course of the Decision-Making Process: 'Accommodating Multiple
Interests'*

Decision-making on the VOC Directive has been dominated at the expert
consultations level in the Commission by national special interests and distributional issues. Although the EU has been considering a VOC directive
since 1991, the Commission has hitherto failed to come up with a viable

draft on which the Council of Ministers could negotiate. Several mutually reinforcing factors have been responsible for this early politicization of the otherwise 'denationalized' processing of problems in this phase: the complexity of the subject matter, the way in which the Commission conducted negotiations, and diverging interests among member states.

The basic problem with the VOC Directive is its complexity. It deals with a large number of industrial sectors, differing in importance and level of development both regionally and from country to country (interview with EU Commission, DG XI, July 1994).[129] The industries concerned are often small and medium-sized enterprises. Because they are economically less robust, it is more necessary to take account of the particular conditions under which such industries operate by providing detailed rules on exemptions.[130]

The Directive has one problem: it deals with an extremely wide range of plants, especially small and medium-sized plants. It tries to go into great detail. It's always a problem doing things like that in Europe because it can suit one country but 'grate' in the others. Or it's a compromise for everyone, and then it 'grates' a bit everywhere (interview with *BDI*, October 1994; interview with *Ministère de l'Environnement*, January 1995).

The complex nature of the subject matter has had a direct impact on the Commission's mode of negotiation and has given rise to strongly diverging constellations of interests not only between member states but also within individual countries. The complexity makes negotiation more difficult for the Commission, which has no clear idea of the sectors to be regulated and is thus unable to ensure consistent problem solving through agenda-setting in expert consultations. The subject matter has expanded continually and become correspondingly more complicated. The focus of discussion shifts from meeting to meeting without sustainable agreement having been reached on the problems dealt with at the preceding session

There's certainly a problem with the Commission's initial strategy on the subject, because at the start we had a number of sectors that was relatively limited, and the number of sectors has grown bit by bit. And what's also surprising is that between Commission meetings there's often been a shift to the other pole. That's to say in March 93 or in June or April 93 there was a meeting on the subject. We had another session a year later. There was yet another in July 94. And every time the basis for the proposed Directive was radically different because from one proposal to the next there had been numerous supplementary annexes, first, and even on fundamentals[131] (interview with *Ministère de l'Environnement*, January 1995).

In view of this point of departure, it is hardly surprising that there has been no breakthrough in consultations at the Commission level. As with the IPC Directive, the main conflict between member states has been about 'subsidiarity versus harmonization' and 'technology-based regime versus quality-oriented regime'. Germany has favoured the technological approach; Britain is for an environmental quality one. France has been steering down the middle, sometimes keeping closer to the Germans, sometimes to the British.

Whilst Commission consultations on the IPC Directive have been substantially influenced by Britain in the role of pace-setter, the VOC measure has seen Germany extremely active in an attempt to swing the Commission's perception of the issue in the direction of their technology-centred approach. Questions of competitiveness have been decisive to the German initiative. German industry has an interest in severe limit values for VOC emissions being made binding on its European competitors since it is already subject to high standards under German law. Industry in the Federal Republic is also keen to open up new markets for German environmental technology (interview with *BMU*, July 1993). German sales would benefit if limit values were made so strict at the European level that compliance would be possible only with the latest technology. 'The Germans ... , and it's doubtless the case with VOC, make a tough policy in saying: "we'll develop *avant-garde* techniques and foreign countries will come to us and buy our licences" ' (interview with *CITEPA*, January 1995). This German interest is clearly apparent, for example, in connection with dry cleaning, where 'the only machines that comply with the Directive' are made in Germany (interview with *Ministère de l'Environnement*, January 1995). In the French view, although no confirmation was forthcoming from the Commission (interviews with EU Commission, DG XI, September 1993; July 1994), there is a '*liaison historique*' between the Commission and a German engineering firm that is commissioned with preparing the technical specifications (interview with *CITEPA*, March 1993). 'It's very tempting to say impose the same limits everywhere, then I could sell machines' (interview with *CNPF*, June 1993). The French claim this makes it comparatively easy for the Germans to introduce their strict standards at the European level.

There are ten sectors of different activities in this Directive of which six had been the subject of preliminary studies to select the best available techniques not entailing excessive cost. In the four remaining sectors, which have not been the subject of a study, the rules have been inspired by German rules[132] (interview with *CITEPA*, March 1993).

The BDI, by contrast, attributes the strong influence of German arrangements to the lack of cooperation on the part of industrial associations in other member states. They had scarcely reacted when approached by the German firm and had made little information and expertise available. 'In effect almost nothing came from the rest of European industry, and from the Germans only a middling amount, but not as much as the principal would have wished. ... Anyway, the fact is that almost nothing came from abroad.

The French really have no grounds for complaint' (interview with *BDI*, October 1994).[133]

The Germans proved correspondingly supportive and gratified when 'their' concept was taken over by the Commission. 'All this was music to the ears of the German delegation, which is strongly in favour of applying the best available technologies which are economically feasible' (ENDS 1991/193, 39). The Germans have thus for the moment succeeded in imposing their view of the issue and their emission-based problem-solving approach at Commission level. Nevertheless, they are not wholeheartedly in favour of Commission plans.

German criticism of the Commission proposal is directed principally at the concrete definition of emission limits, at questions of authorization procedure, and at public information. The Germans consider authorization requirements for small plants to be too strict. Industry fears that the cost of the necessary control technology for small plants will be many times the value of the plant itself, thus flying in the face of the proportionality principle. 'Such requirements are impracticable even on a BAT basis' (*BDI* 1994, 7). The *BDI* also complained that under the provisions of the current draft small plants would be subject to formal authorization procedure, whereas a simplified procedure has hitherto applied in Germany. According to the *BDI*, the result would be 'a flood of authorization proceedings' that would bring the administration to a standstill (*BDI* 1994, 2).

Particular exception is taken to the rules on public participation, hitherto applicable in Germany only within formal procedure.

We'd prefer to see different categories; public access would then be provided only in comprehensive proceedings. Simple authorizations ought to be granted without participatory proceedings, and finally a third category of facilities should need no authorization at all. In the present draft we have the demanding participation level for all units. That has to be changed, and we're confident that it can be changed, because although other member states have accepted it they aren't really interested (interview with *VCI*, March 1993).

Although the Commission has made concessions to German reservations by replacing the term 'permit' by 'allowance', the *BDI* sees no progress in the move, because the equivalent German term for 'allowance', *'Erlaubnis'* is synonymous with *Genehmigung* or 'permit' (*BDI* 1994, 2).

Another German objection concerns the relatively weak limits the draft would impose on large plants. In some measure they are more generous than those under the 2nd Regulation Implementing the Federal Pollution Control Act. In order to safeguard the international competitiveness of industry and to create new markets for their advanced environmental technology, the Germans demand 'that large plants requiring authorization be obliged to respect a much more stringent standard, because they discharge much larger volumes of waste gases; it wouldn't hurt to grant smaller plants some relief. That's our philosophy on the matter' (interview with *BMU*, July 1993).

The German attitude taken in expert consultations at Commission level is therefore ambivalent. Although the technology-oriented approach is essentially in keeping with German regulatory philosophy, the realization of this approach as discussed to date has been opposed by the Germans largely for competition-policy reasons.

Britain's attitude can be similarly characterized, albeit from the opposite direction. Because of their environmental quality orientation, the British criticized the planned introduction of EU-wide emission standards. At the same time they supported the Commission's BAT proposals, which left sufficient leeway for cost-benefit considerations. The British wanted to see only those volatile organic compounds reduced that have an extraordinarily high photochemical reaction potential, and which are therefore particularly responsible for the formation of harmful ozone in the atmosphere. The approach revealed central elements of the traditional British understanding of the issue. On the basis of well-established scientific evidence, the British were attempting to launch a regulative programme oriented on environmental quality, which, not least of all for economic reasons, covered only those substances that caused especially deleterious environmental impacts. Against this background it is not surprising that Britain had already initiated a comprehensive research programme to determine the reaction potential of various substances (ENDS 1991/193, 39). However, since the British have created the preconditions for an emission-based approach with the EPA 1990,[134] they face no special legal and institutional adjustment costs, regardless of what form the Directive finally takes. The cost implications of the Directive for British industry are also relatively slight. This was favoured by the fact that the Commission has taken as a basis for concretizing technical operational requirements both the British guidance notes and the German *TA Luft* (interview with EU Commission, DG XI, September 1993).

In consultations at Commission level, the French position is between those of the British and the Germans. As in discussion on the IPC, the French have sometimes been closer to the Germans and sometimes closer to the British. Although they support German demands for EU-wide emission standards when this is opportune for competition-policy reasons (interview with *Ministère de l'Environnement*, January 1995), when it comes to the BAT criteria, they tend to agree with the British in wanting to include environmental quality considerations. They also feel that limit values ought always to be defined in such a way that a choice is possible between different but equivalent control technologies (interviews with *Ministère de l'Environnement*, January 1995; *CNPF*, January 1995). In this connection the French refer to the subsidiarity principle, which is undermined in their opinion if the stringency of standards allows no choice of technology. 'As far as technology is concerned, from the moment you impose a certain concentration for emissions it's like also imposing the way of tackling it, the means to be used. It's the same thing'[135] (interview with *CITEPA*, March 1993). The French government consequently supports the current Commission draft, under which the prescribed standards give adequate scope for choosing the technological means to realize them. This is in the interests of the French chemical industry, which has profound misgivings about the potential cost of the VOC Directive (interview with *Ministère de l'Environnement*, January 1995). This approach would also permit the French to retain the 'soft' regulatory tools they have hitherto used in this area, which would reduce the cost of legal and institutional adjustment. The French are seeking to reduce pollutant emissions primarily on the basis of voluntary agreements at private law concluded with individual industries, an idea that is at present being discussed in the Commission as a general environmental policy strategy.[136] In these so-called *'engagements de progrès'* or 'commitments to progress' individual industries undertake, for example, to use only products with a low content of organic solvents, or to renounce their use altogether (*IUAPPA* 1991, 145; interview with *CITEPA*, March 1993). Another regulative tool is the 'parafiscal charge'. Emittents are required to pay a charge in proportion to the volume of pollutants released to the environment (*Ministère de l'Environnement* 1991a, 7).[137] Although in this connection the French agree with the Germans about standards being too high for small plants (interviews with *CITEPA*, March 1993; January 1995), they disagree strongly with their demand for more stringent emission limits for larger plants.

The differing positions taken by the countries under review in expert consultations reveal that the concept at issue has 'a bit of everything', in other words, every country has managed to integrate some of its priority concerns in the Commission proposal. The emission and technology-based approach is in line with German regulatory philosophy. British and French interests are reflected in the BAT requirements, which — at least for large plants — leave wide scope for cost-benefit considerations and for choice of technology. The 'classical' member state positions are consequently evident only in tempered form. No country has been fully opposed to regulation of VOC emissions nor has any country been fully satisfied with the draft.

The three factors mentioned above explain why, notwithstanding this apparently favourable negotiatory situation, no decisive breakthrough has yet been achieved in consultations. The first factor, presenting special conditions under which interests are to be reconciled, is the complexity of the subject matter. Regional variation in conditions and national variation in the importance of industrial sectors give each country its own 'problem children' that it watches over with particular solicitude.[138] This heightens the need for detailed arrangements taking especial care of small and medium-sized enterprises. Second, the complex subject matter compounds the difficulty of conducting consistent and purposive negotiations in the Commission. Third, the zero-sum nature of interest distribution diminishes the prospects for consensus; in other words, the benefit of the one is the cost of the other. Thus German industry's stringent BAT rules are to the detriment of its foreign competitors. Less stringent requirements would have the opposite effect.

It is therefore not surprising that some countries, especially Britain, Denmark, and the Netherlands, have advocated an escape clause (interview with DoE, November 1994). On the basis of the subsidiarity principle, the clause would permit any country to derogate from implementation of the Directive if it is able to evidence a programme of its own bringing the same reduction in VOC emissions over the same period, each country having the freedom to determine the ways and means to this end.[139] This proposal, which the Commission incorporated in its new, unofficial draft, has been sharply criticized by German industry, which sees its competitive position at risk, and asks, not without justification, 'why are we making the Directive at all if each of us is going to do it his own way?' (interview with *BDI*, October 1994). It could be said in response that the Directive will at least oblige countries that have so far done nothing to reduce VOC emissions to take some form of action (interview with DoE, November 1994).

Conclusion

Consultations at Commission level on the VOC Directive have shown the gradual changes occurring in the Commission's fundamental view of the issue. Whilst Germany, by taking the initiative, at first succeeded in bringing the Commission's perception of and approach to the problem into line with the German technology-based philosophy, this initial perspective shifted in the course of expert consultations. Other countries, especially Britain and France, which advocated taking greater account of cost-benefit aspects and who argued for quality-related regulatory concepts, managed to introduce certain of these elements, thus modifying the 'German' stance and strategy of the Commission. The resulting Commission proposal contained a bit of everything, but because of the industry-specific zero-sum constellation of interests and the complexity of the subject matter, it has yet to find consensus among member states. This situation made it easy for the Danes and Dutch to add a further element to the Commission proposal ('frame'). They have suggested that agreement could be reached by leaving it to each member state to decide whether to implement the Directive or to introduce a programme of its own to reduce VOC emissions in the same measure and in accordance with European objectives. Whereas it had been possible so far to interpret the Commission's espousal of an emission-oriented procedure as giving the Federal Republic the advantage of the first move, this seems to have been at least neutralized by the Danish/Dutch initiative. Obviously, the difficulties in the uniform regulation of regional diversity suggest an additive rationale (a 'cafeteria' approach) relating to the subsidiarity principle, even if the pervasiveness of the problem justifies Community emission regulation.

3.5   The Commission as Initiator: New Strategies

If we consider the EU clean-air measures we have examined so far, the impression is that in elaborating and initiating directives or regulations the Commission draws mainly on proposals advanced by individual member states or acts in response to such. This policy-making pattern nonetheless admits exceptions. It does occur, albeit not all too often, that the Commission makes 'the first move'. For it cannot always rely on appropriate national regulatory proposals. This can be the case, for example, when obligations that the EU has assumed under international treaties render new regulation necessary. Furthermore, national regulative arrangements transposable to the supranational level are not always to hand for activities with which the Commission seeks to expand and supplement its general range of environmental policy instruments.

It is interesting to note that, in such cases, the Commission does not rely on its own problem-solving capacity alone, but takes up concepts it considers suitable and useful that have been developed in countries outside Europe, especially in the United States. One example is the 1985 Environmental Impact Assessment Directive. An American model has also been under consideration in recent deliberations on emission registers and voluntary agreements with industry on emission abatement intended to strengthen 'autonomy-friendly' forms of control taking due account of the Community dimension. The concept of a European climate tax, however, an extremely controversial and politically highly salient issue from the outset, originated with the Commission.

Environmental Impact Assessment: European Policy on the American Model

In 1985 the Council of Ministers adopted a Directive on Environmental Impact Assessment (EIA) for certain public and private projects.[140] The Commission had relied heavily on the corresponding American arrangements.[141] In 1969 the United States adopted the first legislation in the world providing for the precautionary, cross-medium recording and monitoring of environmental impacts (Weber 1989, 5).[142]

Objective and Content of the Directive
The Directive provides for environmental impact assessment of projects that might have a substantial impact on the environment (Art. 1). The effects of projects on human beings, fauna and flora, soil, water, air, climate, and landscape, and on material and cultural assets are to be investigated across all environmental media. Before constructing a plant or undertaking any other major intervention in Nature or landscape, the entrepreneur is required to provide the licensing authority with information on the site, type, and extent of the project, and on measures for avoiding, abating, or compensating environmental impacts. Furthermore, he must supply the data needed to investigate and evaluate the effects of the project on the environment. Both the information supplied by the project sponsor and the permit application are to be made available for public inspection. Before the project is realized, affected sections of the public are also to be given the opportunity to express their views on it and, once the authority has reached a decision, to be informed on its content and grounds. In implementing the EIA Directive, member states are given discretionary scope in various areas to facilitate its application and integration in national law. This discretion, for example in the choice of projects subject to EIA and in modes of procedure,[143] results in EIA practices varying extremely in level of aspiration[144] (Spindler 1991a, 13; Coenen/Jörissen 1989).

The Decision-Making Process
The Commission initiative to introduce EIA at the European level goes back to the early seventies. Although the Commission had started thinking about the issue in 1974,[145] an official proposal for a directive was forthcoming only in 1980. The reason for this long preparatory phase was that EIA represented a completely new instrument, which was based on a philosophy quite contrary to existing environmental regulation in most member states. The question of how compatible the directive was with different national legal systems was the subject of heated debate.

Despite the long run-up phase in the Commission, negotiations in the Council of Ministers were also extremely protracted. Although there was agreement in most member states that EIA was a useful concept, the Commission proposal initially attracted little support in the Council. Particularly controversial aspects were the type of project to be assessed, assessment procedure, public participation, and the participation of foreign neighbours in internal national administrative and court proceedings under environmental law. After more than five years of negotiations and more than forty meetings of the Council working group on environmental affairs and nine Council meetings, agreement was finally reached by means of appropriate compensation deals (Strübel 1992, 143f.; Cupei 1986). In order to lower member states' legal and institutional adjustment costs, the Commission had markedly broadened margins for action in implementation at the national level.

It is interesting to note that Britain was among the opponents, although the procedural character of the Directive was largely in keeping with British regulatory tradition. Moreover, the EIA instrument was nothing new to Britain, having been in use there since the seventies, albeit on a voluntary and legally non-binding basis.[146] Nevertheless the British faced substantial legal and institutional adjustment costs. They feared a growing shift in political and administrative decision-making powers in favour of the courts. First experience in the United States had shown that the concrete carrying out of EIA had produced a large number of court cases. 'The fears of the government ... were in part that opponents to a development would be provided with the opportunity to seize on some procedural failure as a ground for challenging planning decisions in the courts' (Haigh 1990, 353). Moreover, this conflictual type of practice was fundamentally contrary to British regulatory style, which put great store by consensus and informality. It was also seen as a problem that the proposed Directive would have required a number of fundamental adjustments in legal arrangements. For instance, the number of projects for which environmental impact assessment was to be mandatory went far beyond normal British practice. And the project sponsor's obligation to supply information was not nearly as comprehensive in Britain as in the Commission proposal (ibid.). What is more, the British feared higher planning and administrative costs (Rehbinder 1991, 116).

Britain changed its attitude only when the Commission submitted a modified draft giving greater room for manoeuvre in implementation at the national level and thus reducing the extent of adjustment the British would have to make. The list of projects subject to environmental impact assessment was substantially cut. Member states were also given the right to exempt certain projects without explicit Commission approval. The type and extent of information to be supplied was now to be determined by the member state concerned and no longer at Community level. The thus amended draft Directive 'now [accorded] very closely with existing British development control procedures' (Haigh 1990, 355). British concurrence was also encouraged by a report from the House of Lords Scrutiny Committee, 'which ... played a key role in persuading the government to withdraw its opposition' (ibid., 354). It had supported the Commission proposal from the outset and urged a change in the British attitude. With its vote, the decision-making context in Whitehall altered. 'Those within the government who supported the Directive had their hands greatly strengthened' (ibid.).

At first glance, the lack of opposition on the part of Germany appears quite as surprising as British rejection of the proposal, if one considers the strong contradiction between the procedural nature of EIA and technology-based German regulative law. The fact that there were no comparable arrangements at the time in Germany,[147] might have led one to expect greater resistance to the Directive. But German reserve was by no means indicative of support. It was simply that no-one in Germany had expected the Directive ever to be adopted after the never-ended negotiations. 'Nobody really expected it to come. EIA wasn't taken seriously. There were only one or two voices in the wilderness who said, "For God's sake what've you gone and done?" ' (interview with *DIHT*, March 1993). Another factor in the lack of German opposition may well have been the compensations granted member states by the Commission with respect to national implementation on the Directive.

It was not until the Directive was due for implementation at the national level — at much too late a stage for revision of the Council decision — that the industrial associations protested. Industry regarded the EIA Directive as a superfluous burden that would lead to high control costs and thus to competitive disadvantages (Hübler 1989, 95; *DIHT* 1988, 6). No special EIA arrangements were necessary, it was claimed, since 'the efficient system of German environmental law already contained the provisions envisaged in the Directive' (*DIHT* 1988, 1; interview with steel producer, February 1992). Another objection raised by industry was the lack of evaluation criteria for the adequate implementation of EIA, possibly resulting in 'unreliable procedures' (*DIHT* 1988, 5). It was also claimed that public participation would further intensify these imponderabilities, radically modifying the 'prospect of having a right to authorization where certain conditions are met' (ibid.).

France, true to itself, was neither an enthusiastic proponent nor a bitter opponent of the EIA Directive. The French took an overall favourable stance in the decision-making process, helped by lower legal and institutional costs than those faced by the British and Germans. In the 1976 Nature Conservation Act, the French had established the legal basis for carrying out EIAs. Although the EIA Directive differ in some aspects from French legislation, the wide range of regulatory instruments available in France meant that in this case, too, the French could support the measure without much hesitancy (IPEE 1992, 194).

Implementing the Directive at the National Level
In France and — with some reservations — in Britain, it was possible to implement the EIA Directive at little legal and institution cost. Environmental impact assessments had been carried out in both countries before adoption of the EU Directive. The EU assessment provisions could thus be largely integrated into existing procedures without alterations to public authority powers or comprehensive modifications to procedural law being needed. Whereas the broad margin for action in implementation under the Directive obviated any legal adjustments at all in France, Britain had to expand certain provisions of the 'Town and Country Planning Act', because the EU rules went beyond British arrangements, especially on the information to be provided and the environmental effects to be considered (Haigh 1990, 356). However, by taking recourse to existing arrangements it proved possible 'to minimize the additional cost to project sponsor and public authorities' (Coenen/Jörissen 1989, 141).[148]

Implementation was much more problematic in Germany. As with the Environmental Information Directive or the Eco-Audit Regulation, existing regulatory practice faced comprehensive legal and institutional adjustments. The pressure to adjust in the case of the EIA Directive was caused particularly by the cross-media perspective, which contradicted the medium-specific approach taken in Germany. Administrative processing of a project is medium-specific both vertically and horizontally, and consequently uncoordinated in either legal terms or in practical performance (Pfeiffer 1991, 57). Seen from this perspective, EIA introduces 'something revolutionary into the Federal German legal system' (interview with EIA, October 1993). 'The administrative authorities with their expert staff and departmentalized division of functions and labour are mostly not adapted' to procedures appropriate to EIA (ibid.). Another problem is that procedural rules (as in the EIA directive) play a minor role in German practice with its emphasis on technical standards. Especially the provisions of public participation are difficult to reconcile with the restrictive German tradition.

These differences make it hardly surprising that implementation of the EIA Directive was accompanied in Germany by much 'gnashing of teeth'. Implementation was neither on time nor — at least in the view of environmental organizations — did it do justice to the central objectives of EIA in the sense of assuming an integrated and open outlook on environmental issues. Instead, the process was characterized by efforts to minimize the effects of the Directive on national permitting practice.

Only two years after expiry of the deadline set by the Directive were steps taken in Germany to transpose it. In August 1990 the Environmental Impact Assessment Act (*UVPG*)[149] was adopted. However, this did not mean that industrial plants subject to authorization needed to carry out an

EIA. A further legislative modification was necessary, which came only in 1992 with the amendment of the Regulation on the implementation of authorization procedure under the Federal Pollution Control Act. Since this amendment, EIA has been an integral part of authorization procedure under the law relating to pollution control. There are four new elements in procedure: definition of the frame of assessment, public participation, and the overall description and evaluation of environmental effects (Spindler 1992, 53). Although the European Directive has thus been formally transposed into national law, the administrative directives needed to interpret the *UVPG* are still lacking, so that the implementing authorities and project planners have no guidance on the concrete application of the Act.[150]

Since the Federal Government had failed to transpose the EIA Directive into federal German law by the time-limit, environmental associations claimed that the Directive had direct effect in Germany, and called for the carrying out of an EIA in numerous proceedings. Whilst Federal Environment Minister Töpfer also accepted the direct applicability of the Directive,[151] state authorities refused in many cases to insist on an EIA. Various environmental organizations[152] filed a complaint with the Commission (*KGV Rundbrief* 4/91, 29f.), which thereupon brought action before the European Court of Justice against the Federal Republic for infringement of Community law. Furthermore, many projects carried out without EIA now face protracted procedural delays, because the direct effect of Community law was not respected (*Süddeutsche Zeitung*, 9 January 1995).[153] In addition to belated formal implementation, environmental organizations as well as scientific institutes and associations criticize substantive deficiencies: the inadequate account taken of the cross-media perspective in EIA, the dilution of EIA requirements at the implementation level, and the restricted opportunities for public participation.

These organizations asserted that cross-media, interdisciplinary EIA is impossible. Since EIA was defined as an integral part of administrative authority procedure, specialist legislation and the interests embodied in it had more clout than the EIA — 'the primacy of specialist planning' (Spindler 1991b, 4). A cross-media approach was also hampered by the fact that the *UVPG* did not exclude taking recourse to existing technical standards like the *TA Luft*. But since the latter set limit values only for air, the other media were left out of account (Führ 1992, 9).

It was claimed that reduction of EIA to technical standards watered it down in concrete implementation. 'The further down the regulatory hierarchy you go, the less demanding EIA becomes. The *UVPG* is still comparatively progressive. In the Regulation [implementing the Federal Pollution Control Act], EIA has been largely reduced to the *TA Luft*. If the TA

Luft limits are conformed with, environmental computability is assured. ... This means that the Federal Government has swept everything the EIA had been intended to do out the backdoor not so much with the Environmental Impact Assessment Act but with the implementing Regulation. So its direct effect in licensing procedure has been strongly curtailed (interview with eco-institute, August 1993; Schulz 1990, 1).

Public participation in the EIA procedure was also criticized as inadequate, since not 'everyone' but only affected parties — who are theoretically to be ascertained only in the course of assessment — were entitled to take part. Moreover, in the definition of the context within which important preliminary decisions are to be made, public participation was only facultative and not mandatory (Spindler 1991b, 4f.). Since under the *UVPG* the authorizing authority summarizes the overall environmental impacts of a project, but prepares no separate EIA document, the public has no opportunity to express its opinion on the outcome of the assessment (EC Commission 1993, 4a).

Conclusion

The example of the EIA Directive shows that European environmental measures need not in all cases originate in regulatory competition between member states. In this — comparatively rare — case it was the Commission itself that took the initiative in European policy formation on the American model. Decision making at the European level proved extremely arduous in both the problem-solving phase in the Commission and in negotiations in the Council of Ministers. Although some member states like Britain and France had also taken up the American concept, complex procedural matters concerning differences in national regulatory systems hampered progress.

At first glance, the behavioural patterns exhibited by some member states in the Council were surprising. The British were initially against the measure although its procedural character fitted in with their regulatory tradition. The reason was that the Directive threatened to involve certain legal and institutional adjustments. They gave their approval only when the provisions were modified to become largely compatible with British assessment practice. Looked at superficially, German behaviour was also not what our hypotheses would have given us to expect. After all, the Germans were endorsing a directive likely on implementation to have serious consequences for their licensing practice. But their behaviour was by no means indicative of a sudden change of mind. The Germans 'overlooked' the decision. Because negotiations had gone on for so long, neither German industry nor the German authorities had expected ever to be confronted by EIA. The awakening was all the ruder when EIA came up for implementation. Efforts to integrate the Directive as far as possible in the existing regulatory tradition took two forms: first, implementation was delayed and second the cross-media nature of EIA was given inadequate expression.

France was the only member state that could follow the European decision-making process with equanimity. It was the only country to have put a comprehensive EIA concept into effect as long ago as the seventies. It hence faced only low legal and institutional adjustment costs, thanks to the multiplicity of French regulative instruments.

Environmental Registers and Cooperation with Industry: Innovation on a Voluntary Basis

In its quest for instruments that both respect member state autonomy and ensure effective implementation through greater transparency, the Commission has for some time been working on a novel concept envisaged as complementing the eco-audit and integrated environmental protection provisions. The present Commission strategy, aiming to combine quality orientation and public information, is to be expanded to include two new elements, so-called 'pollution emission registers', and voluntary private-law agreements between the Commission and individual industries on emission reduction. The Commission, which had relied heavily on the American 'toxic release inventory' as a model in elaborating its proposal, is planning a two-stage procedure. First, industry is to be required[154] to publish its emission data in pollutant emission registers, and, second, voluntary emission abatement agreements are to be concluded on this basis with industry. 'In the United States one has this aspect of "regulation followed by voluntary agreement", and we're doubtless also going to copy that' (interview with EU Commission, DG XI, July 1994).

The emission registers, a sort of 'negative hit parade' of the biggest polluters, are intended to exploit 'pressure from below' more fully. The annual publication of lists of firm names and locations classified in terms of emission quantities is intended to offer industry, with its concern about its public image, with an incentive to reduce emissions of polluting substances to avoid ending up as 'top of the pops'. 'Then the public can see where the biggest polluters are. ... In the United States the result has been that people topping the list of emittents were concerned not to be up there the following year' (interview with EU Commission, DG XI, September 1993). Vice versa, such registers permit the public and environmental protection organizations to target individual firms via demonstrations and campaigns to induce more environment friendly behaviour. 'The population ... won't put up with it for very long if their industry misbehaves in comparison to other industries in other countries, because these things are made public and the environmental organizations are certain to get onto it. They'll make an effort' (ibid.).

The conclusion of agreements at private law between industrial associations and the Commission or national governments on reducing industrial emissions is the second novel strategic element put forward by the Commission. To date, however, there have only been vague reflections on how such agreements should look. It is still unclear whether the Commission or the member states should to be party to them (interview with EU Commission, DG XI, July 1994). Strong opposition is expected from national governments if the Commission becomes active in this field. They fear loss of sovereignty, since these agreements would replace national legislative measures. However, a procedure of this type — at the sectoral level and differing from the interventionist approach hitherto preferred — would allow the Commission to take account of both subsidiarity and of the excessive regulation with which they are frequently reproached (interview with EU Commission, DG XI, September 1993).

The not particularly complex subject matter, where member states can estimate potential costs and benefits with relative ease and the populist project of ranking the biggest environmental polluters kept problem solving in the Commission out in the open. Discussions were politicized early on, although the sensitive issue of agreements with industry had been largely excluded. Both governments and industry have hitherto reacted negatively to the plans (interview with EU Commission, DG XI, July 1994; interview with DoE, November 1994).

Member states fear adjustment costs. A number of countries have only recently introduced similar publication rules, and foresee having to revamp their arrangements because of the EU Directive. 'Some of the member states ... have elaborated institutions or systems providing for public information on industrial emissions, and fear that their efforts will be supplemented by another type of effort that would be a real duplication in sort ... without adding anything very important' (interview with EU Commission, DG XI, July 1994).

Although the Commission's plans conformed largely with the general British approach, which gives broad scope to industrial self-regulation and public participation, the British reacted negatively for this very reason. They feared that the publication provisions under the 1990 EPA would have to be amended to comply with the EU proposal. The British feel that the proposal is less comprehensive than their own rules. For, whilst the Commission intends only an annual list of total plant emissions to be produced, there is continuous publication of monthly data in Britain in the 'public registers'.

'So we don't have a problem with the principle of information being available, but simply with the way in which the Commission wants to do it' (interview with DoE, November 1994).

France also disposes of a similar instrument. Since 1988, the Environment Ministry has published a list once a year of the ten biggest air and water pollutant emittents. Like the Commission, the French Environment Ministry pursues a strategy of putting industry under public pressure to reduce pollution. 'The Ministry of the Environment thought that informing the public was an important motor for accelerating progress in environmental matters. Involving the public makes a lot of noise' (interview with EU Commission, DG XI, March 1994). The French position on the Commission initiative is, however, less sceptical than that of the British, since they would face less legal and institutional adjustment (interview with EU Commission, DG XI, July 1994). The same is true of agreements between the Commission and industry. Such agreements between the Environment Ministry and individual sectors of industry are a classical emission abatement instrument in French clean-air policy. Emission values and technical requirements are agreed with the aim of attaining a fixed minimum of polluting emissions (Rengeling 1985, 124).

Whilst the French have little objection to a Directive on pollution registers, Germany categorically rejects the Commission's plans. It fears substantial legal and institutional adjustment costs, since the German command and control philosophy is fundamentally opposed to such arrangements for comprehensive publication of company emission data and the self-regulation of industry via private-law agreements. The Germans also point out that they already have a comparable instrument, the so-called Air Quality Declaration Regulation,[155] which provides for an overall description of air quality in the states of the Federation (interview with EU Commission, DG XI, July 1994). However, this instrument is less comprehensive than the Commission concept, since no information on emissions by individual plants is provided. For the purpose of emission registers, the Germans want to see life-cycle analysis used and not merely consideration of the end product as planned by the Commission. The Commission regards such a procedure as too complicated and demanding. It tends to see the German proposal as more of a delaying tactic than a constructive contribution to the discussion (interview with EU Commission, DG XI, September 1993).

Large sections of European industry also reject the Commission plans, and, together with Germany, demand introduction of the more comprehensive life-cycle analysis concept. Industry feels that merely publishing annual emission figures provides no objective assessment of a company's environmental endeavours. Emission data are, for example, not set in relation to production volume. The pollutant emission registers would also give no indication of the extent to which higher emissions are caused by the production of goods that are far more environmentally friendly in use and disposal than comparable goods (interviews with EU Commission, DG XI, July 1994; *BDI*, October 1994). Life-cycle analysis would make it possible to judge the environmental impact of a good in the entire production process and not only in the final phase of production.

The new strategic elements with which the Commission hopes to improve implementation of EU provisions continue a trend that emerged with the strategic re-orientation under the Fifth Action Programme. Introducing pollutant emission registers is intended to increase public access and influence. Like the registers, the private-law agreements place greater stress on industrial self-regulation. The Commission hopes this will produce more effective implementation than interventionist limit values. It is interesting to note that the model taken by the Commission, in the absence of member state initiative, was the American 'toxic release inventory' concept. Although the Commission is seeking to make greater use of control forms respecting autonomy and Community interests, its emission register concept has been attracting little applause. Legal and institutional adjustment costs would be too high for some countries. Whilst in France and especially Britain it would be necessary to amend recently introduced instruments of a similar type, Germany resisted for another reason. Publication of emission data and industrial self-regulation were difficult to reconcile with the German command and control philosophy. In order to avoid European regulation in this direction, Germany, together with the industrial lobby, urged adoption of a more comprehensive concept — which the Commission considers to have no prospect of success because of the complex subject-matter — namely life-cycle analysis (interview with EU Commission, DG XI, July 1994).

CO_2/Energy Tax: A European Tax?

Since the late eighties there has been discussion at the EU level on introducing a Community CO_2/energy tax as a possible instrument for substantially reducing emissions of the greenhouse poison carbon dioxide. In contrast to many other activities, the initiative in this case has come from the Commission, which is actively promoting this novel concept of a European tax. Because of the extensive changes such a new tool would bring, the decision-making process has exhibited two specific features. The Council of Ministers became involved already in the problem-solving phase at Commission level. This early politicization is attributable to the increased need for consensus on such a measure. The particular need for coordination is also evident in the internal coordination among different Directorates General within the Commission. Whilst legislative initiatives are normally subject only to negative coordination at this stage, the positive variety was to be observed in this case between the Environment and Energy DGs, since the two worked actively on elaborating the draft Directive, making consensus finding within the Commission more difficult.

The Background to the Commission Initiative

There are various reasons for the Commission's particular interest in a European environmental tax. It is keen to strengthen its international negotiating position, wishes to further political and economic integration within the hitherto poorly 'Communitized' energy policy sector, and finally hopes to reduce implementation deficiencies at the national level by introducing new, economic control instruments. Since no member state submitted a regulatory proposal apt, in the Commission's view, to afford a suitable approach to the issue, the Commission took the initiative itself.

Although scientists are still divided on the exact causes of global climatic changes and the so-called greenhouse effect, there appears to be far-reaching agreement on one point, namely that carbon dioxide emissions, primarily from the burning of fossil fuels, are a major contributor to this environmental problem that affects the entire eco-system of the Earth. Although Third World countries are at present the main victims of climatic change, responsible for drought and floods,[156] the industrialized world is principally responsible for the world-wide increases in CO_2 emissions. EU countries contribute 12.9 per cent, the United States 23 per cent, and Japan 5 per cent (see table 17). Although warming of the Earth's atmosphere is a world-wide problem, the EU as one of the strongest economic powers, bears a special responsibility as a potential motor for a global climate protection policy.

The Commission is seeking to use this international responsibility of the EU to strengthen its mandate for negotiating at the international level. At the international Earth conferences in Geneva (1990) and Rio (1992), the Commission assumed a pacemaker role. The concept of '"environmental leadership", introduced and actively promoted by Environment Commissioner Ripa di Meana, became the basis of the Commission's proposals in the field of global environmental policy during the phase of the elaboration of the greenhouse strategy' (Jachtenfuchs 1994, 201). As promoter of efficient climatic protection policy, the EU is, however, credible only if 'it offers a convincing package of measures itself' (Hey 1992, 83). But — as the Berlin conference in March 1995 showed — there is great discrepancy between proposed measures and actual pollutant reductions. Although the EU called in Berlin for CO_2 emissions to be kept constant from the year 2000 at the 1990 level, it became clear that the Union was highly unlikely to attain the targets agreed in Rio. There participant countries had undertaken to reduce CO_2 emissions to the level of 1990 by the year 2000. Meanwhile, the EU Commission expects a 5 to 8 per cent rise in emissions within this period (*FAZ*, 10 March 1995).

Table 17: CO_2 Emissions Caused World-Wide by Energy Consumption — in millions of tonnes

	1971	1975	1980	1985	1988
Canada	94	109	124	115	124
USA	1,209	1,240	1,369	1,339	1,433
Japan	217	252	261	253	272
Australia	48	56	63	66	71
New Zealand	4	5	5	7	7
Austria	15	15	17	16	16
Belgium	36	36	37	30	32
Denmark	17	16	18	18	18
Finland	15	16	19	17	18
France	126	126	139	109	103
Britain	187	170	167	159	163
W.Germany	208	198	219	200	198
Ireland	6	6	7	7	8
Italy	92	97	106	101	108
Netherlands	44	46	50	48	51
Norway	7	7	9	8	9
Portugal	6	7	8	8	10
Spain	35	46	55	54	57
Sweden	27	26	24	22	21
Switzerland	12	11	12	12	13
OECD	2,427	2,522	2,756	2,648	2,793
World-wide	4,380	4,811	5,528	5,802	6,256

Source: International Energy Agency (1989)

A European environmental tax would not only give the Commission the opportunity to enhance its international influence, but would also provide the possibility of using international constraints to expand its poorly developed competence in energy policy. In the energy sector no noteworthy policy or economic activities have yet been undertaken at the supranational level. 'For most energy sources there is neither a common market nor a common policy. The energy sector is one of the sensitive key sectors where transfers of sovereignty have so far been unacceptable' (Hey 1992, 17; see Dantith 1987, 155).

At the beginning of the seventies, the Commission attempted first steps towards a Community energy policy, but under the impression of the 1973 oil price crisis, the Council of Ministers rejected the draft proposals the Commission submitted (Hawdon 1988, 105f.). Following further unsuccessful sallies into the field, the Commission has since 1981 referred only to a 'common energy strategy', providing general objectives, the attainment of which has been left fully in the hands of individual member states (Hey 1992, 18).

There has been just as little economic integration in the energy sector as policy integration. According to a Commission study, it is hindered by various market barriers resulting from divergence in the following areas: technical standardization for plants and products, energy tariffs, regulation of the import and export of gas and electricity, cost and pricing structures determined by specific financing requirements and subsidization, and oil industry environmental standards and conditions (EC Commission 1988). In view of this heterogeneity and the impediments accompanying it, there is transnational cooperation at best in gas supplies, and then only between primary distributors. The final consumer is usual supplied by state monopolies (Hey 1992, 56f.).

Since there is little Community policy in the energy sector and weak market integration, it is not surprising that a Community environmental energy policy should meet with opposition and difficulties. The differences in energy regulation from country to country in the Union (with regard to norms and environmental standards etc.) and the lobby of the powerful monopolist state and quasigovernmental actors in the energy sector, required energy environmental measures that allowed wide scope for a variety of regulation, unlike the other approaches usual in the clean-air field. This 'principle of gradational integration' (Hey 1992, 25) finds tangible expression in the margin for manoeuvre granted member states in setting limit values. Neither the EU provisions on the lead content of petrol, on the sulphur content of fuel oil, nor those on the reduction of SO_2 and NO_2 for old plant set uniform limits, but left it up to member states to fix varyingly stringent limit values within a certain margin ('target sharing').

Disregarding this instance of 'optimal harmonization' (Hey/Jahns-Böhm 1989, 98f.), the regulatory approach taken in EU energy environmental policy to date has been largely in line with other common environmental policy. The control mode is determined by the adoption of requirements and the setting of limit values. The use of 'soft' market control instruments, which would include a CO_2/energy tax, is a novelty at the supranational level. For this reason the discussion about a climate protection tax is

... more than a debate on a specific policy proposal among others. It is embedded in a broader discussion on the reorientation of EC environmental policy. The tax proposal and the place it occupied in this debate is thus an indicator of a frame shift from classic environmental policy to sustainability within the Commission and partly also in the Council (Jachtenfuchs 1994, 170).

A further motive for Commission interest is the dissatisfaction with the extent to which European environmental policy has been implemented. In advocating 'soft' control via voluntary agreements or economic and fiscal instruments, the DG XI is consciously moving away from the classical environmental policy approach, the strongly regulative nature of which has produced a high degree of inflexibility, bureaucratic effort, and barriers to advances in environmental technology. Within the DG XI, introduction of the CO_2/energy tax has thus become a 'test case for the introduction of economic and fiscal instruments' (Jachtenfuchs 1994, 212).

'Problem Solving' with the Participation of the Council of Ministers
Because of the profound changes a European tax would cause at both the supranational and national levels, the policy formation process was politicized at a comparatively early date. As an expression of this development, the Council of Ministers became involved already in the problem-solving stage at Commission level. In addition to early politicization, there was another circumstance making it more difficult for the Commission to produce a draft proposal capable of finding consensus: the problem of positive coordination between the Environment and Energy Directorates General, both of which were involved in elaborating the Directive.

European-level discussions on combating the greenhouse effect began in 1988, when the Commission submitted a first communication to the Council on alternative modes of action. Since then, several communications and proposals have been exchanged between Commission and Council, giving progressive form to the topic of a European environmental tax (Hey/Brendle 1994, 7). A decisive step was the 'Dublin Declaration' of the European Council of June 1990, in which the global responsibility of the EU and the necessity to limit CO_2 emissions were particularly stressed. Against the background of this declaration and the international pressure to which the EU was subject in view of the 1990 Geneva Conference, the Council of Ministers resolved in the autumn of 1990 to stabilize CO_2 emissions in member states at the 1990 level by the year 2000 (interview with EU Commission, DG XI, September 1994). At the same time, the Commission was called upon to prepare the measures necessary to attain this goal. Since then, 'the debate shifted from *whether* this should be done to *how* [emphasis in orig.] it could be achieved' (Jachtenfuchs 1994, 207).

In the autumn of 1991, the Commission responded to this demand and sent a communication to the Council of Ministers on 'a Community strategy to limit carbon dioxide emissions and to improve energy efficiency'[157] (EC Commission 1991). On the basis of these studies, the Commission came to the conclusion that conventional methods like more stringent energy-saving standards for household appliances and motor vehicles, voluntary undertakings by industry to save energy, and greater R&D efforts in the field of energy technologies would not suffice to stabilize EU-wide CO_2 emissions. To achieve long-term price signals and changes in energy consumption behaviour, the introduction of a combined CO_2 and energy tax was necessary.

In a number of follow-up meetings the Energy and Environment Councils discussed the general criteria for such a European tax. Even at this stage the positions of member states on the proposal became clear. Four basic attitudes can be distinguished. The first group of countries, which was in principle favourable towards a European environmental tax, included Denmark, Germany, the Netherlands, Belgium, Luxembourg, and Italy (interview with EU Commission, DG XI, September 1994). In Germany the introduction of a climate tax at the national level has been under discussion for some years.

The eco-tax discussion in Germany had its boom-time in 1989. When the Federal Environment Office submitted its concepts, when the Social Democrats changed course on ecological taxes, and in principle all parties drew up generally worded policy papers going in this direction (interview with EURES, August 1993).

In the run-up to the discussion, however, there was 'a huge row between the Economics and Environment Ministries'.

A struggle about competence, in which the Economics Ministry took the more progressive position ... The Environment Ministry was acting in the context of a more regulatory approach, where the state has control over what is levied and how it is allocated, in other words a purely financial instrument. Whereas the Economics Ministry was more interested in introducing market-mechanism control, that is one that acts via prices. The conflict was then resolved by saying in effect 'Well, let the EC handle it' (interview with EURES, August 1993; interview with Greenpeace, October 1994).[158]

In the wrangle on framing an energy tax or charge, the buck was not only passed to the EU to put an end to altercation at the ministerial level. Other background factors are revealed by statements made by the Environment Minister on the subject of climatic protection, 'which were made in an very interesting order':

Töpfer first stated that we absolutely had to do something to reduce our CO_2 emissions by 25 to 30 per cent. Then he said, all right, it could also be a CO_2 tax. And the next step was that we make a tax but not in the national context but in the EC context. And then the topic was dead for the next three or four years ... , although we had already had a bill ready in the drawer two and a half years ago (interview with *UBA*, November 1992).

The reasons why 'the EC was made use of to appease and hedge' (interviews with Robin Wood, November 1992; Greenpeace, October 1994), were largely economic.

One outcome of the recession is that we won't see the government's implementation of its CO_2 resolution on reducing emissions by 25 to 30 per cent by 2005 and the attached CO_2 charge because of substantial intrusion by economics. We're waiting for an EC decision. Because of the economic situation we can't go ahead by ourselves, we have to wait for an EC decision (interview with *BMU*, July 1993; see interview with *BDI*, March 1993).

A second group of countries, the southern member states, fears that a climate tax would slow economic development. They therefore make their approval contingent on appropriate compensation and derogations. A third standpoint is taken by the French, whose interests are strongly determined by the high share of nuclear power generation in France and who also give only conditional support to the project. They want to see the tax levied only on CO_2 and not on energy. In contrast to France, as usual 'ambiguous, not opposed and not in favour of this directive' (interview with EU Commission, DG XI, January 1994), Britain exhibited the most vehement opposition of all member states. Apart from economic objections, it was the potential loss of national autonomy — a fourth position — that aroused British resistance to a European tax.

These basic member-state positions were reflected in both the Environment and the Energy Councils. In a series of Council meetings in the autumn of 1991 it became clear that, if the Commission wanted to draft a directive acceptable to all, it would have to take crucial member state interests into account. Britain and the southern member states in particular insisted for reasons of competitiveness that a European arrangement could only be found if the EU's major trading partners (especially the United States and Japan) initiated comparable measures. The British Environment Secretary claimed that, though reducing CO_2 emissions was imperative, raising energy prices could be effective only over the long term and through coordination at the international level (Jachtenfuchs 1994, 253f.). Although Denmark, the Netherlands and Germany urged a single-handed European effort, the Commission was in no position to ignore the reservations of other countries (Jachtenfuchs 1994, 257). Finally, the Commission could not be sure of French support unless its nuclear power interests were taken into due account (ibid., 256).

Discussion at the Commission Level

Interestingly enough, the opposing positions taken by member states were reflected at Commission level, where the Directorates General XI (Environment) and XVII (Energy) were particularly involved in preparing the draft Directive. Individual countries made a massive attempt to bring 'their' commission officials into line with national policy (interview with EU Commission, DG XI, September 1994). Decision making at Commission level was made difficult not only by the relatively substantial national pressurizing, but also by the necessity to coordinate positions between two Directorates General.

The Energy DG, like the Environment DG, considered stabilization of CO_2 emissions a crucial condition for combating the greenhouse effect. But while the DG XI favoured a tax with a 75 per cent energy component and a 25 per cent carbon dioxide component, the Energy Commissioner Cardoso e Cunha advocated giving stronger weighting to the CO_2 component with a correspondingly lower burden on energy prices (Jachtenfuchs 1994, 214). A compromise was finally found in the form of a 50/50 provision as proposed in the Commission Communication of September 1991. Furthermore, the DG XVII feared, with its eye on Europe's competitors the United States and Japan, that the tax would put European Union member states at a competitive disadvantage ('first-mover disadvantage'), and should therefore be introduced only if the United States and Japan took similar measures.[159] Also for reasons of competitiveness, energy-intensive industries were to be exempted from the tax (ibid.).

These compromises, reached after more than a year's discussion within the Commission, are attributable not only to conflicts of interest between Directorates General and to national pressurizing. They also reflect massive lobbying by European industry, which did everything in its power to prevent adoption of a Community-wide climate tax. Immediately before the Commission presented its strategy paper,[160] various European industrial associations[161] staged a joint press conference to express their opposition to such a tax. The most important argument advanced against taxing energy was the fear that 'a "unilateral" EC-wide tax on energy consumption or CO_2 emissions would cause severe economic damage without any certainty of achieving the desired environmental objective' (Jachtenfuchs 1994, 267). For competition reasons, the European Union, since it caused 'only' 13 per cent of global CO_2 emissions, ought therefore to wait for the United States to make a similar move, and until such time support measures to save energy in the countries of central and eastern Europe.[162] Industry fears more than distortions in competition. The Association of the Mechanical, Electronics and Metalwork Industries of the EU and EFTA (ORGALIME) suspects that the various national measures taken to implement the tax will hinder the European internal market by raising additional barriers to trade.

Instead of a climate protection tax, the coal industry, the sector most affected, suggests CO_2 emissions could be far more effectively reduced by introducing new technologies. The umbrella organization of the motor industry, ACEA, held out the prospect of compensation, offering to reduce CO_2 emissions from motor vehicles voluntarily by 10 per cent over the period from 1993 to 2005. Further reductions could be obtained, so they claimed, by transport policy measures and the use of non-fossil fuels. From the point of view of industry, the utility of such voluntary agreements is that they cost less and require less bureaucratic effort than central regulation (Jachtenfuchs 1994, 268f; interview with *BDI*, March 1993). The strong resistance from industrial associations, added to national objections and conflicts of interest within the Commission, found expression in numerous derogations and compromises, which the Commission integrated into its proposal for a climate protection tax, and did much to water down the project.

The Commission Proposal

In June 1992, the Commission published its 'Proposal for a Council Directive introducing a tax on carbon dioxide emissions and energy'.[163] It provided for introduction of a CO_2/energy tax and tax incentives for investments saving energy or reducing CO_2 emissions. The tax was to be levied mainly on fuel products: hard coal, lignite, and peat and their products, natural gas and oil. Also to be included were certain ethyl and methyl alcohols where used as fuels, as well as electricity and thermal energy. Exempted from the tax would be renewable energy sources (wind, sun, biomass, biofuels, etc.) and energy resources used in industry as raw materials.

In accordance with the subsidiarity principle, member states would be required to regulate collection in their own responsibility. Revenues would go to the member states. The basis of assessment would be first energy content and second the specific quantities of CO_2 released on combustion of the product. Since electricity consumption is also to be taxed and double taxation is to be avoided, fuels used to generate electrical power would be exempted from the tax.

In order to take account of the different interests brought into play by member states and European industry, the Commission integrated a number of compromises in the draft proposal to permit its adoption in the Council of Ministers. The compromise clauses include the following:

- To ensure the overall fiscal neutrality of the CO_2/energy tax and consistency of the tax burden, member states are called upon to provide compensation. Surplus income is to be offset by lowering other public fiscal

charges (taxes or social insurance contributions) or by granting tax incentives).
- To avoid impairing Community competitiveness, application of the tax is to be made conditional on other OECD countries introducing a similar tax or measure with the same financial impact (principle of conditionality)
- In order to ease the burden further on energy-intensive industries and those heavily involved in international trade, derogations are to apply, reducing the tax rate or providing full exemption. Instead, these sectors are to undertake voluntary energy savings. However, the Commission has yet to elaborate a precise concept on this issue (interview with EU Commission, DG XI, September 1994).
- Also for reasons of economic competitiveness, the draft provides for the gradual introduction of the tax. From an initial rate of US$ 3 per barrel of oil[164] in 1993, the rate is to be raised annually to the year 2000, when the full amount of the equivalent of US$ 10 per barrel is to be charged. Despite this graduated risc, many ecological studies commissioned by the Commission indicate that the envisaged tax rate would not be sufficient to stabilize CO_2 emissions by 2000 at the 1990 level, as defined in the Dublin Declaration (see Capros 1991; DRI 1991; Morse 1991).[165]
- A combined CO_2 and energy tax is envisaged. Tax rates are to be based half on the quantity of CO_2 emissions from the energy source concerned and half on its calorific value (50/50 rule). The relatively high weighting of the CO_2 component, which was adopted largely to counter French reservations, would cause substantial power price differentials among member states. Since energy source structures can vary widely from country to country, 'the CO_2 component produces marked burden differentials ... Countries with a high proportion of (CO_2-free) nuclear power generation (Belgium, France) would have a low tax burden, while countries with a high proportion of fossil-fuel generated power (Denmark, Greece, Britain) would bear a correspondingly higher burden' (Hillebrand 1992, 8).
- In response to southern member state misgivings, compensation payments are envisaged under a 'burden sharing' scheme. The EU is to provide development funding to offset the global economic costs of the tax and other measures within the CO_2 strategy package for the less developed EU member states.

Although the Commission has tried to take account of the major concerns of member states and industry in the proposed Directive, negotiations in the Council of Ministers were far from successful. Notwithstanding numerous compromises, some countries refused to approve the draft. Concessions also failed to dispel misgivings about the extremely comprehensive changes that a European environmental tax would bring at both the supranational and national levels. Finding a consensus was further complicated by the unanimity principle applicable in tax matters. According to a member of the Commission staff, there were 'no prizes to be won' in tackling the CO_2 issue. 'It really gets down to the sum and substance of the matter, to the whole of energy policy, to the whole of structural policy' (interview with EU Commission, DG XI, March 1993).

Council Negotiations on the Commission Draft
Between 1992 and 1994, a number of countries attempted during their presidential term to bring agreement closer. Despite these endeavours, which progressively diluted the Commission proposal, by October 1994, when not even a 'very restrictive and symbolic tax' (Hey/Brendle 1994, 9) could find approval in the Environment Council, the project of a European environmental tax seemed condemned to failure. Apparently, the concessions made by the Commission in its official proposal had not sufficed to bridge the substantial gaps that the concept of a European tax had revealed between conflicting national interests.

A first, Greek compromise proposal was given the thumbs down by Germany and Britain. The Greeks have suggested extending the concept of burden sharing, proposing introduction of a 'burden-sharing index', which would have included both the national per capita emission of CO_2 and GNP. Economically weak member states would be liable to the tax only if their index score attained 85 per cent of the Community average. This concept, which was particularly in the interest of late industrializing countries, was strongly opposed by Britain, which would not have been among the beneficiaries of the derogation. But Germany also showed little interest in such an arrangement. This would have reduced to absurdity its aim of avoiding competitive disadvantages for German industry through a Community-wide provision. (Hey/Brendle 1994, 8).

Much hope was placed in the German presidency (first half of 1994) to bring about an agreement. Germany was keen to establish European consensus on a solution, since it feared that without it the domestic pressure to go it alone would become stronger.[166] But any such hopes quickly dissipated. In July 1992, the Council of Economics and Finance Ministers (ECOFIN), which has to approve all tax issues, rejected such a tax on principle. Instead, they proposed that existing taxes on energy sources such as coal, petrol, diesel fuel, and fuel oil should be raised (interview with EU Commission, DG XI, September 1994). Despite these substantial amendments to the original Commission proposal, Germany failed to achieve agreement at the Environment Council meeting in October 1994. The British still objected to any form of European tax. But the southern member states, too, blocked progress, since they felt the envisaged compensation and derogations were inadequate (*Süddeutsche Zeitung,* 5 October 1994). Only the French, spared potential taxation by their nuclear power, greeted the project (interview with EU Commission DG XI, September 1994). The Commission initiative to reduce CO_2 emissions by means of a European Union climate tax had thus for the moment come to naught.

However, the Commission is planning a new Directive giving member states broad leeway, within a certain framework, in implementing an energy tax (*Süddeutsche Zeitung,* 10 May 1995). The idea is, for example, to introduce taxation definitions for all non-renewable energy resources, leaving it up to the individual country to decide case by case which it chooses to tax, 'i.e., each countries does its own thing. This can range from introducing zero rates in southern countries or Britain. The whole system is then differentiated again for each definition, thus for France "electricity: zero", for Britain "coal: zero", for Germany "here and here: everything" ' (interview with *BDI*, October 1994). Whether this strategy will bring the Commission greater member state approval for a European tax remains to be seen.

Conclusion

The decision-making process on the introduction of a European tax to reduce CO_2 emissions differs in some aspects from that on other directives. This is especially attributable to the substantial national and supranational effects such an instrument would have. Because the energy sector has a low Community-wide profile and energy environmental policy is correspondingly cautious, reserved, and classically interventionist, the Commission plan met with scepticism among member states particularly jealous of their autonomy.

The particular effects of such an innovative measure already became apparent in problem-solving at the Commission level. Already at this stage in the decision-making process, the Commission brought in the Energy and Environment Councils. Early politicization of policy formation was attributable first to the far-reaching impact the measure would have. But involvement of the Council of Ministers can also be interpreted as a strategic move on the part of the Commission to sound out member state attitudes and to integrate them in its proposal. This makes it more difficult for member states, whose objections have already been taken into account in the draft Directive, to legitimate their negative attitude in the Council of Ministers.

However, there were singular aspects not only in the relationship between Council and Commission. Within the Commission, too, decision-making proved comparatively difficult. The reason was that, unusually, more than one directorate general was involved in elaborating the proposal. Positive coordination between the Energy and Environment Directorates General had to be established and differences overcome. This was all the more difficult with internal decision-making hampered by the massive political pressure brought to bear by member states.

Nevertheless, the Commission displayed every possible measure of flexibility and willingness to compromise in an effort to win the minds of the member states. The incentives were and apparently still are great: the possibility of enhancing international negotiating clout, intensifying supranational policy and economic integration, and reducing implementation deficiencies in the environmental field by re-orienting the existing classical regulatory philosophy towards the deployment of market-mechanism instruments.

Notwithstanding the considerable efforts undertaken by the Commission and a number of member states to bring about agreement, the proposal for a European climate tax has for the moment failed. Evidently, the comprehensive compromises offered to reconcile conflicts of interest between member states due not least of all to substantial structural differences in energy supplies, were insufficient to permit agreement on a common denominator. There are strong indications that, if Council of Ministers approval for such a far-reaching measure is to be obtained, member states will have to be granted still greater margins for action than provided for in the already heavily diluted Commission proposal.

Notes

[1] Lorenzetti's impressive frescos can be admired in the Palazzo Pubblico in Siena.
[2] 80/779/EEC; OJ L/229
[3] 85/203 EEC; OJ L/87
[4] 84/360/EEC; OJ L/188, S. 20-25.
[5] 88/609/EEC; OJ L/336, S. 1-13.
[6] The listed pollutants are: SO_2, NO_x, CO, organic substances and hydrocarbons, heavy metals, dust, asbestos, chlorine and fluorine (European Communities 1984, 25).
[7] For the purpose of the directive, emission a limit value is the content and/or mass of pollutants in emissions from plants that may not be exceeded over certain periods (European Communities 1984, 21).
[8] The RCEP has been in existence since 1970. Its members are appointed on the proposal of the Prime Minister. The terms of reference for the RCEP are extremely broad, embracing both domestic and international environmental policy, the anticipation of future environmental risks, and the setting of research priorities. The members of the RCEP (at present fourteen) are personalities from industry, science and politics ('The Great and the Good'). The broad spectrum represented is intended

to ensure that the RCEP remains in touch with the public interest and does not serve to articulate particularist interests. '[The members] serve as individuals and not as representatives of organisations or professions' (RCEP 1992, 2).

9   See chapter 2.1 for greater detail on internal political developments in Britain.

10 All subsequent daughter directives are to be enacted under the roof of the planned Directive on Integrated Pollution Control (IPC) which will replace the framework directives on air and water.

11 In 1980, CEGB research spending was about £1.5 million, or about 45 per cent of total British expenditure on research into the consequences and causes of acid rain (Boehmer-Christiansen/Skea 1991, 208).

12 Industry's main concern was with small plants of between 50 MW and 100 MW, operated by large industrial undertakings for in-house power supplies. The CBI argued that because of the low capacity of such plants, potential emission reductions bore no relation to the costs that abatement would cause (interview with multinational pharmaceuticals manufacturer, January 1993).

13 Although the DoE is one of the largest government departments with far-reaching competence (environmental protection, local authorities, housing), it has relatively few resources in the clean-air field.

14 See below.

15 The largely identical statements by the House of Lords Committee and the RCEP is perhaps attributable to the overlap in membership between the two bodies (Haigh 1990, 185).

16 Because of the close organizational links with the unions, the Labour Party took an extremely ambivalent attitude towards the issue of acid rain. While criticizing the rigid position of the government, it was nevertheless in Labour's interest not to endanger the situation of the British coal mining industry with exaggerated SO_2 reductions. During the 1984 miners' strike some Labour MPs were reported to have declared that anyone raising the issue of acid rain was to be regarded as a class traitor (Boehmer-Christiansen/Skea 1991, 215).

17 Financed fifty-fifty by the CEGB and the NCB.

18 It should, however, be noted that because of its generation capacity at the time, the CEGB needed no new power stations for the time being (interview with former CEGB employee, September 1993).

19 84 per cent of respondents were in favour of fitting FGD to British power stations even if this would entail a 5 to 10 per cent increase in the price of electricity. As early as 1986 59 per cent of respondents in a DoE survey had described such measures as a 'good idea'. Furthermore, 89 per cent had declared themselves to be 'very concerned' or 'fairly concerned' about possible damage in Britain, the corresponding figure for damage in other countries being 82 per cent. (ENDS-Report 1987/155, 3).

20 Privatization of energy supplies is dealt with below.

21 The fifteen largest environmental organizations have a total membership of over five million, representing some ten per cent of the British population (interview with DoE, January 1993).

22 See chapter 2.

23 The intellectual backing for this strategy was supplied in particular by the Conservative think tank, the Centre for Policy Studies, which played a crucial role in the ideological underpinning of Thatcherism (Rüdig 1991, 166).

24 The timing was well chosen: The NUM, which in 1974 had been instrumental in bringing down the Heath government, had emerged emaciated from an unsuccessful strike in 1984/85. A decisive reason for this defeat was that the CEGB had previously — to a certain extent as a contingency measure — stockpiled coal on a large scale. The effects hoped for from industrial action — power cuts and bottle-necks in power supplies — had thus failed to materialize (Boehmer-Christiansen/Skea 1991, 214f.).

25 One important reason was that, following the not very successful privatization of British Gas and British Petrol, the Conservative government felt obliged to legitimate its policy (Rüdig 1991, 167).

26 For Germany and France the figures were 40 per cent for 1980 and 60 per cent for both 1998 and 2003.

27 *Ni la loi sur la pollution atmosphérique de 1961, ni les décrets d'application de 1963 et 1974, ni les arrêtés ministériels pris en application de cette réglementation ne contiennent un* objectif précis [emphasis added] *de qualité de l'air à atteindre; cette absence est volontaire car il incombe aux* politiques régionales de définir ces objectifs [emphasis added] *en fonction de ce qu'elles estiment raisonnable à atteindre. ... La législation française ne contient pas non plus de standard d'émission, qui limiteraient d'une façon générale les émissions des installations de combustion.'*

28 *L'introduction de ces valeurs-limites s'opposait à la philosophie d'intervention développée jusqu' à là par l'administration française. Jusque dans les années 1980, les autorités françaises considéraient que ce type de normes allait conduire accorder des 'droits à polluer' dans les zones où la qualité de l'air se trouvait en dessous de ces normes. En conséquence, le Ministère de l`Environnement se refusait d'en fixer.'*

29 The decree in question (91/1122) also provides for the extension of the alarm zone to mobile sources. This means that in the event of air quality limits being exceeded owing to road traffic pollution, a limited driving ban can be declared (*Ministère de l'Environnement* 1991a, 19).

30 *[Les normes] ont apporté un poids supplémentaire aux autorités françaises chargées de cette politique, ... , qui ont pu renforcer leurs interventions auprès des émetteurs de pollution, notamment dans les régions où la norme était dépassée.'*

31 '*... le Ministère de l`Environnement français a tenté* d'anticiper la décision du Conseil [emphasis added]*, et a repris les principales contraintes de cette Directive dans une proposition de circulaire, applicable aux installations de combustion soumises à la législation des installations classées.''*

32 With the exception of Lille and the '*Petite Couronne de Paris*'.

33 These power stations had a double burden to bear. To begin with, they were in the weaker position vis-à-vis nuclear energy, which was strongly promoted by the state, a position that threatened to deteriorate still further under the strict standards imposed by the directive and the consequent investment in control technology.

34 The establishment of such state 'agencies' has been described by some commentators as a characteristic of or a concession to a non-existent industrial environmental policy (interview with Knoepfel, May 1993).

35 See below on the domestic policy discussion.

36 The distribution of reduction shares among the different sectors was initially controversial. Other sectors in particular complained that the share of the energy industry had been set too low. 'That was because at that time other sectors have been more putting in new plants with BAT requirements in it. So their emissions were

naturally reduced' (interview with British electricity utility, September 1993). Since 1993, however, the positions have been reversing, since the flue-gas desulphurization installations in the energy sector, to some extent still under construction, will make their mark.

37 Firms cannot organize trade-offs between different plants at will. There is a ceiling to upward divergence, the 'cap' varying in terms of the local environmental situation. Furthermore, such changes must be reported to the Inspectorate beforehand (interview with British electricity utility, September 1993).

38 Besides wanting to create a single controlling authority, the RCEP had other aspects in mind. It was of the opinion that environmental interests were not given due attention in the framework of the HSE, which maintained close ties with industry: 'The Health and Safety Commission and its Executive are industry-oriented bodies, set up to protect workers; they do not have an environmental competence, nor would it be appropriate, in view of their purpose, for them to acquire one. The Alkali Inspectorate should be transferred forthwith from the Health and Safety Executive to the Department of the Environment' (RCEP 1976, 107).

39 See chapter 3.1.2.

40 Conservative losses in comparison to the 1984 election were particularly high. They won only 32 seats with 34.7 per cent of the vote, while in 1984 they had won 40.8 per cent of the vote and 45 seats. However, the Green Party was not the only beneficiary of Conservative losses. Labour, too, gained a few percentage points, and with 40.1 per cent of the vote and 45 seats had the best result (compared with 36.5 per cent and 32 seats in the 1984 election) (Source: The Times Guide to the European Parliament, 1989, p. 86f.).

41 In the seventies, Tom Burke had been director of FoE and in the eighties chairman of the 'Green Alliance' (Grove-White 1992, 113).

42 With the poll tax, central government wanted to place local authority spending, which since the beginning of the eighties had repeatedly been the target of central government cutback attempts, more strongly under the control of the citizens, and thus to achieve further reduction in local authority expenditures. Prior to the poll tax, local authority budgetary autonomy had been possible because the local authorities received revenues from a fixed and comprehensive tax on property, or 'rates'. Fifty per cent of rates revenue came from firms, the remainder being levied on private individuals in the form of a monthly income tax. These revenues covered a third to a half of local authority expenditures, the rest being compensated by central government block grants (Bramley 1991, 284). The purpose of the tax reform in introducing nationally uniform corporation tax and an annual block grant fixed in advance in terms of respective local needs was to make all increases in local authority spending dependent on the poll tax, which was to be levied on all voters, i.e., not only by the well-off but also by the socially weaker members of the community. Since the poll tax was to produce about 25 per cent of local income, increasing local authority spending by 1 per cent would in principle increase the poll tax by 4 per cent. With this tool the government wanted additional local authority spending to take direct effect as additional costs for the voter, thus ensuring improved public control of local spending (Crick/van Klaveren 1991, 400).

43 The council tax introduced in April 1993 is a mixture of property and poll tax. In principle it is levied on all residents over the age of 18, but liability is subject to certain conditions involving individual assets and property. The government thus departed from its original intention of establishing a direct link between local spending and costs for the voter (Farrington/Lee 1992, 44f.).

44 The Customer's Charter, which is part of the Citizen's Charter, sets further standards on information, openness, and responsibility for public authorities, which, if they live up to them, are rewarded on request by a government seal of approval, the co-called Charter Mark (interview with local authority, September 1993).

45 The amalgamation of the authorities with responsibility for water pollution control, the Regional Water Authorities, was not foreseen for the moment, since partial reorganization in this area was planned because of the privatization of certain functions. In 1989 this process produced the National Rivers Authority, which was responsible for water pollution control. It is intended to merge this authority with the HMIP in the course of the nineties.

46 The Campaign had repeatedly introduced a private member's bill in the House calling among other things for free access to environmental information (Weidner 1987, 93). (In Britain individual Members of Parliament can introduce bills, which, if they attract a majority of votes can become directly applicable law).

47 The EPA 1990 supersedes numerous old laws such as the Alkali Act 1906, large sections of the Health and Safety at Work Act 1974 and the Control of Pollution Act 1974 (NSCA 1992, 42).

48 Part 1 of the EPA has been in force since 1 April 1991 for all new plants as well as large combustion plants (existing and new). All other processes, regardless of whether under IPC or local control, progressively came within the ambit of the new statutory provisions (last on 1 April 1992). All licensing proceedings are unlikely to have been completed before 1995. During the interim the earlier rules continue to apply for existing plants (Gibson 1991, 25).

49 So-called 'presumptive limits' agreed between the AI and industry.

50 Although this sort of procedure was already provided under the bpm principle, the lack of emission standards meant that it had no bite (interview with IEHO, March 1992).

51 The lack of limit values under bpm meant that 'BPM ceased to be a sword used by those who wanted to maintain and improve environmental standards, and became a shield for those who were unwilling or unable to do any more' (environmental spokesman for the Labour Party, quoted in ENDS 1990/181, 22).

52 On the procedure for preparing these 'notes' see below.

53 The NRA was set up only under Water Act 1989 to assume the regulatory function of the former Regional Water Authorities. The non-regulative function of these authorities (drinking water supplies and sewage disposal) were privatized by the Water Act 1989. Their regulatory duties were assigned to the NRA. The government was keen to see rapid privatization, because otherwise the state would have had to bear the high cost of bring drinking water quality up a standard in compliance with EU limit values. With the establishment of the NRA, it was thus no longer possible to wait until IPC had been introduced, which resulted in the present jurisdictional overlap (interview with IEHO, March 1992; Gibson 1991, 23f.).

54 Because of the large number of NRA employees, many IPC inspectors fear 'that their voice will become a minority voice in a body that has to do largely with water' (interview with AMA, November 1994).

55 British moves towards deregulation are also evident at the European level. Under strong pressure from Prime Minister Major and Federal Chancellor Kohl, an expert group was established to prepare proposals for the Commission on deregulation in the environmental field with the aim of furthering the competitiveness of European

industry. The proposals met with strong opposition in the EP and were dropped (The European 9-15 June 1995).

56   The IPC registers are kept at HMIP and local authorities to facilitate public access to the data.

57   The privatized power supply utilities had a particular problem in this respect. They wished to prevent publication of their *anticipated* pollutant emissions, which under the national reduction plan for implementing the Large Combustion Plant Directive also formed part of the operating permit. They justified their objection on the grounds that these figures permitted fuel suppliers to find out exactly what fuels the power stations were going to use in the future, thus allowing them to dictate prices. However, both the complaint to HMIP and the appeal to the Environment Secretary as the last instance were of no avail (interview with British electricity utility, September 1993; interview with HMIP, September 1991).

58   In 1994 HMIP issued a consultative document developing general decisional criteria for determining BPEO based on simple mathematical models (HMIP 1994).

59   In 1993 the BSI conducted a pilot study to test its quality assurance standard, in which some 100 firms participated. The response of the firms was largely positive; proposals for a few slight modifications are at present being looked into by the BSI. In the meanwhile many more firms have expressed their interest in introducing the BSI standard (interview with BSI, February 1994).

60   The following quote gives a graphic insight into new local authority powers: 'For years we have been battling in this borough with the smell that comes from a maggot farm ... thousands or millions of maggots for fishermen. And it breeds them of decaying organic material. They stink, exactly. We have been battling with this over the Public Health Acts for years. But now it's a process that requires authorization. There is a set standard to comply with and so the controls will be more effective' (interview with Bexley local authority, March 1992).

61   NSCA News Release of 18 January 1993.

62   Greater use is made of the public registers by industrial undertakings than by the public. The 1993 NSCA survey showed that 7 authorities reported frequent inspection of data by firms, 183 reported infrequent inspection, and 97 none at all.

63   *Before* adoption of the Environmental Protection Act 1990 the British public had access to environmental information on consumer protection, waste water regulation, and in the planning field (ENDS 1989/168, 28).

64   The Act, '*No. 78-753 du 17 juillet 1978 portant diverses mesures d'amélioration des relations entre l'administration et le public et diverses dispositions d'ordre administratif, social et fiscal'* initially applied only to non-personal data, but in 1979 it was amended to apply to personal data in the event of personal concern (Winter 1990, 176).

65   The definition or concretization of the concepts 'incomplete' ('*inachevé, préalable'*) or 'preparatory' ('*préparatoire'*) document is also the responsibility of the *CADA*. The borderline between 'incomplete/preparatory' documents and those that are actually open to inspection is a hazy. There are also documents (in authorization proceedings) that have to be made public in a formal public enquiry (*enquête publique*), in so far as this is required by the rules of procedure (Winter 1990, 194).

66   See Wegener (1992).

67   The Rosenheim *Landratsamt* takes first prize for charging DM 109 for the information that it would furnish no information (*Die Zeit*, 10 March 95, 35).

68   See chapter 3.5.
69   COM (93) 0423; C 311, 17/11/1993.
70   The Commission lays a great deal of store by having the EU set quality standards rather than leaving it to individual member states, since they fear that the latter will fix the quality standards halfway between the Sun and the Moon' (interview with EU Commission, DG XI, September 1993).
71   This concept was originally to be complemented by a so-called environment register (see above), but on objections being raised by industry, this first of all shelved and is now to be the subject of a separate directive.
72   The term is defined in the Commission draft as 'the latest stage in the development of activities, processes and their methods of operation which indicates the practical suitability of particular techniques as the basis of emission limit values for preventing or, where that is not practicable, minimizing emissions to the environment as a whole, without predetermining any specific technology or other techniques' (Art. 2, para. 10).
73   First difficulties have already arisen in the formulation and application of the IPC Directive. Definition of the best available techniques in the various sectors 'is distributed on a national basis. So the Spanish have described the best available techniques for blast furnaces. We were [at that level] in 1945.' A lot of effort will still be needed, and it will take a few years yet before 'levels are harmonized in individual process engineering fields' (interview with steel producer, March 1993).
74   There are two options under discussion. Southern member states in particular advocate balancing the economic costs and ecological benefits for individual plants. Northern countries would prefer to assess economic proportionality in terms of the impacts on the given industrial sector regardless of special local aspects.
75   In the opinion of the *BDI*, the French position can be explained by the fact that 'the people in authority and the *DRIRE* representatives can negotiate individual conditions with industry'. Cooperation and coordination there is 'much more constructive and "chummy"' than in Germany (interview with *BDI*, March 1993).
76   First reading in Parliament under the cooperation procedure. The first reading was originally planned for May 1994, by was taken off the agenda at the last moment. The reason was intervention by the Agriculture Committee, which had yet to be heard. Since the IPC Directive also contains certain provisions relating to intensive livestock farming, the Committee had to be heard (interviews with EU Commission, DG XI, June 1994; July 1994).
77   The national emission limits provided for in the Commission proposal should, in the view of the EP, apply only until corresponding European standards have been adopted.
78   The German Christian Democrat MEP Karl-Heinz Florenz proved a particularly committed advocate of the technology-based philosophy in the Environment Committee (interview with DoE, November 1994).
79   The environmental registers had been dropped from the draft proposal by the Commission on the urging of industry (see above).
80   This high priority is due to the circumstance that the Commission's IPC draft proposal would require substantial re-orientation of existing regulatory practices in Germany. 'This directive is a real slap in the face. We won't give way as easily as that. As it is we can't go along with it under any circumstances' (interview with BMU, July 1993).

81 Coordination problems between the Federal government and the states meant that the proposal was relatively poorly drafted and forwarded to the COREPER working group only a few days before the relevant meeting (interview with Council of Ministers, Environment Committee, January 1995).

82 'One of the main problems was that the Germans deleted the article that said "you have to coordinate the decision on the different releases". And we understand the reason for that is that at the federal level air is controlled and at the *Länder* level water is controlled and the *Länder* were unhappy in giving away that permission' (interview with DoE, November 1994).

83 'We understand the obligation on the operator is more or less directly imported from German legislation' (interview with DoE, November 1995; interview with Council of Ministers, Environment Committee, January 1995).

84 Regulation (EEC) No. 1836/93 of the Council of 29 June on voluntary participation by industrial companies in a Community eco-management and audit scheme (see EC Official Journal No. L 168/1).

85 The concept of the environmental audit underlying the official Commission draft proposal of 1990 is, however, no invention of Brussels (interview with steel producer, March 1993), but is based largely on the position paper of the International Chamber of Commerce (ICC) on the content and implementation of 'l'Environmental Audits' (ICC 1989). The paper, published in 1989, is designed to guide industrial undertakings in establishing in-house environmental management, and is part of an ICC effort to propagate self-regulative action by industry in environmental protection.

86 Particularly important is the British Standard 7750 — 'Specification for Environmental Management Systems'.

87 Whereas environmental impact assessment (see chapter 3.5) ensures environmental precaution in the planning phase, the environmental audit system, and the 'logical extension and consistent continuation of the EIA idea' provides for the evaluation of the environmental impact of the management and organization of a firm in day-to-day operation (interview with environmental impact assessment association, October 1993).

88 Auditing measures include the investigation and evaluation of the management system, the recording of relevant information, the evaluation, preparation and documentation of the audit results, and the elimination of flaws in the corporate environmental system by the competent management.

89 The statement of participation may appear on the environmental statements of the firm, in brochures, reports, and documents, and in the company letterhead. To provide firms with an incentive to participate in the audit system, an earlier Commission proposal provided for the possibility of awarding a mark of approval (eco-audit logo). However, this mark was strongly criticized by both industry and environmental organizations: 'Industry may well have been primarily concerned by competitive considerations; thus German firms were worried that the mark would be much more easily obtainable in other EU countries, while the impression would be given that all companies had been investigated in accordance with identical standards'. Environmentalist complained that the environmental audit logo would differ from the environmental marks hitherto in use in that 'it has nothing to do with a particular product. A product of the company concerned may cause serious environmental problems although the firm — with individual sites — satisfies the requirements of the auditing procedure' (Führ 1993, 26). Because of this criticism,

the logo was first transformed into a less conspicuous statement of participation that was nevertheless still provided with a graphic symbol. In the meanwhile, however, the idea of a label has gained ground. The Commission is at present working on the drafting of a directive to develop a Community-wide eco-label 'which is intended to: — promote the design, production, marketing and use of products which have a reduced environmental impact during their entire life cycle, and — provide consumers with better information on the environmental impact of products without, however, compromising product or workers' safety or significantly affecting the properties which make a product fit for use' (EU Commission 1992, Art. 1).

90 German consultants apparently have greater inhibitions and difficulties in finding their way on the European market: 'The German assessor is German and stays German and can only speak German.' (interview with EIA association, October 1993).

91 A comprehensive empirical study commissioned by the Federal Environment Office in 1989 came to the conclusion that 'in most cases environmental risks are dealt with in preference by technical means (77.5 per cent); organizational (14.5 per cent) or personnel measures (8.0 per cent) are relatively insignificant' (Antes et al. 1992, 379).

92 Like the industrial associations, environmental associations feel that concrete criteria and yardsticks for appraising environmental protection management are lacking (see *DNR* s.a.). However, the realization that 'when you look at the emperor he's naked and hasn't anything on at all' gives the environmental associations occasion to fear not that requirements will diverge but that they will become blurred and reduced to a minimum (interview with European Environment Bureau, March 1993).

93 The efforts by the BSI, which started back in 1990, to develop a British standard for an environmental management system as quickly as possible is only one example for the standardization contest between member states. The British in particular 'are doing their utmost to avoid having to adopt German standards' (interview with *DIN*, May 1993).

94 In Germany, Environment Minister Töpfer and the German Institute for Standardization (*DIN*) reached agreement 'on taking account of environmental interests in standardization' only in October 1992. The basis for cooperation between the Federal Ministry of the Environment and the *DIN* is the Environmental Protection Coordination Office especially set up for this purposes and the likewise newly established standardization committee on the Foundations for Environmental Protection (*NAGUS*), which is responsible for standardization in environmental protection at the national, European and international levels (see Spindler 1993b).

95 It is interesting to note that individual firms made this move much earlier than the body representing them, the Confederation of German Industry (*BDI*). The *BDI* was at times rebuked for hindering environmental protection activities in firms through its 'destructive' criticism of the important and necessary Eco-Audit Regulation (IBM Deutschland, Rhotert; quoted in Spindler 1993b, 87).

96 The original plan was to charge one chamber with accreditation for the entire country. However, this having been rejected on legal grounds as an inadmissible infringement of the autonomy of chambers of commerce, industry modified its model, now proposing that accreditation be carried out decentrally by all chambers of industry and commerce and by chambers of handicrafts (interview with *BMU*, October 1994).

97 *Zentralverband des deutschen Handwerks.*

98 *Bundesverband der Freien Berufe.*

99 The setting up of such an *Umweltgutachterausschuß* had been projected from the outset in the model proposed by the Federal Environment Ministry.

100 *Presque tout est commun pour le principe essentiel entre la loi des installations classées et l'approche intégrée. ... Donc la France a assisté avec une certaine satisfaction à l'élaboration de [cette directive].'*

101 *Il y a déjà une organisation administrative qui prévoit la coordination des autorités et il y a déjà l'accès du public aux procédures.'*

102 *Il y aura de petites règles nouvelles qui ne sont pas d'un grand coup pour la France.*'

103 *qui a été ni pour pousser le réglement ni pour s'opposer à ce réglement.'*

104 At the EU level there are at present quality standards only for SO_2 and suspended particulate matter, NO_x, and lead.

105 COM 94/109.

106 Besides SO_2, suspended particulates, NO_x, lead, and ozone, already dealt with in EU directives, this classification would require EU regulation of CO_2, cadmium, arsenic, nickel, fluoride, benzene, and certain hydrocarbons (European Communities, 1994, 29).

107 The Commission refers especially to studies of the WHO and other national and international specialist bodies.

108 *On avait essayé de réfléchir tout seul et puis au bout d'un moment on s'est dit qu'on irait voir ce qui se fait ailleurs. On est allé voir la Commission en leur disant: Alors, si il y a eu des réflexions du même style en Europe sur ces éléments-là qu'on puisse aller voir les gens qui ont réfléchi là-dessus et discuter avec eux.*

109 German misgivings about such a concept has been clearly apparent in the reluctant implementation of the Environmental Information Directive (see above).

110 *Si il y a un aspect commercial à la base des propositions de réglementations plus strictes? Je réponds 'oui, oui, oui' à votre question. ... Puisque l'Allemagne a des concentrations d'industries, il faut utiliser des seuils sévères et des machines pour obtenir ces seuils. Une fois qu'on a les machines, la tentation est grande de dire que le reste de l'Europe devrait utiliser ces machines lui aussi. ... C'est pour cela que certaines directives ont une forte inspiration germanique.*

111 SO_2 emissions caused by the use of liquid fuels constitute about 40 per cent of total SO_2 releases. Half come from the so-called 'heavy fuel oils' used in industry and ships (interview with EU Commission, DG XI, September 1994).

112 'Integrated Approach for Sulphur and Sulphur Dioxide, Limits in European Refining Industry' (Arthur D. Little, October 1991).

113 An additional obstacle in negotiations at the Commission level arose during discussion of the legal basis for the Directive. The Commission had initially planned to base the Directive on Art. 100a, which relates to measures which have as their object the establishment and functioning of the internal market, and which provides for the Council of Ministers to act by a qualified majority. After internal consultations, the Commission came to the conclusion that Article 130s would prove a more appropriate basis, since protection of the environment is a central objective of the planned measure. But because this article requires unanimity where questions of energy supplies are concerned, yet another change was made in the legal basis. The Commission feared that some member states could block negotiations by pleading the unanimity principle, and so after protracted internal discussions returned to Article 100a (interview with EU Commission, DG XI, September 1994).

114 The almost daily appearance of oil industry representatives at the Commission provoked one Commission representative to remark acidly that 'one could say this directive was made by the oil industry' (interview with EU Commission, DG XI, September 1994).

115 EC O.J. No. C 122 of 18 May 1990. The resolution states three principles on which the Commission was to act: primary among these was the avoidance of wastes by using the best available technology. Only in cases where this was not possible should the extent be investigated to which wastes can the reintroduced into the production process by recycling. If this option proved impossible, wastes were to be disposed of in the most environmentally friendly manner possible (interview with EU Commission, DG XI, March 1993). However, according to the Commission, regulating emissions from waste incineration plants was only one side of the coin. A consistent waste disposal policy needed to take account both of incineration and landfill. The Commission argued that both are final disposal methods and therefore belong together. Regulation of incineration thus increases the attractiveness of landfill, which — if not subjected to similar control requirements and with adequate capacity — would offer a much cheaper alternative. Whereas the Directive on the incineration of hazardous waste was adopted relatively rapidly by the Council of Ministers, member states showed little interest in the issue of landfill (interview with EU Commission, DG XI, September 1993).

116 The Germans had already been the driving force behind the Council Resolution calling on the Commission to prepare a proposal.

117 In preparing the proposed Directive, the Commission relied on the know-how of a national expert from the Tübingen *Regierungspräsidium*, which in general prescribes relatively extensive and stringent environmental technology controls in this field (interview with EU Commission, DG XI, March 1993).

118 Because of this technological development, it proved possible in Germany to reduce the limit values defined for dioxin under the 17th Regulation Implementing the Federal Pollution Control Act fivefold in comparison to those laid down in 1986 in the *TA Luft* (interview with EU Commission, DG XI, March 1993).

119 Since the Directive is to come into force only on 1st January 1997 and CEN will by that time have produced an adequate measuring method, the guide value provision has become irrelevant.

120 French waste disposal policy was moving away from landfill to incineration (interview with *conseil régional*, March 1994; interview with *Ministère de l'Environnement*, March 1994).

121 The Dutch had originally demanded an even stricter limit. However, in bilateral negotiations, Britain persuaded Germany to advocate a certain easing of the standard (interview with DoE, September 1993).

122 It was only in connection with sewage discharge that the British advanced their quality arguments, to the effect that the environment can absorb a certain amount of harmful substances without deleterious effects: 'The English say, for example, we have such a huge, beautiful body of water at our front door. Let's dilute it a bit' (interview with EU Commission, DG XI, March 1993).

123 In 1995 there is already technology on the market undercutting by half the limit values in the Directive, and costing only 10 per cent more than conventional control installations (interview with Tübingen *Regierungspräsidium*, June 1995).

124 In the transport field the EU adopted a framework Directive in 1991 providing for the approximation of legal provisions on hydrocarbon emissions. This Directive

requires the Commission to prepare further proposals for directives with the aim of restricting losses from evaporation and vaporization at all stages of fuel storage (European Communities 1993, 2). A two-phase procedure is planned. In a first phase, concerned with the storage and distribution of fuels for petrol, the Commission has already elaborated an official proposal for a directive, which is now being discussed in the Council of Ministers. It provides for VOC emissions in this phase to be reduced by 80 to 85 per cent by 2000 (European Communities 1993, 3). A proposal for the second phase, addressing the avoidance of emissions during the fuelling of motor vehicles, is to be ready in the course of 1995 (*CITEPA* 1992/106, II.70). The EU initiatives were strongly influenced by Germany, which had announced that it intended to regulate VOC emissions by motor vehicles by itself at the national level. The EU thereupon declared that they intended to take action on the issue, thus preventing the German initiative (interview with *UBA*, November 1992).

125 Although the Geneva Protocol provided for a total reduction in VOC emissions of only 30 per cent, the Commission envisages a 50 per cent reduction for industrial plants, since the measures planned by the EU for motor vehicles are unlikely to be enough to achieve the level agreed in Geneva (interview with EU Commission, DG XI, July 1994).

126 'BAT means the latest stage of development of activities, processes and their methods of operation which indicate the practical suitability of particular techniques ... , without predetermining any specific technology or other techniques' (European Communities 1993, 4).

127 A distinction is made between a simplified A plan and a more comprehensive B plan. Plan B is more elaborate in that it documents the path of every single VOC in the entire production process, whereas Plan A looks only at the final stage in production (European Commission 1993, Annex I).

128 Measurements are to be made regularly. Since in many production processes, volatile organic compounds are released only at irregular intervals, an eight-hourly average is taken as indicator. It is, however, planned to embody measurement procedure in standardized CEN requirements. National rules are to apply until such supranational provisions have been elaborated.

129 'In Holland, for example, dry-cleaning is done a lot by big companies. In a country like France it's still done a lot by small companies' (interview with *CNPF*, January 1995).

130 Differing conditions arise simply from the fact that ozone is formed by two sorts of pollutant: hydrocarbons and nitrogen oxides. Endless combinations of these two substances are conceivable with the same potential for ozone formation (so-called isoflats). This situation makes an EU-wide regime difficult, because the occurrence of the two substances varies strongly throughout Europe (interview with EU Commission, DG XI, July 1994).

131 *Il y eu certainement un problème de stratégie initiale de la Commission sur le sujet, puisqu'au départ on a un nombre de secteurs qui était relativement limité et petit à petit ce nombre de secteurs a accru. Et ce qui surprend en plus, C'est qu'entre chaque réunion de la Commission on a changé fréquemment fusé des pôles. C'est à dire qu'en mars '93 ou en juin ou en avril '93 il y a eu une réunion sur le sujet. On a eu une autre session un an après. Et il y a eu une autre au mois de juillet '94. Et chaque fois la base du projet directive était radicalement différente, puisque d'un projet à l'autre il y avait de nombreux annexes supplémentaires, premièrement, et, d'autre part, même sur le fond.*

132 *Dans cette directive il y a dix secteurs d'activités différentes dont six étaient objets d'études préalables pour choisir les meilleures techniques disponibles à coûts économiques acceptables. Dans les autres quatre secteurs qui n'ont pas fait l'objet d'étude, les réglementations se sont inspirées des réglementations allemandes.*

133 Another reason why the Environment Ministry in particular was keen to make national standards binding at the European level was to defend its negotiating position 'in the constant struggle against the Economics Minister, the Transport Ministry, and all the other Ministries' (interview with *BMU*, July 1993).

134 The British EPA 1990 assigned control of VOC emissions to local authorities. Permitting procedure and controlled substances are defined in Environmental Protection (Prescribed Processes and Substances) Regulations. HMIP issues guidance notes, providing concrete details of authorization conditions, and specifying the technological control requirements under the BATNEEC principle (Gibson 1991).

135 *Au point de vue technique, à partir du moment où on impose une concentration à l'émission, C'est comme si on imposait également la façon de s'y prendre les moyens à utiliser. C'est pareil.'*

136 See chapter. 3.5.2.

137 For the purpose of reducing VOC emissions, the French Environment Ministry had suggested levying a charge at a rate of 765 francs per tonne emitted, which was vetoed by Prime Minister Balladur because he considered the cost to industry would be too high. 'But you know that VOCs are very varied, that their impacts on ozone formation are very variable. We were facing a difficult problem. And then the Prime Minister ... cut short discussion by stating that 180 francs were to be levied per tonne for all pollutants and not 765 for VOCs' (interview with *CITEPA*, January 1995).

138 The Germans and the French keep a special eye on the motor industry and the Italians on the leather industry, while the Greeks want to safeguard the interests of their textile industry.

139 National control as well as financial incentives or voluntary industry agreements could be envisaged.

140 85/337/EEC.

141 There are at present moves at the EU level to supplement the existing EIA concept. Also taking up an American model, the Commission is working on plans for so-called 'Strategic Environmental Assessment', which would expand existing EIA of projects to include EIA for planning and programmes. The Commission has been thinking about such strategic assessment since 1990, since it considers that project assessment comes too late in the day. (interview with EU Commission, DG XI, February 1994). At the time, however, the *ballon d'essai* was shot down by member states. Industry in particular showed little enthusiasm, since it feared licensing would be unnecessarily delayed. A new attempt is now to be made on the basis of scientific studies to push through planning and programme EIA.

142 The 'National Environmental Policy Act' (NEPA).

143 Projects coming within the ambit of the Directive are listed in Annexes I and II. With regard to those named in Annex II, member states may decide at their discretion whether the expected environmental impacts of the project justify assessment or not. Moreover, member states may exempt individual projects entirely or partly from environmental impact assessment. In so far as implementation of environmental impact assessment is possible 'through integration in existing or other procedures', and there is room for manoeuvre on the legal procedure for detailed

implementation, the effective shape and form of EIA is left largely to the member states (see Bleckmann 1985, 88ff.).

144 In the opinion of an environmental consultancy firm, the broad leeway for interpreting the EIA Directive leads to strong differences in EIA reports. 'When the Germans do an EIA it costs two million marks and the report is 300 pages long. When the Italians do an EIA, the result is two pages stating that everything's fine and looks good. Period.' (interview with German environmental consultancy firm, August 1993).

145 The Commission takes as its basis the requirement of the 1973 First Action Programme of the European Union that, within the context of permitting proceedings, environmental impacts are to be investigated and taken into account as early as possible.

146 Only in the case of particularly risky projects does the 'Town and Country Planning Act' give authorities the right of demanding comprehensive information from the project sponsor for the purpose of evaluating the project (Coenen/Jörissen 1989, 136f.).

147 Another Interior Ministry bill inspired by the American model from 1973 for an 'Act on the Assessment of the Environmental Impacts of Public Works' never even got as far as the *Bundestag*. The 'Principles for the Assessment of the Environmental Impacts of Federal Public Works' adopted by the Federal Government in 1975 have also had little effect worth mentioning outside the field of traffic route planning (Otto-Zimmermann 1989, 110ff.).

148 These changes were implemented by a series of regulations (see DoE 1986; 1988). In particular, the Town and Country Planning (Assessment of Environmental Effects) Regulations and the Highways (Assessment of Environmental Effects) Regulations of 1988 were appropriately amended. To some extent, however, completely new legislation had to be passed, for example on reforestation and drainage and on transmission line construction (Rehbinder 1991, 116f.). Various DoE circulars fill in the details of the regulations, especially on consultation procedure (NSCA Handbook 1992, 23).

149 Act Implementing the Directive of the Council of 27 June 1985 on the environmental impact assessment of certain public and private projects (85/337/EEC) of 12 February 1990 (*BGBl.* I, Nr. 6, p 205).

150 *Süddeutsche Zeitung,* 9 January 1995. Only the Federal Ministry of the Environment draft of a general administrative directive implementing the Act on Environmental Impact Assessment of June 1991 has 'already been distributed and put into practice on a non-mandatory basis', although still incomplete (Commission of the European Community 1993, 7a).

151 The Lower Saxony Environment Minister Griefhahn, for example, stopped authorization proceedings on the planned nuclear waste disposal site 'Schacht Konrad' on the grounds that no EIA had been carried out. Before the Federal Constitutional Court in Karlsruhe, Federal Environment Minister Töpfer's response was that the EC EIA Directive had had direct effect for some time in relation to the final disposal project, and that frequent and intensive discussion on aspects of EIA had already taken place with Lower Saxony and the outcome taken into account in the application (quoted in *KGV-Rundbrief* 3/91, 3).

152 The '*Verein zur Förderung der Umweltverträglichkeitsprüfung e.V.*', the '*Bund für Umwelt und Naturschutz Deutschland*' and the '*Deutsche Naturschutzring*'.

153 The ECJ recently stopped construction of a federal highway in Bavaria, and the Rhineland-Palatinate Higher Administrative Court voided planning approval for a

section of the 'Eiffel Autobahn'. In both instances the direct effect of the EIA Directive had been ignored. According to experts, these rulings are likely to have repercussions for a many large projects carried out on similar premises (*Süddeutsche Zeitung* 9 January 1995).

154 The Commission had originally planned to introduce the registers on a voluntary basis. But the lack of enthusiasm on the part of industry has induced it to envisage a mandatory procedure. The Commission feels that 'the stick has to be there to force surrender of the emission data', since industry is otherwise not willing to supply all relevant information (interview with the EU Commission, DG XI, July 1994).

155 *Immissionserklärungsverordnung.*

156 On the ecological, economic, and political consequences of climatic change see Hey (1993, 5ff.).

157 SEC(91) 1744 final, 14 October 1991. The proposals are based on 24 studies commissioned by the Commission over the years.

158 Although Industry generally prefers market economy instruments, in this case it backed the Environment Ministry: 'We don't want a tax. A tax is something the government collects and that it can distribute the way it wants. If anything is to be done, then a system of charges ought to be introduced, which like the waste water charge system is the best thing because it functions as an incentive' (interview with steel producer, March 1993; interview with power utility, July 1993)

159 During the election year 1992, the American government had been very reserved about introducing an energy tax. President Clinton meanwhile supports the goal of stabilizing CO_2 emissions and advocates adoption of an energy tax. However, unlike the European Commission draft, the American measure would be a pure energy tax, 'not discriminating coal against nuclear energy and it ends where the EC tax starts. ... The US will certainly not fulfil the OECD-Clause, making action in the EC more difficult' (Hey 1993, 15f.). In contrast to the United States, Japan continues to hesitate on the issue. During the eighties, energy prices in Japan were much higher than in the EU (50 per cent) and the United States (200 per cent). Although Japanese prices have since fallen, the Japanese government seems 'to be reluctant to increase taxation again' (ibid., 15).

160 Of 14 October 1991. See EC-Commission (1991a).

161 The European Chemical Industry Association (CEFIC), the Association of European Automobile Manufactures (ACEA), the European Association of Metals (EUROMETAUX), the European Cement Association, the European Petroleum Industry Association (EUROPIA) and the European Federation of Industrial Energy Consumers (IFIEC-Europe).

162 On the course of the press conference see Agence Europe 25 September 91, 13.

163 COM(92) 226 final, 30 June 1992.

164 The initial rate is calculated on the basis of $ 3 mark-up per boe (barrel of oil equivalent). This represents a rate of ECU 17.70 per toe (tonne of oil equivalent). The tax amounts to ECU 2.81 per tonne of carbon dioxide released, and ECU 0.21 per gigajoule of energy generated. Since present excise on oil products does not always relate to energy content but to other physical units of measurement, the following converted rates apply for oils products: for leaded and unleaded engine petrol ECU 13.46 per 1000 litres, for diesel fuel and light fuel oil ECU 15.42 per 1000 litres, for kerosene and I.P. fuel ECU 14.4 per 1000 litres, for heavy fuel oil ECU 17.21/t, for other heavy distillates ECU 15.36/t, and for refinery and liquid gas ECU 0.39/GJ.

165 The Commission expects that, without Community measures, CO_2 emissions will rise by between 11 and 23 per cent depending on economic growth (EC Commission 1989b). The tax would therefore bring an approximate decline in CO_2 emissions of 6 per cent (DRI 1991). However, this figure is far below target, so that accompanying measures would be needed to raise energy efficiency. If the stabilization target is to be attained solely by means of the taxation instrument, the required mark-up would be 18 per boe (not $ 3) (Morse 1991).

166 A noted study carried out for Greenpeace by the German Institute for Economic Research (*DIW*) concluded that an ecological tax reform undertaken in Germany alone need not have only negative repercussions, but an overall positive effect on the German economy (DIW 1994).

4 The Regulatory Contest and its Consequences: A Subtle Redefinition of the State

The nature of the state in European Union countries has been experiencing changes that have been partly spectacular, partly unobtrusive and incremental. The focus of public attention has been on the enlargement of the Union with the associated accession negotiations and on the Maastricht II debate, because they transform the quality of the European edifice directly, drastically and for everyone to see. At the same time, behind the main negotiatory scenes, European legislation has been bringing about innumerable, subtle, and complex changes that are politically less visible, but nonetheless of immense significance for national state structures. Our interest has centred on the many, unspectacular but — in their cumulative effects on the state at the national and supranational levels — extremely important transformations in the formation and implementation of common policies.

Such changes are by no means the result of unidirectional influence exerted by Brussels on member states. The latter, as we have seen, engage in a regulatory contest to shape European policy, to champion their interests and push through their policy concepts. All member states are interested to a greater or lesser degree in putting their stamp on Community policy. In the area we have been investigating, clean-air policy/industrial emissions, the big three member states Britain, France, and Germany have been the major protagonists in a regulatory contest to introduce their philosophies and practices in combating industrial emissions at the European level, whilst other countries in the Union have acted as coalitionists, negotiating specific exchange deals for their concessions. The motivation for the three countries' endeavours to influence European legislation has been institutional/legal in nature — to minimize the cost of adjusting to European legislation — and economic — to avoid the regulatory costs harming the competitive position of domestic industry. A necessary precondition for a country to gain the upper hand in the contest and bring Community strategy into line with its own ideas is congruence between national strategies and solutions and those of the European Commission.

In the course of coordinating the multiplicity of European interests, informal coordination patterns develop — embedded in the institutional structures of Europe — that produce specific policy contents and require member states to undertake specific adjustments in their policy practices. As our account and the empirical examples of various directives and regulations show, these coordination patterns are: the use of the 'strategy of the first move', 'problem solving', and the related pattern of 'negative coordination, negotiation, and compensation'.

The sequence of coordination patterns and their course from problem definition to adoption of the measure can be shown in four different path profiles, systematized from the perspective of the 'first mover' in the question: In making the first move, what success does the initiator country have in shaping the measure concerned? In the case of a 'clear homerun' the first mover succeeds in pushing through its regulatory proposal in all phases up to policy formation with respect to both the basic regulatory approach and substantive provisions. The second path, that of the 'saddled homerun', is taken where the initiator's proposal is complemented and expanded by additional requirements that are similar from the point of view of the fundamental approach. Here, too, the first mover attains its regulatory goal, which is, however, amplified and augmented by the Commission and other member states until a comprehensive piece of legislation on the issue is adopted. In the 'moderated homerun' scenario, the principles of the regulatory approach are not modified in the first two coordination phases, but the substantive demands on member states are scaled down in the final phase of negative coordination/negotiation/compensation so that countries opposed to the measure can see their way to giving their approval. In the case of the 'thwarted homerun' the initiator comes up against the decided opposition of another high-regulating country and might have to accept far-reaching changes already in the problem-solving phase that call in question the fundamental regulatory approach. The reason may be that either another high-regulating country prefers a different approach and wants to see it receive equal consideration, which results in the typical policy mix or leads to a new, third solution being worked out jointly. Early opposition is also likely where clear distributional dimensions become apparent, inducing the potential loser to mark resistance *qua* negative coordination, thus initiating negotiations.

Once the Commission has given the go-ahead, the course that policy coordination takes depends largely on the nature of the subject matter to be regulated. If it is a technically and legally very complex matter difficult for the public to grasp, it has low political visibility. Thus, there is little chance of a clear homerun if an issue with easily comprehensible redistributory implications is to be decided, where the public can be easily mobilized and where negotiating processes set in early.

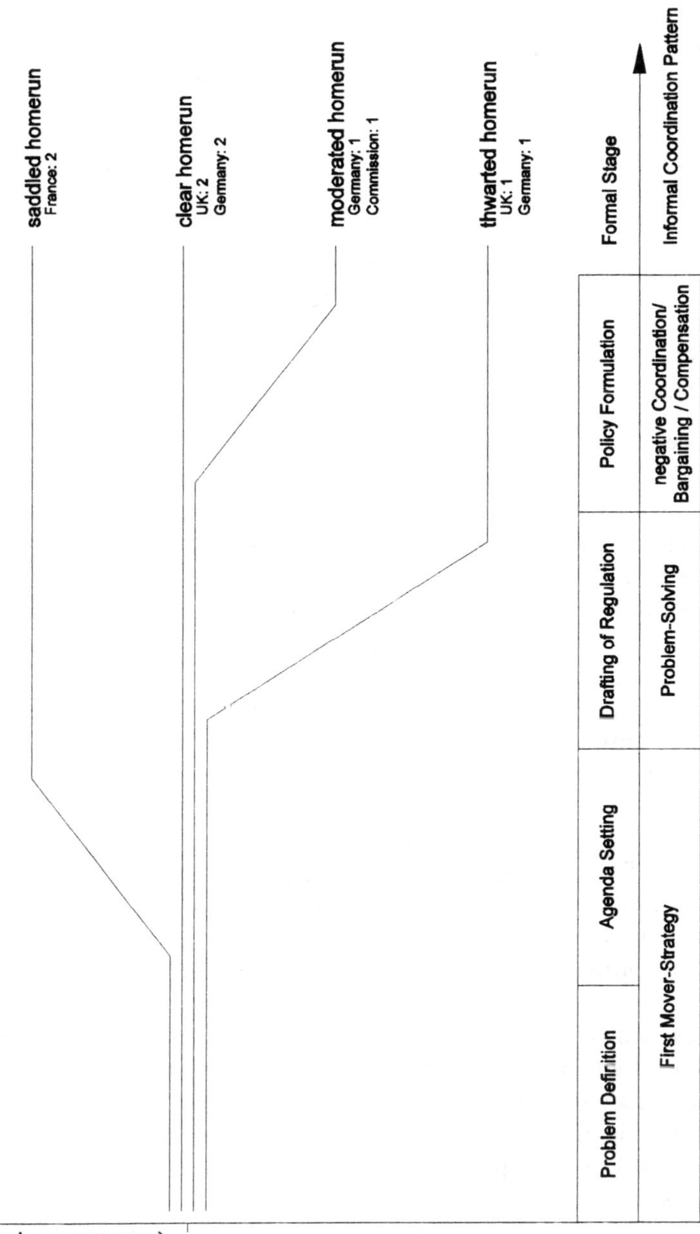

Figure 2:   Patterns of Coordination: Long-term Perspective

Table 18: Success/Failure as Policy Initiator

Policy Initiator/ Path Pattern	UK	France	Germany	Commission
clear homerun	-Eco-Audit- -Access to Information		-Framework Directive -Hazardous Waste	
moderated homerun			-Large Com- bustion Plants	-EIA
saddled homerun		-Product Stan- dards for SO$_2$* -Air Quality Framework*		
thwarted homeruned (policy-mix)	-IPC*		-VOC*	

* not yet decided

The most numerous category of the major clean-air/industrial emission measures investigated is that of the 'clear homerun', including the 1984 Industrial Plant Directive, that on the Incineration of Hazardous Waste, the Information Directive, and the Eco-Audit Directive. The two major contestants, Britain and Germany, each initiated two in this category and pushed them through to adoption. One measure was a German moderated homerun, the Large Combustion Plant Directive. The other moderated homerun, the EIA-Directive, was initiated by the Commission. Two measures, the *Chaîne pétrolière* and the new Framework Directive on Air Quality were French homeruns reinforced and expanded by the Commission, that is 'saddled homeruns'. Two further pieces of legislation — still to be adopted — the Integrated Pollution Control Directive and the VOC Directive, are measures that were controversial at an early stage, where a compromise was found on regulatory approach or where a policy mix was proposed. Two proposals, concerning the eco-register and the CO$_2$/energy tax, are Commission initiatives, which met with early opposition from all member states. They are not mentioned in the table because at this point of time it is still too early to classify them as clear, moderated, saddled or thwarted homeruns.

Our investigation of European regulation reveals that EU policy is a patchwork where — of the twelve measures under review — four clear runs could be scored by different players, in two cases substantive concessions had to be made on instruments, and where — even on the moderated homeruns — considerable concessions had to be made on aspiration and range if not on basic approach, and finally where quite new principles of regulation were introduced by the Commission. Different players put their respective stamps on entire directives, or on parts of directives. The outcome is a motley pattern of European instruments. The development of the patchwork may be put down to regulatory competition on the one hand, and on the other to the tendency of the Commission to keep a balance of member states' 'first moves' and the necessity to strike package deals.

Having thus far considered the patterns of European policy processes, we now turn to the impacts these processes have on the nature of the state in the three countries under review. The patchwork of policy measures that emerges from regulatory competition and its dynamics calls for varying degrees of policy adjustment in the different member states. There are three dimensions to consider: reactions to European requirements, forms of influence exerted in modifying state arrangements, and specific impacts on state practices. Forms of influence and impacts are then summarized in relation to the various dimensions of the state.

Taking reaction profiles first: it is possible to make European regulatory policy without provoking any adjustment in state arrangements worth mentioning if, for example, 'old structures' can be adapted to 'new' rules. Merely formal and not substantive adjustment is, however, also to be found, where only legal transposition into national legislation occurs. Finally, real change in member states can take place in structures, processes and policy contents. European influence can be defined as processes more or less directly attributable to European policy.

It is to be distinguished from the more indirect — and frequent   influence manifested in the anticipation of European policy measures. Countries often jump the gun at the national level to reduce legal and institutional adjustment costs and/or to be in a position to exert an active influence on Community policy.[1] This accelerates national policy formation and can result in a *politique de la surenchère'*, with those member states that entertain regulatory ambitions seeking to outbid one another. Finally, convergence of national and supranational policy developments can be observed, generated independently but exhibiting substantive congruence.

These influences cause tangible changes in state arrangements markedly transforming existing practices, in replacing organizational structures, problem-solving traditions and strategies by new structures and procedures. Old patterns of the state are removed or substituted. A second, much more frequent form is additive. New measures and institutions are added to existing national arrangements.

Forms of mutual influence and fundamental effects on the state in member countries can be shown in five dimensions. Relating to aspects of polity, politics and policy, these dimensions are the basic distribution of formal power structures in the organizational constitution of the state; the control instruments used by the state to attain policy goals; the related fundamental problem-solving tradition; the structure and form of interaction between the state and associations and individual clients; and, finally, the aggregation and mediation of societal interests by political parties.

The core of the state is affected by changes occurring in the fundamental political and organizational structures and in the distribution of powers among state actors. Membership of the EU has brought shifts in competence in the policy area under review. As policy formation has shifted towards the Community level, powers of decision have shifted — quite trivially — from the individual countries to Brussels. National governments are the affected actors. Thus, in the case of Germany, federal government (and *Bundesrat*) regulations on emission limits have to remain within the frame set by the EU. In the federally constituted Germany, this shift in competence is also at the expense of the states, where some primary powers of decision — for instance in waste disposal — have 'migrated' to the supranational level.

However, it would be superficial to interpret this shift simply as a zero-sum process (Wessels 1992), because member states retain substantial scope for action. The individual country or 'national state', 'although it lost its national "sovereignty" to a system of supranational institutions, is always and everywhere present as the guardian of national identity and the representative of national interests. As a national state it can no longer achieve anything, but as a national state nothing can be achieved against it either' (Grande 1993, 64). One reason is that it is responsible for implementing European legislation. Only member states have the necessary organizational and personnel capacities and expertise. In Germany this applies to the states, which are responsible for implementing legislation. As criticism of Brussels bureaucracy has grown, this scope has in recent years been enlarged by adoption of the subsidiarity principle. The Commission is more and more concerned in its legislative proposals to apply a strategy that is both 'autonomy-friendly and compatible with Community goals' (Scharpf 1993b) and which combines general objectives with the establishment of transparency and 'mobilization from below'.

On the other hand it cannot be denied that only the individual 'national states' with their traditional structures are able to provide the political and social infrastructure in which democracy becomes living reality beyond formal democratic institutions. Only the national state disposes of operational party organizations, trade, professional and other associations, citizen action groups, and social movements constituting the indispensable basis for work in democratic institutions and for democracy itself. The political acceptability of European policy depends on the activities of such organizations 'on the spot'. 'Enhancement of the legitimacy for European decision-making processes can ... not be based solely on direct legitimation by representative institutions; processes of legitimation through multiphase interest formation must also be taken into account' (Lepsius 1992, 187; Grimm 1994).

A further political development shows that a shift in policy formation to the Community level need not mean a loss in the decision-making clout of actors in national networks. Our study shows that the regions and local authorities find new interlocutors at the supranational level. They can turn to them to strengthen their own position within the national political system, seeking backing from such supranational actors as the Commission and the Court of Justice for their policy objectives. This potential for 'coalition formation' is exploited particularly by local territorial corporations in the centralist countries. Thus British local authorities willingly establish good network contacts with Brussels institutions because their political and financial position in the British political system has been weakened by the reforms of the eighties. European influence has been directly instrumental in strengthening the formal administrative decision-making powers of British local authorities. It was essentially European pressure — in addition to converging domestic changes (such as privatization) — that induced the British to revise their environmental legislation. But the EPA enlarged local authority licensing powers, thus increasing their competence in what was a bad period for them.

French regional authorities and their elected representatives have — since enactment of the decentralization laws — also been turning with greater self-confidence to supranational actors, 'and complaining in Europe about Paris' (interview with EC Commission DG XI, March 1993). The changes that have taken place at the regional level over the past ten years in France are consciously seen in relation to the *espace politique européen*. 'The close correlation existing between the emergence of European public policy and local public policy relatively independent of national policy must be emphasized ... , quite clearly ... a new public sphere is under construction' (Mueller 1992, 292).

A second important dimension in which changes in state arrangements induced by Europe have taken place is that of state control instruments and legal rules. Significant changes have occurred through direct European influence, through anticipation, and through convergence in the use of national and European instruments. Most marked are the transformations in the use of policy instruments in Britain.

For Britain, the predominant influence at the instrumental level has been a growing reliance on legal rules and regulations. After several years of determined opposition, the British finally accepted under protest a European regulatory practice providing for statutory emission limits and requiring the use of state-of-the-art technology, which they proceeded to implement systematically and conscientiously.

Another instrumental innovation in administrative practice, almost revolutionary in view of the typical 'culture of secrecy' dear to British government, is the opening up of administration (including industrial licensing procedures) to the interested public. This is an instance of convergence between a domestic development and a European trend. Since the eighties, the British have been pushing a policy of openness to enhance the accountability of individual public authorities to the citizen (Citizens' Charter) and generally to 'roll back' the state. Demands for access to information were being made at the same time in European directives. Once Britain had embodied these elements of administrative transparency in its own legislation, it became a fervent advocate for strengthening openness and 'mobilization from below' at the European level, since this was (now) in line with its own objectives and practices. These two changes modified British modes of regulation. Amicable negotiations between inspector and operator behind closed doors and the subsequent setting of plant-specific clean-air measures were no longer possible.

The pressure for change the EU has exerted on Germany in regard to state control instruments is diametrically opposite from that to which Britain has been exposed. In the eighties, Germany succeeded in imposing its own 'command and control' regulatory style at the European level, and therefore initially had no adjustment problems in the field of stationary emission sources. It has, however, determinedly opposed the control elements that in recent years have been developed in cooperation with the Commission — under British influence — and introduced at the European level. The Federal Republic rejects the return to a quality-based approach that gives the responsibility for setting of emission limit values to national governments, and which abstains from imposing the mandatory use of the best available technology. It distrusts the hopes put in industrial self-regulation based on eco-management and eco-auditing. It regards environmental impact assessment as an alien — Anglo-American — legal element incompatible with the German regulatory system as laid down in the Federal Pollution Control Act, and which disrupts the accustomed course of German authorization practice. It is sceptical about efforts towards informational openness and optimum transparency in public authorities and industry.

Whereas in Britain control instruments in licensing practice were directly replaced by instruments attributable to the (European) influence of the Large Combustion Plant Directive, in Germany we find an additive effect. With the introduction by Community legislation of new control instruments and legal rules, many competing arrangements exist side by side without clarity on how they interrelate. One example is the uncertain relationship between authorization of a plant pursuant to the Federal Pollution Control Act and environmental impact assessment.

In contrast to the two other countries, France — a tortoise to two hares — required little effort in integrating European innovations because it needed only to activate existing domestic instruments. One exception was the statutory regulation of emission limits at the national level, where France was obliged to adjust to European legislation. Like Britain, France was required in this respect to introduce greater legal as opposed to administrative regulation, an obligation attributable to German influence. To this end, France adopted a regulation in 1993 setting limit values for all industrial plants and processes on the model of the German *TA Luft*. Its implementation has, however, hitherto foundered on the opposition of industry.

Disregarding these necessary adjustments, which differed in extent and quality from country to country, all three tried to satisfy European regulatory requirements with existing legal regulatory instruments. These efforts, for example Germany's attempt to implement the Large Combustion Plant Directive by means of the *TA Luft* or France's endeavour to do the same with a mere administrative directive, were scotched by the ECJ on the grounds that such arrangements were not statutory in nature.

Re-orientations — closely related to changes in policy instruments — also occurred at a third level of state activities, in the problem-solving philosophy prevailing in the given policy area. Britain was under strong pressure to abandon a problem-solving approach too rigidly attached to unequivocal scientific evidence, which permitted state intervention only when the harmful impact of pollutants on human health had been proved. Cost-benefit considerations had to be set aside in favour of regular, emission-based intervention applicable regardless of regional ambient air quality.

Germany, by contrast, had to move away from its far too rigid source-oriented, end-of-the-pipe technological approach, and accept a more flexible approach with more diverse instruments permitting a range of approaches to solving given problems. Its narrow 'technology' orientation had to give way to a broader 'technique' understanding, which, in the British sense, includes the deployment and training of personnel as well as the operational organization of an enterprise. The concept of environmental protection has thus been broadened. France — again multifaceted in its problem-solving philosophies — needed only to activate those elements of its multiple problem-solving approaches required to meet European demands.

In all three countries, change at the level of problem-solving culture occurred to only a very limited degree through substitution. Problem-solving philosophies were added rather than replaced, expanding the national spectrum in Britain and Germany.

Changes in the dimension of the state relating to the interface between state and society, i.e., to interaction between the state and associations, individual clients, and the general public find expression in both policy formation (including self-regulation by associations) and policy implementation. National governments and executives have lost influence in policy formation because European legislation has displaced national legislation in many areas. It is thus only logical that national associations turn with growing frequency directly to the Commission, which drafts this legislation. Progressive market integration alone has multiplied the number of associations present in Brussels and has vastly complicated the European 'association landscape' (Kohler-Koch 1992). National and regional associations and individual enterprises often address the Commission directly. The complexity of influencing and lobbying processes is compounded by the lack of institutionalized and systematized access for associations to the Commission, which regulates the issue in a relatively arbitrary and haphazard manner (Mazey/Richardson 1993).

Interest representation structures have also changed with regard to the self-regulation of associations in so far as such regulation has passed to European bodies. For example, the DIN Institute acts for the Federal Republic in the European standardization institutions CEN and CENELEC. As a result, the trade unions, like small and medium-sized industries and consumer and environmental associations are underrepresented, because for lack of resources they do not participate intensively in the work of the DIN (Voelzkow/Eichener 1990). On the urging of the Commission and with particular support from Britain, there has been a marked shift in accent in the division of labour between the state and industry. More and more reliance is being placed on industrial self-regulation and eco-management for the purpose of emission abatement, while direct public-authority intervention is diminishing.

Adjustment to Brussels' regulatory instrument of 'prescribed transparency' has had substantial consequences on interaction between administration and public. Especially in Germany — which 'suffered' this development in contrast to the active role played by Britain — it brought a marked change in relations between the administration/operators on the one hand and the public on the other. The Federal Republic is under particularly strong pressure to change its practice of administrative interest mediation.

On a final level, that of interaction between the state and political parties and their function in interest aggregation, European legislation and political institutional structures exert indirect influence of considerable significance for national party systems. National political parties have been losing influence for two reasons.

First, competence in policy formation has been shifting to Brussels, and national parties have less say in policy-making within the the national framework. On the other hand, the European parties as 'subunits' of which national parties could operate in the political and social organizational process are weak entities. The reason, not to put too fine a point on it, is that the EU is not a fully fledged political system with interlinked, mutually answerable political institutions in which the European parties have a corresponding function of interest aggregation in the election of a parliamentary majority, which would then produce a government responsible to this majority. Because there are no such institutional translation mechanisms to lend effect to party political positions via a government answerable to parliament, European parties are under no pressure to act in a coherent manner calculated to concentrate political forces. At the European level there are accordingly no government platforms backed by party political majorities to guide action. The particularist interests of individual member states and individual actors within member states act on the legislative activities of European institutions without European parties exerting any effective programmatically integrative influence.

At the European level, policy is therefore segmented and fragmentary in nature. Parties have an aggregating and integrating function only at the national level, i.e., they provide a certain coherence in the implementation of European legislation in so far as a margin for action has been provided. Thus, under the influence of progressive market integration, policy-making in Europe is increasingly multifarious, but very heterogeneous and fragmented. The consequence for member states is that more and more areas of regulative policy are escaping the reach of the coherence-generating function of party democracy.

Thus, as we have shown in various dimensions of one regulatory policy area, the state in Europe is undergoing an ineluctable, often imperceptible process of change little noticed by the general public.

Notes

1 From the empirical perspective, this form of influence is the most difficult to ascertain. The frequency of this anticipatory form of handling European policy is repeatedly stressed by experts and office holders interview.

Appendix: Actor Groups Interviewed

A   Respondents in the Federal Republic of Germany:

Parliament and Parties:
Ausschuß für Umwelt, Naturschutz und Reaktorsicherheit, March 1993
Interparlamentarische Arbeitsgemeinschaft von Luftverschmutzung, March 1993
Bündnis 90/Die Grünen, March 1993
CDU, March 1993
FDP, March 1993
SPD, March 1993

Federal Authorities:
Bundesumweltministerium, March and July 1993, March, September and October 1994
Umweltbundesamt, November 1992 and May 1993

State Authorities:
Factory inspectorates, March and July 1993
Landesanstalt für Immissionsschutz, February 1993
Regierungspräsidium, April 1993 and February 1995
Umweltministerium NRW, February 1993

Local authorities:
Environment offices, August 1993 (2 interviews)

Local authority consultancy organizations:
Deutsches Institut für Urbanistik, November 1992
Kommunale Umwelt-Aktion, February 1993

Industrial polluters:
Chemical company, February 1993
Utility, July 1993
Steel producer, February and March 1993 and March 1994

Environmental protection equipment industry:
July and August 1993

Insurance industry:
Insurance company, August 1993
Haftpflichtverband, August 1993

Evironmental consultant:
September 1993

Associations:
Bundesdeutscher Arbeitskreis für Umweltbewußtes Management (B.A.U.M.), January 1993
Bundesverband der Deutschen Industrie (BDI), March 1993, March and October 1994
Deutsches Institut für Normung (DIN), May 1993
Kommission zur Reinhaltung der Luft im VDI und DIN, January 1993
Technischer Überwachungsverein (TÜV), February 1993
Verband der Chemischen Industrie (VCI), March 1993
Verein dt. Eisen-, Hüttenleute und Wirtschaftsvereinigung Stahl, April 1993

Chambers of industry and commerce:
Deutscher Industrie- und Handelstag (DIHT), March 1993
International Chamber of Commerce (ICC), July 1993

Industrial relations:
Deutscher Gewerkschaftsbund (DGB), February 1993

Science and research:
EURES-Institut, August 1993
Medizinisches Institut für Umwelthygiene und Silikoseforschung, February 1993
Öko-Institut, August 1993
Rat von Sachverständigen für Umweltfragen, March 1993
Unabhängiges Institut für Umweltfragen, November 1992
Wissenschaftszentrum Berlin (Weidner), November 1992

Environmental protection organizations:
Arbeitsgemeinschaft für Umweltfragen (AGU), March 1993
Bund für Umwelt und Naturschutz Deutschland (BUND), March 1993
Bundesverband Bürgerinitiativen Umweltschutz (BBU), March 1993
Deutscher Naturschutzring (DNR), February 1993
Greenpeace, November 1992 and October 1994
Interessengemeinschaft gegen Luftverschmutzung, February 1993
Robin Wood, November 1993
UVP-Förderverein, October 1993 and March 1994

B   Respondents in Britain:

Parliament and parties:
Labour Party, September 1992
Environment Shadow Minister, January 1993

Central authorities:
Her Majesty's Inspectorate of Pollution (HMIP), October 1991 (3 interviews), November 1992 (4 interviews), September 1993 (3 interviews), November 1994

Department of the Environment, November 1992, January 1993 (2 interviews), March 1993, September 1993 (3 interviews), November 1994 (2 interviews)

Regional authorities:
County, September 1992

Local authorities:
Association of Metropolitan Authorities (AMA), January 1993 (3 interviews), November 1994 (2 interviews)
Civic Office, September 1991 and September 1993
Corporation of London Health Department, September 1991
Institution of Environmental Health Officers (IEHO), October 1991
Local Government Management Board (LGMB), January 1993
National Society for Clean Air (NSCA), June 1993

Industrial polluters:
Central Electricity Generating Board (CEGB), September 1993 (2 interviews)
Multinational chemical company, January 1993

Consultancy firm:
September 1993

Associations:
British Standard Institution (BSI), January 1993
Confederation of British Industry (CBI), September 1991, September 1992 and November 1994

Science and research:
Boehmer-Christiansen, S., January 1992, July and September 1993
Haigh, N., December 1991
Royal Commission on Environmental Pollution (RCEP), September 1992, January 1993 and November 1994

Environmental protection organization:
Greenpeace, January 1993

C   Respondents in France:

Parliament and parties:
Génération Ecologie, June 1993
Les Verts, June 1993
Les Verts, July 1993
Rassemblement pour la République (RPR), June 1993

Central authorities:
Agence de l'Environnement et de la Maîtrise de l'Energie (ADEME), March 1993
Institut Français de l'Environnement, March 1994
Ministère de l'Infrastructure, June 1993 and March 199

Secrétariat Général du Comité Interministériel pour les Questions de Coopération Economique Européenne (SGCI), March 1993
Ministère de l'Environnement, March 1993, June 1993, March 1994 (2 interviews) and January 1995 (3 interviews)

Regional authorities:
Factory inspectorates (DRIRE), November 1992, June 1993 (2 interviews), February and March 1994
Prefecture, June 1993
Conseil régional, July 1993

Local authorities:
AIRPARIF, March 1993
AREMA, June 1993
ASPA, July 1993
Environment office, November 1992

Industrial polluters:
Utility, June 1993
Refuse incineration plant operator, September 1993
Paper manufacturer, September 1993

Associations:
Association pour la Prévention de la Pollution Atmosphérique (APPA), March and June 1993
Centre Interprofessionel Technique d'Etudes de la Pollution Atmosphérique, March 1993 and January 1995
Confédération Nationale du Patronat Français (CNPF), June 1993 and January 1995

Science and research:
Knoepfel, P., May 1993
Roqueplo, Ph., June 1993

Environmental protection organizations:
Alsace Nature, July 1993
Amis de la Terre, June 1993
Nord Nature, June 1993

D   Respondents at the European Level:

EU Commission DG XI, March 1993 (3 interviews), September 1993 (6 interviews), December 1993, January 1994, March 1994 (4 interviews), July 1994 (2 interviews), September 1994 (6 interviews)
Council of Ministers, Environment Committee, October 1994
Liason Office of the State of North Rhine-Westphalia, March 1993
European Environment Bureau, March 1993

References

Adams, W.J./Stoffaes, C. 1986. French Industrial Policy. Washington.

Agence Europe 16.05.1992. No. 5731, 11.

Agence Europe 25.09.1991. No. 5574, 13.

Aldrich, H.E./Pfeffer, J. 1976. *Environment of Organizations.* In: A. Inkeles, J. Coleman and N. Smelser, Eds., Annual Review of Sociology, No. 2, 79-105.

Antes, R./Steger, U./Tiebler, P. 1992. *Umweltorientiertes Unternehmensverhalten — Ergebnisse aus einem Forschungsprojekt.* In: U. Steger, Ed., Handbuch des Umweltmanagements. Munich. 375-393.

Ashby, E/Anderson, M. 1981. The Politics of Clean Air. Oxford.

Ashford, D.E. 1982. Policy and Politics in France. Living with Uncertainty. Philadelphia.

Bach, M. 1992. *Eine leise Revolution durch Verwaltungsverfahren. Bürokratische Integrationsprozesse in der Europäischen Gemeinschaft.* In: Zeitschrift für Soziologie, Vol. 21, No. 1, 16-30.

Backhaus, L. 1980. *Gefahren und Gefahrenbeurteilungen in der Rechtsordnung Frankreichs.* In: R. Lukes, Ed., Gefahren und Gefahrenbeurteilungen im Recht, Bd. 2. Cologne.

Badie, B./Birnbaum, P. 1983. The Sociology of the State. Chicago.

Bechmann, A. 1984. Leben wollen. Cologne.

Beerwerth, G. 1993. *Grünes Frankreich. Die französischen Umweltparteien vor den Parlamentswahlen.* In: Blätter für deutsche und internationale Politik, No. 3, 347-354.

Bennett, G. 1979. *Pollution Control in England and Wales. A Review.* In: Environmental Policy and Law, No. 5, 93-99.

Bennett, G., Ed., 1991. Air Pollution Control in the European Community. London.

Berg, A. 1992. Frankreichs Institutionen im Atomkonflikt. Rahmenbedingungen und Interessenlagen der nichtmilitärischen Atompolitik. Munich.

Beyme, K. von 1991. Das politische System der Bundesrepublik Deutschland nach der Vereinigung. Munich.

Bieber, R. 1984. *Europäisches Parlament.* In: W. Woyke, Ed., Europäische Gemeinschaft. Problemfelder-Institutionen-Politik. Munich. 339-346.

Bleckmann, A. 1985. *Die Umweltverträglichkeitsprüfung von Großvorhaben im europäischen Gemeinschaftsrecht.* In: Wirtschaft und Verwaltung, 86ff.

Blowers, A. 1987. *Transition or Transformation? Environmental Policy under Thatcher.* In: Public Administration, 65, 277-294.

Boehmer-Christiansen, S. 1988. *Pollution Control or Umweltschutz?* In: Energy and Environment, Vol. 2, No. 1, 6-10.

Boehmer-Christiansen, S./Skea, J. 1991. Acid Politics: Environmental and Energy Policies in Britain and Germany. London.

Bongaerts, J.C. 1989. *Die Entwicklung der europäischen Umweltpolitik*. In: WSI-Mitteilungen, 10, 575-584.

Bramley, G. 1991. *Kommunalfinanzen. Alternative Modelle und britische Städte*. In: B. Blanke, Ed., Staat und Stadt. Systematische, vergleichende und problemorientierte Analysen "dezentraler" Politik. PVS-Sonderheft 22, 283-307.

Brasser, L.J./Mulder, W.C. 1989. Man and his Ecosystem. Proceedings of the 8th World Clean Air Congress. Amsterdam.

Braudel, F./Labrousse, E. 1980. Histoire économique de la France: L'Ere industrielle et la Société d'aujourd'hui: siècle 1880-1980. Paris.

Bresser, A.H.M./Salomons, W. 1990. Acidic Precipitation. International Overview and Assessment. New York.

Bridge, J. 1981. *National Legal Tradition and Community Law: Legislative Drafting and Judicial Interpretation in England and the European Community*. In: Journal of Common Market Studies, 4, 351-376.

Brodhag, C. 1990. *Pollutions atmosphériques: mettre en place de nouvelles régulations politiques*. In: Annales des Mines, Nov. 1990, 107-110.

Brunsson, N. 1989. The Organization of Hypocrisy. Talk, Decisions and Actions in Organizations. Chichester.

Budge, I./ McKay, D. 1988. The Changing British Political System: Into the 1990s. Essex.

Bugge, H.C. 1976. La pollution industrielle: Problèmes juridiques et administratifs. Paris.

Bundesministerium des Inneren 1974. Umweltgesetze. Textsammlung der in Kraft getretenen Gesetze. Umweltschutzmaßnahmen des Bundesministeriums des Inneren. Bonn.

Bundesministerium für Umwelt, Naturschutz und Reaktorsicherheit 1990. Umwelt '90. Umweltpolitik, Ziele und Lösungen, Essen-Werden.

Bundesministerium für Umwelt, Naturschutz und Reaktorsicherheit 1992. Referentenentwurf. Gesetz zur Umsetzung der Richtlinie des Rates vom 7. Juni 1990 über den freien Zugang zu Informationen über die Umwelt (90/313/EWG), Dec. 1992. Bonn.

Bundesministerium für Umwelt, Naturschutz und Reaktorsicherheit 1993. Eckpunkte für ein Gesetz über die Zulassung von Umweltgutachtern und über die Registrierung geprüfter Standorte — Umweltgutachter und Standortregistrierungsgesetz (USG) -, 23. November 1993. Bonn.

Bundesverband der Deutschen Industrie (BDI) 1992. Stellungnahme des BDI zur Umsetzung der Richtlinie 90/313/EWG des Rates vom 7. Juni 1990 über den freien Zugang zu Informationen über die Umwelt — Umweltinformationsgesetz, Sept. 1992. Cologne.

Bundesverband der Deutschen Industrie (BDI) 1992a. Stellungnahme zum Vorschlag der EG-Kommission für eine Verordnung (EWG) des Rates, "die die freiwillige Beteiligung gewerblicher Unternehmen an einem gemeinschaftsweiten Öko-Audit-System ermöglicht". Cologne.

Bundesverband der Deutschen Industrie (BDI) 1993. Stellungnahme zu dem Vorschlag der Kommission "Vorschlag für eine Verordnung (EWG) des Rates über die freiwillige Beteiligung gewerblicher Unternehmen an einer gemeinschaftlichen Umweltmanagement- und Betriebsprüfungsregelung" ("Öko-Audit"), 18 Dec. 1992, Cologne.

Bundesverband der Deutschen Industrie (BDI) 1994. Vorläufige Stellungnahme des BDI zum Arbeitsentwurf der Europäischen Kommission vom Oktober 1993 für eine Richtlinie des Rates über die Begrenzung von Emissionen organischer Verbindungen, die beim Einsatz von organischen Lösungsmitteln in bestimmten Prozessen und industriellen Anlagen entstehen, Jan. 1994, Cologne.

Bungarten, H.H. 1978. Umweltpolitik in Westeuropa. EG, internationale Organisationen und nationale Umweltpolitiken. Bonn.

Burmeister, J.H./Winter, G. 1990. *Akteneinsicht in der Bundesrepublik*. In: G. Winter, Ed., Öffentlichkeit von Umweltinformationen. Europäische und nordamerikanische Rechte und Erfahrungen. Baden-Baden.

Capros, P. et al. 1991. Impact of an Energy and Carbon Tax on CO_2 Emission. Report for the Commission of the European Communities Directorate General XII. Athen.

Carden, S. 1992. Local Authority Control and Responsibilities. Paper presented at the CBI conference "Integrated Pollution Control — The Lessons Learnt". April 1992, London.

Chandler, A. 1962. Strategy and Structure. Chapters in the History of the Industrial Enterprise. Cambridge.

CITEPA 1992. Etudes Documentaires, No. 104 and 106/1992.

CITEPA 1993. Etudes Documentaires, No. 107 and 110/1993.

Coenen, R./Jörissen, J. 1989. Umweltverträglichkeitsprüfung in der Europäischen Gemeinschaft. Derzeitiger Stand der Umsetzung der EG-Richtlinie in zehn Staaten der EG. Berlin.

Coleman, J.S. 1972. Policy Research in the Social Sciences. Morristown N.J.

Communauté Urbaine de la Ville de Strasbourg (CUS) 1993. Charte de l'Environnement. Strasbourg.

Constantinesco, V./Hübner, U. 1974. Einführung in das französische Recht. Munich.

Converse, P. 1964. *The Nature of Belief Systems in Mass Politics*. In: D. Apter, Ed., Ideology and Discontent. New York. 206-261.

Corbett, R. 1990. *The Growing Influence of the European Parliament on EC Legislation*. In: Parliamentary Affairs, No. 3, 26-28.

Crick, M./van Klaveren, A. 1991. *Mrs Thatcher's Greatest Blunder*. In: Contemporary Record, Vol. 5, No. 3, 397-416.

Cupei, J. 1986. Umweltverträglichkeitsprüfung: ein Beitrag zur Strukturierung der Diskussion. Cologne.

Czada, R. 1993. *Konfliktbewältigung und politische Reform in vernetzten Entscheidungsstrukturen. Das Beispiel der kerntechnischen Sicherheitsregelung*. In: R. Czada and M. G. Schmidt, Eds., Verhandlungsdemokratien. Interessenvermittlung, Regierbarkeit. Festschrift für Gerhard Lehmbruch. Opladen.

Czada, R./Lehmbruch, G. 1990. *Parteienwettbewerb, Sozialstaatspostulat und gesellschaftlicher Wertewandel. Zur Selektivität der Institutionen politischer Willensbildung*. In: U. Bermbach, B. Blanke and C. Böhnert, Eds., Spaltungen der Gesellschaft und die Zukunft des Sozialstaats: Beiträge eines Symposiums aus Anlaß des 60. Geburtstags von Hans-Hermann Hartwich. Opladen. 55-84.

Damaska, M.R. 1986. The Faces of Justice and State Authority. A Comparative Approach to the Legal Process. New Haven.

Dantith, T./Williams, S.F. 1987. The Legal Integration of Energy Markets. Berlin.

Deetjen, H. 1992. La législation des installations classées pour la protection de l'environnement, Manuskript. Strasbourg.

Delandre, J.-R. 1991. *La surveillance de la pollution de l'air en France. Histoire, évolution, structures*. In: Pollution Atmosphérique, 131, 376-379.

Department of the Environment 1982. Air Pollution Control. Pollution Paper No. 18. London.

Department of the Environment 1986. Implementation of the European Directive on Environmental Assessment — Consultation Paper, 22. April 1986. London.

Department of the Environment 1986a. Air Pollution Control in Great Britain, Review and Proposals. London.

Department of the Environment 1988. Draft Circular from DoE and the Welsh Office — Environmental Assessment: Implementation of the EC Directive, 26.1.1988. London.

Department of the Environment 1988a. Integrated Pollution Control, A Consultation Paper. London.

Department of the Environment 1988b. Air Pollution Control in Great Britain. Works proposed to be scheduled for prior authorization, A Consultation Paper. London.

Department of the Environment 1988c. Air Pollution Control in Great Britain. Follow-up Consultation Paper issued in December 1986. London.

Department of the Environment 1989. Integrated Pollution Control and Local Authority Air Pollution Controls. Public Access to Information. London.

Department of the Environment 1991: Integrated Pollution Control. A Practical Guide. London.

Department of the Environment 1992. Digest of Environmental Protection and Water Statistics. London.

Despax, M./Coulet, W. 1982. The Law and Practice Relating to Pollution Control in France, 2nd ed. London.

Détrie, J.P. 1984. *Les pluies acides et l'évolution de la réglementation de la pollution atmosphérique*. In: Pollution Atmosphérique, 102, 114-121.

Deutscher Industrie- und Handelstag (DIHT) 1988. Stellungnahme des DIHT zum Referentenentwurf des Bundesministers für Umwelt, Naturschutz und Reaktorsicherheit für ein Gesetz zur Umsetzung der Richtlinie des Rates vom 27. Juni 1985 über die Umweltverträglichkeitsprüfung bei bestimmten öffentlichen und privaten Projekten (85/337/EWG).

Deutscher Industrie- und Handelstag (DIHT) 1992. Stellungnahme des DIHT zum Referentenentwurf eines Gesetzes zur Umsetzung der Richtlinie des Rates vom 7. Juni 1990 über den freien Zugang zu Informationen über die Umwelt (90/313/EWG), Sept. 1992.

Deutscher Naturschutzring (DNR) undated. Stellungnahme des Deutschen Naturschutzring (DNR) zum Vorschlag für eine Verordnung (EWG) des Rates über die freiwillige Beteiligung gewerblicher Unternehmen an einem gemeinschaftlichen Öko-Audit-System. Bremen.

Deutscher-Naturschutzring (DNR) undated. Gutachten zur Umsetzung der EG-Richtlinie über den freien Zugang zu Informationen über die Umwelt. Bremen.

Deutsches Institut für Wirtschaftsforschung (DIW) 1993. *Umweltschutz und Standortqualität in der Bundesrepublik Deutschland*. In: Wochenbericht 16, 199-206.

Dierkes, M./Fietkau, H.-J. 1987. Umweltbewußtsein — Umweltverhalten. Gutachten für den Rat von Sachverständigen für Umweltfragen. Bonn.

Doherty, J. 1989. *Britain in a Changing International System*. In: A. Cochrane and J. Anderson, Eds., Restructuring Britain, Politics in Transition. London, 7-33.

Döhler, M. 1990. Gesundheitspolitik nach der Wende: Policy-Netzwerke und ordnungspolitischer Strategiewechsel in den USA, Großbritannien und der Bundesrepublik. Berlin.

Drexler, A.-M. 1980. Umweltpolitik am Bodensee. Constance.

DRI 1991. The Economic Impact of a Package of EC Measures to Control CO_2-Emissions. Final Report for DG XI, Nov. 1991. Brussels.

Dubois, P. 1987. *L'investissement des entreprises: diagnostic macro-économique*. In: Revue d'Economie Industrielle, 40/41, 13-25.

Duhamel, O. 1991. Le pouvoir politique en France. Paris.

Dunleavy, P. 1993. Introduction: Stability, Crisis or Decline? In: Patrick Dunleavy, Andrew Gamble, Ian Holliday and Gillian Peele, Eds., Developments in British Politics. London.

Dyson, K. 1980. The State Tradition in Western Europe. A Study of an Idea and Institution. Oxford.

Ebers, M. 1989. Die Europäische Gemeinschaft. Ihre institutionelle Struktur, Verwaltungsorganisation, Verwaltungsverfahren und Finanzen. In: Wirtschaftswissenschaftliches Studium, 7, 321-370.

EC-Commission 1991a. A Community Strategy to Limit Carbon Dioxide Emissions and to Improve Energy Efficiency. SEC (91) 1744, 14.10.1991. Brussels.

EC-Commission 1991b. A Community Strategy to Limit Carbon Dioxide Emissions and to Improve Energy Efficiency, 07.10.1991. Brussels.

EC-Commission 1991c. A Community Strategy to Limit Carbon Dioxide Emissions and to Improve Energy Efficiency, 31.05.1991. Brussels.

EC-Commission 1992. Communication from the Commission. A Community Strategy to Limit Carbon Dioxide Emissions and to Improve Energy Efficiency, COM(92) 246 final, 1.6.1992. Brussels.

EC-Commission 1993. Proposal for a Council Directive on the Incineration of Hazardous Waste. Brussels.

EG-Kommission 1988. Der Binnenmarkt für Energie. Arbeitsdokument der Kommission, KOM(88) 238. Brussels.

EG-Kommission 1989a. L'énergie et l'environnement. Communication de la Commission au Conseil. Brussels.

EG-Kommission 1989b. Energy in Europe. Major Themes on Energy. Brussels.

EG-Kommission 1993. Bericht der Kommission über die Durchführung der Richtlinie 85/337/EWG über die Umweltverträglichkeitsprüfung bei bestimmten öffentlichen und privaten Projekten und Anhang über Deutschland, Vol. 3. Brussels.

EG-Kommission 1993a. Gemeinschaftsrecht im Bereich des Umweltschutzes, Vol. 2, Luft. Brussels.

EG-Ministerrat 1986. Entschließung des Rates über neue energiepolitische Ziele der Gemeinschaft für 1995 und die Konvergenz der Politik der Mitgliedstaaten (86/C/241/01). In: Amtsblatt der Europäischen Gemeinschaften, 25.9.1986, No. C 241/1. Brussels.

Ehlermann, C.D. 1990. Die institutionelle Entwicklung der EG unter der Einheitlichen Europäischen Akte. In: Außenpolitik, Vol. 41, No. 2, 136-146.

Ehrmann, H. 1976. Das politische System Frankreichs. Eine Einführung. Munich.

Eichener, V. 1992. Social Dumping or Innovative Regulation? Processes and Outcomes of European Decision-Making in the Sector of Health and Safety at Work Harmonization. Florence.

Environmental Data Services Ltd. 1987. ENDS-Reports No. 147, 148, 155. London.

Environmental Data Services Ltd. 1988. ENDS-Report No. 161. London.

Environmental Data Services Ltd. 1989. ENDS-Reports No. 168, 176, 178. London.

Environmental Data Services Ltd. 1990. ENDS-Reports No. 180, 181, 182. London.

Environmental Data Services Ltd. 1991. ENDS-Reports No. 193, 194. London.

Environmental Data Services Ltd. 1992. ENDS-Reports No. 205, 207. London.

Ernst, J. 1989. Kommunale Einwirkungsmöglichkeiten auf Genehmigungsverfahren nach dem Bundesimmissionsschutzgesetz (BImSchG). Unpublished paper. Darmstadt.

EU-Commission 1992. Council Regulation (EEC) No. 880/92 of 23 march 1992 on a Community Eco-Label Award Scheme. In: Official Journal of the European Communities, L 99/1.

EU-Commission 1993. Report from the Commission of the Implementation of Directive (85/337/EEC) on the Assessment of the Effects of Certain Public and Private Projects on the Environment, Com(93) 28 final — Vol. 13. Brussels.

Europäische Gemeinschaften 1973. *Erklärung des Rates der Europäischen Gemeinschaften und der im Rat Vereinigten Vertreter der Regierungen der Mitgliedstaaten vom 22. November 1973 über ein Gemeinsames Aktionsprogramm der Europäischen Gemeinschaften für den Umweltschutz.* In: Amtsblatt der Europäischen Gemeinschaften, No. C 112, 1-33.

Europäische Gemeinschaften 1977. *Entschließung des Rates der Europäischen Gemeinschaften und der im Rat Vereinigten Vertreter der Regierungen der Mitgliedstaaten vom 17. Mai 1977 zur Fortschreibung und Durchführung der Umweltpolitk und des Aktionsprogramms der Europäischen Gemeinschaften für den Umweltschutz.* In: Amtsblatt der Europäischen Gemeinschaften, No. C 139, 1-46.

Europäische Gemeinschaften 1983. *Entschließung des Rates der Europäischen Gemeinschaften und der im Rat Vereinigten Vertreter der Regierungen der Mitgliedstaaten vom 7. Februar 1983 zur Fortschreibung und Durchführung der Umweltpolitk und des Aktionsprogramms der Europäischen Gemeinschaften für den Umweltschutz (1982-1986).* In: Amtsblatt der Europäischen Gemeinschaften, No. C 46, 1-46.

Europäische Gemeinschaften 1984. *Richtlinie des Rates vom 28. Juni 1984 zur Bekämpfung von Luftverunreinigungen durch Industrieanlagen.* In: Amtsblatt der Europäischen Gemeinschaften No. L 188, 20-25.

Europäische Gemeinschaften 1985. Richtlinie des Rates vom 27. Juni 1985 über die Umweltverträglichkeitsprüfung bei bestimmten öffentlichen und privaten Projekten No. L 158, 40.

Europäische Gemeinschaften 1987. *Entschließung des Rates der Europäischen Gemeinschaften und der im Rat Vereinigten Vertreter der Regierungen der Mitgliedstaaten vom 19. Oktober 1987 zur Fortschreibung und Durchführung der Umweltpolitk und des Aktionsprogramms der Europäischen Gemeinschaften für den Umweltschutz (1987-1992).* In: Amtsblatt der Europäischen Gemeinschaften, No. C 328, 1-44.

Europäische Gemeinschaften 1988. *Richtlinie des Rates vom 24. November 1988 zur Begrenzung von Schadstoffemissionen von Großfeuerungsanlagen in die Luft.* In: Amtsblatt der Europäischen Gemeinschaften No. L 336, 1-13.

Europäische Gemeinschaften 1990. *Richtlinie des Rates vom 7. Juni 1990 über den freien Zugang zu Informationen über die Umwelt.* In: Amtsblatt der Europäischen Gemeinschaften, No. L 175, 56.

Europäische Gemeinschaften 1991. Der Rat der Europäischen Gemeinschaft. Luxembourg.

Europäische Gemeinschaften 1992. Die Europäische Union. Luxembourg.

Europäische Gemeinschaften 1992a. *Fünftes Aktionsprogramm.* In: EG-Bulletin No. 3, 48-54.

Europäische Gemeinschaften 1993. Explanatory Memorandum. Explanation of the Purpose of the Council Directive on the Limitation of the Emissions of Organic Compounds Due to the Use of Organic Solvents in Certain Processes and Industrial Installations, Oct. 1993. Brussels.

Farrington, C./Lee, M. 1992. Council Tax: Your Guide 1992. London.

Faudry-Brenac, E. 1981. *Crise et lutte antipollution dans l'industrie.* In: Futuribles, May 1981, 51-64.

Feick, J./Hucke, J. 1980. *Umweltpolitik. Zur Reichweite und Behandlung eines politischen Thema.* In: PVS Sonderheft 11, 168ff.

Fiebig, K.-H./Krause, U./Martinsen, R. 1986. Organisation des Kommunalen Umweltschutzes. Berlin.

Frankel, M. 1978. The Social Audit Pollution Handbook. London.

Fromont, M. 1983. *Die französischen Dezentralisierungsgesetze.* In: Die öffentliche Verwaltung, 10, 398-402.

Führ, M. 1989. Sanierung von Industrieanlagen. Düsseldorf.

Führ, M. 1991. *"Dauer von Genehmigungsverfahren" kein Standortfaktor für die Industrie. Zwei von drei Unternehmen sehen positiven Einfluß der Öffentlichkeitsbeteiligung.* In: KGV-Rundbrief No. 4, 25ff.

Führ, M. 1993. Betriebsorganisation als Element proaktiven Umweltschutzes. Zur Diskussion über die EG-Verordnung zum "Öko-Audit", Beitrag für das Jahrbuch des Umwelt- und Technikrechts 1993, Schriftenreihe des Instituts für Umwelt- und Technikrecht an der Universität Trier.

Gabriel, O.W., Ed., 1992. Die EG-Staaten im Vergleich. Strukturen, Prozesse, Politikinhalte. Opladen.

Gamble, A. 1988. The Free Economy and the Strong State. The Politics of Thatcherism. London.

Gaussens, P. 1991. *Quel rôle pour l'APPA.* In: Pollution Atmosphérique, No. 131, 368-375.

Gibson, J. 1991. *The Integration of Pollution Control.* In: Journal of Law and Society, 18, No. 1, 18-31.

Giraud, H. 1991. *La grande misère des commissaires-enquêteurs. Chargés de consulter le public sur les problèmes de l'administration ou des élus locaux, ces "honnêtes hommes" ont des rémunérations indécentes.* In: Le Monde, 28./29.04.1991.

Grande, E. 1993. *Die neue Architektur des Staates.* In: R. Czada and M.G. Schmidt, Eds., Verhandlungsdemokratie, Interessenvermittlung, Regierbarkeit, Festschrift für Gerhard Lehmbruch. Opladen. 73-81.

Greenpeace 1993. *Kreisverwaltung bloßgestellt. Der Kreis Paderborn wollte Greenpeacern Einsicht in Umweltdaten verwehren — vergeblich.* In: Greenpeace Nachrichten, Nov 1993.

Grefen, K. undated. Technische Regelsetzung zur Luftreinhaltung (DIN, VDI, CEN, ISO). Studienbausteine "Luftreinhaltung" an der RWTH. Aachen.

Gregory, F.E.C. 1983. Dilemmas of Government. Britain and the European Community. Oxford.

Grimm, D. 1995. Braucht Europa eine Verfassung? Munich.

Große, E.U./Lüger, H.-H. 1989. Frankreich verstehen. Eine Einführung mit Vergleichen zur Bundesrepublik. Darmstadt.

Grosser, A./Goguel, F. 1980. Politik in Frankreich. Stuttgart.

Gunningham, N. 1974. Pollution, Social Interest and the Law. London.

Hadesbeck, M. 1991. Modernisierungspolitik in Frankreich. Mit kulturellem Elan gegen die blockierte Gesellschaft. Münster.

Haigh, N. 1990. EEC Environmental Policy and Britain. Harlow.

Hall, P. 1986. Governing the Economy. The Politics of State Intervention in Britain and France. Cambridge.

Hansmann, K. 1992. Bundes-Immissionsschutzgesetz und ergänzende Vorschriften: Textausgabe mit Einführung und Anmerkungen. Baden-Baden.

Harbrecht, W. 1978. Die Europäische Gemeinschaft. Stuttgart.

Hartkopf, G./Bohne, E. 1983. Umweltpolitik: Erste Grundlagen, Analysen und Perspektiven. Opladen.

Haussmann-Grassel, B. 1985. Bürgerbeteiligung an gebundenen Verwaltungsentscheidungen. Darmstadt.

Hawdon, D. 1988. *Energy Policy.* In: P. Coffey 1988. Main Economic Policy Areas of the EEC Towards 1992. The Challenge to the Community's Economic Policies when the 'Real' Common Market is Created by the End of 1992. Dordrecht.

Henn, K.-P. 1993. *Das Öko-Audit wird Element im Unternehmensalltag. 1. Deutscher Umweltschutz-Auditing Roundtable.* In: UVP-Report, 2, 89.

Her Majesty's Alkali and Clean Air Inspectorate 1974. Annual Report for 1974. London.

Héritier, A. 1993. *Regulative Politik in der Europäischen Gemeinschaft: Die Verflechtung nationalstaatlicher Rationalitäten in der Luftreinhaltepolitik — Ein Vergleich zwischen Großbritannien und der Bundesrepublik Deutschland.* In: W. Seibel, Ed., Festschrift für Thomas Ellwein. Baden-Baden.

Héritier, A. 1993a. *Policynetzwerkanalyse als Untersuchungsinstrument im europäischen Kontext: Folgerungen aus einer empirischen Studie regulativer Politik.* In: A. Héritier, Ed., Policy-Analyse. Kritik und Neuorientierung. PVS Sonderheft 24, 432-447.

Herz, O. 1989. *La politique française de prévention de la pollution atmosphérique.* In: L.J. Brasser and W.C. Mulder, Eds., Man and his Ecosystem, Proceedings of the 8th World Clean Air Congress. Amsterdam.

Hesse, J. J., Ed., 1978. Politikverflechtung im föderativen Staat: Studien zum Planungs- und Finanzierungsverbund zwischen Bund, Ländern und Gemeinden, Baden-Baden.

Hey, Ch. 1992. The proposed EC-Energy Tax and GCC Interests. An European Point of View. Paper prepared for the symposium on EUROPE — GCC: Economic Relations and Environment organized by the Parliamentary Association for Euro-Arab Cooperation. Brussels.

Hey, Ch./Brendle, U. 1992. Umweltverbände und EG. Handlungsmöglichkeiten der Umweltverbände für die Verbesserung des Umweltbewußtseins und der Umweltpolitik in der europäischen Gemeinschaft. Freiburg.

Hey, C./Brendle, U. 1994. Towards a New Renaissance: A New Development Model. Brussels.

Hey, Ch./Jahns-Böhm, J. 1989. Ökologie und freier Binnenmarkt. Die Gefahren des neuen Harmonisierungsansatzes, das Prinzip der Gleichwertigkeit und Chancen für verbesserte Umweltstandards in der EG. Studie für das Europäische Umweltbüro. Freiburg.

Hill, M. 1983. *The Role of The British Alkali and Clean Air Inspectorate in Air Pollution.* In: P.B. Downing and K. Hanf, Eds., International Comparisons in Implementing Pollution Laws. Boston. 87-106.

Hillebrand, B. 1992. *Auswirkungen der EG-CO2/Energiesteuer auf die Energiepreise in den Ländern der EG.* In: RWI-Papiere, No. 33. Essen.

Hlubek, W. 1992. *Braunkohleverstromung in den alten und neuen Bundesländern.* In: StromTHEMEN Extra, 57, 1-8.

HMIP 1994. Environmental, Economic and BPEO Assessment Principles for Integrated Pollution Control. Consultation Document. London: HMIP.

Hohmann, H. 1989. *Die Entwicklung der internationalen Umweltpolitik und des Umweltrechts durch internationale und europäische Organisationen.* In: Aus Politik und Zeitgeschichte, B 47-48, 29-45.

Hoppe, W. 1984. Die wirtschaftliche Vertretbarkeit im Umweltschutzrecht. Schriftenreihe Recht, Technik, Wirtschaft. Cologne.

Hrbek, R. 1984. *Rat der Europäischen Gemeinschaften (Ministerrat).* In: W. Woyke, Ed., Europäische Gemeinschaft. Problemfelder-Institutionen-Politik. Munich. 392-397.

Hübler, K.-H. 1989. *Anmerkungen zu den Einwänden gegen die UVP.* In: K.-H. Hübler and K. Otto-Zimmermann, Eds., 1989. UVP-Umweltverträglichkeitsprüfung: Gesetzgebung, Sachverstand, Positionen, Lösungsansätze. Taunusstein.

Igl, G. 1976. Die rechtliche Behandlung der industriellen Luftverunreinigung in Frankreich und in der BRD. Berlin.

Institut National de la Statistique et des Etudes Economiques (INSEE) 1988. Annuaire Statistique de la France. Paris.

Institut pour une Politique Européenne de l'Environnement (IPEE), Ed., 1992. L'application de la législation communautaire environnement en France. Paris.

International Energy Agency (IEA) 1989. Energy Policies and Programmes of IEA Countries, 1988 Review. Paris (OECD).

International Union of Air Pollution Prevention Associations (IUAPPA) 1991. Clean Air Around the World. National and International Approaches to Air Pollution control. Brighton.

Internationale Handelskammer (ICC), Ed., 1989. Umweltschutz-Audits, Positionspapier. Paris.

Ismayr, W. 1992. Der deutsche Bundestag: Funktionen, Willensbildung, Reformansätze. Opladen: Leske und Budrich.

Jachtenfuchs, M. 1994. International Policy-Making as a Learning Process: The European Community and the Greenhouse Effect. Florence.

Jaedicke, W./Kern, K./Wollmann, H. 1990. "Kommunale Aktionsverwaltung" in Stadterneuerung und Umweltschutz. Cologne.

Jänicke, M. 1978. Umweltpolitik. Opladen.

Jänicke, M./Mönch, H. 1988. *Ökologischer und wirtschaftlicher Wandel im Industrieländervergleich.* In: M.G. Schmidt, Ed., Staatstätigkeit. Opladen. 389-405.

Jegouzo, Y. 1990. Le guide de l'environnement. Paris.

Jesse, E. 1992. *Wahlsysteme und Wahlrecht.* In: O.W. Gabriel, Ed., Die EG-Staaten im Vergleich. Strukturen, Prozesse, Politikinhalte. Opladen. 172-191.

Johnson, S.P./Corcelle, G. 1989. The Environmental Policy of the European Communities. London.

Jordan, A. 1993. *Integrated Pollution Control and the Evolving Style and Structure of Environmental Regulation in the UK.* In: Environmental Politics, Vol. 2, No. 3, 405-428.

Jordan, G./Richardson, J. 1982. *The British Policy Style or The Logic of Negotiation.* In: J. Richardson, Ed., Policy Styles in Western Europe. London. 80-109.

Karwiese, D. 1989. *Neue Struktur der französischen Verwaltungsgerichtsbarkeit durch Einführung von Berufungsgerichten.* In: Die öffentliche Verwaltung, 16, 713-719.

Keiter, H./Staupe, J. 1991. *Organisation des Umweltschutzes.* In: J.Vogl, A. Heigl and K. Schäfer, Eds., Handbuch des Umweltschutzes. Landsberg/Lech.

Kellow, A. 1988. *Promoting Elegance in Policy Theory. Simplifying Lowi's Arenas of Power.* In: Policy Studies Journal, Vol. 16, No. 4, 713-728.

Kempf, U. 1980. Das politische System Frankreichs. Eine Einführung. Opladen.

Kenis, P./Schneider, V. 1987. *The EC as an International Corporate Actor: Two Case Studies in Economic Diplomacy.* In: European Journal of Political Research, 15, 437-459.

Kessler, P. 1984. *Der staatliche Umweltschutz als organisatorisches Problem.* In: Die öffentliche Verwaltung, 37, 285-293.

Kimmel, A. 1988. *Die Präsidentschafts- und Parlamentswahlen 1988 und die verfassungspolitische Entwicklung in der V. Republik.* In: Frankreich Jahrbuch 1988. Opladen. 49-62.

Kirchner, E.J. 1992. Decision-Making in the European Community: The Council Presidency and European Integration. Manchester.

Klöpfer, M. et al. 1991. Umweltgesetzbuch — Allgemeiner Teil. Berlin.

Knoepfel, P. 1991. Remarques evaluatives d'un observateur étranger sur la lutte contre la pollution atmosphérique en France, Paper, Institut de Hautes Etudes en Administration Publique (IDHEAP). Lausanne.

Knoepfel, P./Kissling-Näf, I. 1993. *Transformation öffentlicher Politiken durch Verräumlichung — Betrachtungen zum gewandelten Verhältnis zwischen Raum und Politik.* In: A. Héritier, Ed., Policy-Analyse. Kritik und Neuorientierung, PVS Sonderheft 24, 267-288.

Knoepfel, P./Larrue, C. 1985. *Distribution spaciale et mise en oeuvre d'une politique publique: le cas de la pollution atmosphérique.* In: Politique et Management Public, 2, 45-69.

Knoepfel, P./Weidner, H. 1985. Luftreinhaltepolitik (stationäre Quellen) im internationalen Vergleich. Band 2 (Bundesrepublik Deutschland), Band 3 (England), Band 4 (Frankreich). Berlin.

Knoepfel, P./Weidner, H. 1986. *Explaining Differences in the Performance of Clean Air Policies: an International and Interregional Comparative Study.* In: Policy and Politics, Vol. 14, No. 1, 71-91.

Kohler-Koch, B. 1990. *Vertikale Machtverteilung und organisierte Wirtschaftsinteressen in der Europäischen Gemeinschaft.* In: U. von Alemann et al., Eds., Die Kraft der Region: Nordrhein-Westfalen in Europa, Bonn. 221-235.

Kohler-Koch, B. 1992. *Die Rolle organisierter Interessen im westeuropäischen Intergrationsprozeß.* In: M. Kreile, Ed., Die Integration Europas, PVS Sonderheft 23, 81-119.

Kohler-Koch, B. 1993. *Interessen und Integration. Die Rolle organisierter Interessen im westeuropäischen Integrationsprozeß.* In: Politische Vierteljahresschrift 34/2, 181-191.

Kommunale Gleichstellungsstelle für Verwaltungsvereinfachung (KGSt) 1985. Organisation des Umweltschutzes, KGSt-Bericht 5. Cologne.

Krasner, S.D. 1988. Sovereignty: *An Institutional Perspective.* In: Comparative Political Studies, Vol. 21, No. 1, 66-94.

Kukawka, P. 1993. *Dezentralisierung in Frankreich: Bilanz und Perspektiven.* In: Aus Politik und Zeitgeschichte, B32, 17-23.

Küppers, P. 1992. *Einsicht in Umweltdaten. Ab 1.1.1993 auch ohne bundesdeutsche Regelung möglich.* In: KGV-Rundbrief No. 4, 9.

Küppers, P. 1993. *Demokratisieren und Verkürzen von Genehmigungsverfahren. Ist das möglich?* In: KGV-Rundbrief No. 1+2, 61-68.

Küppers, P. 1993a. *Einsicht in Umweltdaten. EG-Richtlinie wird unterlaufen.* In: KGV-Rundbrief No. 1+2, 11-12.

Lahl, U. 1986. *Wendepunkt für die etablierte Politik.* In: Der Gemeinderat. 24-26.

Lahl, U. 1988. *Nicht machbar, Herr Nachbar.* In: O. Huter, W. Schneider and B. Schütt, Eds., Umweltschutz für uns. Cologne. 35-53.

Lane, J.E./Ersson, S.O. 1991. Politics and Society in Western Europe. London.

Larrue, C. 1992. *Environmental Monitoring in France.* In: H. Weidner, R. Zieschank and P. Knoepfel, Eds., Umwelt — Information. Berichterstattung und Informationssysteme in zwölf Ländern. Berlin. 286-313.

Larrue, C./Prud'homme, R. 1990. La Politique de l'Environnement. Permet-elle de mieux protéger l'environnement en France? Paris.

Laufer, H. 1985. Das föderative System der Bundesrepublik Deutschland, 5. ed. Munich.

Laumann, E.O. et al. 1978. *Community Structure as Interorganizational Linkages.* In: Annual Review of Sociology, 4, 455-484.

Lavoux, T. 1990. *La mise en oeuvre des directives "air" dans les états membres de la communauté*. In: Annales des Mines, Nov. 1990, 113-115.

Lebas, Y. 1981. *Die Internationalisierung der französischen Wirtschaft*. In: R. Lasserre, W. Neumann and R. Picht, Eds., Deutschland — Frankreich: Bausteine zum Systemvergleich, Bd. 2. Gerlingen.

Leggewie, C./de Miller, R. 1978. Der Walfisch. Ökologiebewegungen in Frankreich. Berlin.

Legrand, H./Biren, J.M./Jonan, M. 1987. *Politique Générale de Protection de l'Environnement en France en Matière de Pollution Atmosphérique*. In: Pollution Atomosphérique, 29, 113, 10-17.

Lehmbruch, G. 1976. Parteienwettbewerb im Bundesstaat. Stuttgart.

Lehmbruch, G. 1994. A Developmental Analysis of the German State, Unpublished paper. Constance.

Lepsius, M.R. 1992. *Zwischen Nationalstaatlichkeit und westeuropäischer Integration*. In: B. Kohler-Koch, Ed., Staat und Demokratie in Europa. Opladen. 180-192.

Macrory, R. 1985. *Environmentalism and the Courts*. In: T. O'Riordan and R.K. Turner, Eds., Progress in Resource Management and Environmental Planning. Chichester.

Macrory, R. 1988. *Environmental Policy in Britain — The Challenge Ahead*. In: European Environmental Yearbook. Institute for Environmental Studies. London. 670ff.

Majone, G. 1989. *Regulating Europe: Problems and Prospects*. In: T. Ellwein et al., Eds., Jahrbuch zur Staats- und Verwaltungswissenschaft. Baden-Baden. 159-177.

Majone, G. 1989a. Evidence, Argument and Persuasion in the Policy Process. New Haven.

Majone, G. 1993. *Wann ist Policy-Deliberation wichtig?*. In: A. Héritier, Ed., Policy-Analyse. Kritik und Neuorientierung, PVS Sonderheft 24, 97-115.

Marin, B., Ed., 1990. Generalized Political Exchange. Antagonist Cooperation and Integrated Policy Circuits. Frankfurt.

Maxeiner, D. 1993. *"Verdammt, er grinst"*. *Der Ausstieg aus der Autogesellschaft kommt*. In: Der Spiegel, No. 30, 92-94.

Mayntz, R. 1987. *Politische Steuerung und gesellschaftliche Steuerungsprobleme — Anmerkungen zu einem theoretischen Paradigma*. In: Th. Ellwein et al., Eds., Jahrbuch zur Staats- und Verwaltungswissenschaft, Bd. 1. Baden-Baden. 89-110.

Mayntz, R. 1993. *Policy-Netzwerke und die Logik von Verhandlungssystemen*. In: A. Héritier, Ed., Policy-Analyse. Kritik und Neuorientierung, PVS Sonderheft 24, 39-56.

Mayntz, R. et al. 1978. Vollzugsprobleme der Umweltpolitik, (Materialien zur Umweltforschung, Rat von Sachverständigen für Umweltfragen). Stuttgart.

Mayntz, R. et al. 1982. Regulative Politik und administrative Kultur. Cologne.

Mazey, S./Richardson, J. 1993. *EC Policy-Making: An Emerging European Policy Style?* In: L. Duncan and Ph. Lowe, Eds., European Integration and Environmental Policy. Bellhaven.

Mazey, S./Richardson, J. 1993a. *Pressure Groups in the EC*. In: J. Lodge, Ed., The EC and the Challenge of the Future. London.

McLoughlin, J. 1982. The Law and Practice Relating to Pollution Control in the United Kingdom. London.

Mény, Y. 1987. *The Socialist Decentralisation*. In: G. Ross et al., Eds., The Mitterrand Experiment. Continuity and Change in Modern France. Plymouth.

Mény, Y. 1988. *France: Rapport National*. In: H. Siedentopf and J. Ziller, Ed., Making European Policies Work. L'Europe des Administrations? Bd. 2. London.

Mény, Y. 1993. Government and Politics in Western Europe. Britain, France, Italy, West Germany. Oxford.

Menysch, D./Uterwedde, H. 1982. Frankreich. Wirtschaft-Gesellschaft-Politik. Opladen.

Merkel, W. 1993. Integration and Democracy in the European Community. The Contours of a Dilemma, Centro de Estudios Avanzados Ciencias Sociales (Working Paper No. 42). Madrid.

Meurin, G. 1992. *Umweltschutz-Audits contra Öko-Audit-System der EG*. In: Zeitschrift für angewandte Umweltforschung, 3, 300-303.

Meurin, G. 1992a. *Umweltschutz-Audits: Überprüfungen des betrieblichen Umweltschutzes sollen Schwachstellen aufdecken. Wirkungsvolles Management-Instrument. Leitfaden der Internationalen Handelskammer*. In: Blick durch die Wirtschaft, 24. April 1992.

Minister für Umwelt, Raumordnung und Landwirtschaft des Landes Nordrhein-Westfalen 1986. Wer ist zuständig? Aufgaben der Fachbehörden im Umweltschutz. Münster-Hiltrup.

Ministère de l'Environnement 1990. Plan National pour l'Environnement. Paris.

Ministère de l'Environnement 1991. Etat de l'Environnement. Paris.

Ministère de l'Environnement 1991a. La Qualité de l'Air en France. Année 1990. Bilan d'Application des Directives Européennes, Direction de l'Eau et de la Prévention des Pollutions et des Risques (DEPPR). Paris.

Ministère de l'Industrie et de l'Aménagement du Territoire 1991. Bilans de l'Energie 1970 à 1989. Paris.

Morand-Deviller, J. 1987. Le Droit de l'Environnement. Paris.

Morse, M. 1991. A Survey of recent CO_2 Tax Impact Studies. Paper presented at the BAEE seminar on applicability and effectivieness of a CO_2 tax to reduce CO_2 emissions. Den Haag.

Mousel, M./Herz, O. 1990. *La politique française de prévention de la pollution atmosphérique*. In: Annales des Mines, Nov. 1990, 69-73.

Mueller, P. 1992. *Entre le local et l'Europe. La crise du modèle français de politiques publiques*. In: Revue française de science politique, 2, 275-297.

Müller, E. 1986. Innenwelt der Umweltpolitik: sozial-liberale Umweltpolitik — (Ohn)macht durch Organisation? Opladen.

Müller-Rommel, F. 1992. *Erfolgsbedingungen Grüner Parteien in Westeuropa*. In: Politische Vierteljahresschrift, 2, 189-217.

National Society for Clean Air and Environmental Protection 1992. Pollution Handbook. Brighton.

National Society for Clean Air and Environmental Protection 1992a. *Implementation of the Environmental Protection Act 1990, Part 1*. In: Clean Air, Vol. 22, No. 2, 57-62.

Nettl, J.P. 1968. *The State as a Conceptual Variable*. In: World Politics, 7, 559-592.

Neumann, W./Uterwedde, H. 1986. Industriepolitik. Ein deutsch-französischer Vergleich. Opladen.

O'Riordan, T. 1979. *The Role of Environmental Quality Objectives in the Politics of Pollution Control*. In: T. O'Riordan and R. d`Arge, Eds., Progress in Resource Management and Environmental Planning, Vol. 1. Chichester. 221-258.

O'Riordan, T. 1988. *The Politics of Environmental Regulation in Great Britain*. In: Environment, Vol. 30, No. 8, 5-22.

Offe, C. 1990. Die Aufgabe von staatlichen Aufgaben. "Thatcherismus" und die populistische Kritik der Staatstätigkeit, Zentrum für Sozialpolitik, Arbeitspapier 9. Bremen.

Olier, J.Ph./Duprilot, V./Hertz, O. 1989. *Le programme français de lutte contre les emissions d'hydrocarbures dans l'atmosphère*. In: L.J. Brasser and W.C. Mulder, Eds., Man and his Ecosystem, Proceedings of the 8th World Clean Air Congress. Amsterdam.

Olier, J.Ph./Jarrault, P. 1988. *La politique française de limitation des pollutions atmos-phériques et le rôle de l'agence pour la qualité de l'air*. In: Administration, 139, 1-12.

Olier, J.Ph./Milhau, A./Hamelin, M. 1989. *Les strategies de lutte contre la pollution at-mosphérique: la solution française*. In: L.J. Brasser and W.C. Mulder, Eds., Man and his Ecosystem, Proceedings of the 8th World Clean Air Congress. Amsterdam.

Oliver, D. 1991. Government in the United Kingdom. The Search for Accountability, Ef-fectiveness and Citizenship. Milton Keynes.

Otto-Zimmermann, K. 1989. *Aus der Chronik des UVP-Gesetzesvorhaben 1972-1974*. In: K.-H. Hübler and K. Otto-Zimmermann, Eds., UVP-Umweltverträglichkeitsprüfung: Gesetzgebung, Sachstand, Positionen, Lösungsansätze. Taunusstein.

Pappi, F.-U. 1993. *Policy-Netze: Erscheinungsform moderner Politiksteuerung oder me-thodischer Ansatz*. In: A. Héritier, Ed., Policy-Analyse. Kritik und Neuorientierung, PVS Sonderheft 24. Opladen. 84-94.

Parodi, M. 1971. L'Economie et la société française de 1945 à 1970. Paris.

Peretti, J. 1990. *La politique énergétique et la pollution de l'air*. In: Annales des Mines, Nov. 1990, 111-112.

Perry, A.H. 1981. Environmental Hazards in the British Isles. London.

Peters, G.P. 1992. *Bureaucratic Politics and the Institutions of the European Community*. In: A.M. Sbragia, Ed., Europolitics. Institutions and Policy-Making in the "New" European Community, Washington D.C. 75-122.

Pezet, M. 1984. *Agence pour la qualité de l'air, son rôle, ses modalités d'action, ses ac-tions*. In: Annales des Mines, August 1984, 63-72.

Pfeiffer, R. 1991. Probleme bei der Umsetzung der EG-Richtlinie 85/337 über die Um-weltverträglichkeitsprüfung in das deutsche Recht. Trostberg.

Posse, A.U. 1986. Föderative Politikverflechtung in der Umweltpolitik. Munich.

Prieur, M. 1984. Droit de l'Environnement. Paris.

Prittwitz, V. 1984. Umweltaußenpolitik. Grenzüberschreitende Luftverschmutzung in Eu-ropa. Frankfurt/M.

Prittwitz, V. 1990. Das Katastrophenparadox: Elemente einer Theorie der Umweltpolitik. Opladen.

Rehbinder, E. 1991. Das Vorsorgeprinzip im internationalen Vergleich. Düsseldorf.

Rehbinder, E./Stewart, R. 1985. Environmental Protection Policy. Berlin.

Rengeling, H.-W. 1985. Der Stand der Technik bei der Genehmigung umweltge-fährdender Anlagen: Zur Bekämpfung der Luftverunreinigung im Recht der EG-Mitgliedstaaten und im Europäischen Gemeinschaftsrecht. Cologne.

Rest, A. 1986. Luftverschmutzung und Haftung in Europa. Kehl.

Rhodes, R.A.W. 1991. *Now Nobody Understands the System*. In: P. Norton, Ed., New Directions in British Politics. Essays on the evolving constitution. Aldershot. 83-112.

Romi, R. 1990. L'administration de l'environnement. La Garenne-Colombes.

Roqueplo, P. 1986. *"Der Saure Regen: ein Unfall in Zeitlupe"*. In: Soziale Welt, 37, 402-426.

Roqueplo, P. 1988. Pluies acides: menaces pour l'Europe. Paris.

Royal Commission on Environmental Pollution 1976. 5th Report. Air Pollution Control. An Integrated Approach. London.

Royal Commission on Environmental Pollution 1988. 12th Report. Best Practicable Envi-ronmental Option. London.

Royal Commission on Environmental Pollution 1992. A Guide to the Commission and its Activities. London.

Rüdig, W. 1991. *Die (gescheiterte) Privatisierung der Kernenergie: Krise des Thatcherismus?* In: R. Sturm, Ed., Thatcherismus — eine Bilanz nach zehn Jahren. Bochum. 155-178.

Sabatier, P.A. 1988. *An Advocacy Coalition Framework of Policy Change and the Role of Policy-Oriented Learning Therein.* In: Policy Sciences, 21, 129-168.

Sabatier, P.A. 1993. *Advocacy-Koalitionen, Policy-Wandel und Policy-Lernen: Eine Alternative zur Phasenheuristik.* In: A. Héritier, Ed., Policy-Analyse. Kritik und Neuorientierung, PVS Sonderheft 24, 116-148.

Sachverständigenrat für Umweltfragen (SRU) 1983. Waldschäden und Luftverunreinigungen (Sondergutachten). Stuttgart.

Sachverständigenrat für Umweltfragen (SRU) 1987. Stellungnahme zur Umsetzung der EG-Richtlinie über die Umweltverträglichkeitsprüfung in das nationale Recht. Stuttgart.

Sachverständigenrat für Umweltfragen (SRU) 1988. Umweltgutachten 1987. Stuttgart.

Salzwedel, J./Preusker, W. 1983. Umweltschutzrecht und -verwaltung in der Bundesrepublik Deutschland. Cologne.

Samuel, L. 1978. Guide pratique de l'écologiste. Paris.

Samuel, P. 1989. Quel air respirons-nous? Rapport sur la surveillance de la qualité de l'air en France, Les Amis de la Terre. Paris.

Sandnes, H./Styve, H. 1992. Calculated Budgets for Airborne Acidifying Components in Europe, 1985, 1987, 1988, 1989, 1990 and 1991, EMEP/MSC-W Report 1/1992.

Sauvy, A. 1965. Histoire économique de la France entre les deux guerres. Paris.

Scarrow, H. 1972. *The Impact of the British Domestic Air Pollution Legislation.* In: British Journal of Political Science, 2, 261-282.

Schäfer, K. 1977. *Entwicklung des Umweltrechts.* In: Vogl, J., Heigl, A. and Schäfer, K., Eds., Handbuch des Umweltschutzes. Losebl.-Ausg., 2. ed. Landsberg/Lech.

Scharpf, F.W. 1985. *Die Politikverflechtungs-Falle: Europäische Integration und deutscher Föderalismus im Vergleich.* In: Politische Vierteljahresschrift, 4, 323-356.

Scharpf, F.W. 1988. *Inflation und Arbeitslosigkeit in Westeuropa. Eine spieltheoretische Interpretation.* In: Politische Vierteljahresschrift, 1, 6-41.

Scharpf, F.W. 1993a. *Versuch über Demokratie im verhandelnden Staat.* In: R. Czada and M.G. Schmidt, Eds., Verhandlungsdemokratie, Interessenvermittlung, Regierbarkeit, Festschrift für Gerhard Lehmbruch. Opladen. 25-50.

Scharpf, F.W. 1993b. Autonomieschonend und gemeinschaftsverträglich. Zur Logik der europäischen Mehrebenenpolitik, Diskussionspapier 9, Max Planck-Institut für Gesellschaftsforschung, Cologne.

Scharpf, F.W. 1993c. *Positive oder negative Koordination in Verhandlungssystemen.* In: A. Héritier, Ed., Policy-Analyse. Kritik und Neuorientierung, PVS Sonderheft 24, 57-83.

Scharpf, F.W./Brockmann, M., Eds., 1983. Institutionelle Bedingungen der Arbeitsmarkt- und Beschäftigungspolitik. Frankfurt/M.

Scharpf, F.W./Reissert, B./Schnabel, F. 1976. Politikverflechtung: Theorie und Empirie des kooperativen Föderalismus in der Bundesrepublik, Kronberg/Taunus.

Scherzberg, A. 1994. *Freedom of Information — deutsch gewendet: Das neue Umweltinformationsgesetz.* In: Deutsches Verwaltungsblatt, July 1994, 733-745.

Scheuing, D. 1989. *Umweltschutz auf der Grundlage der Einheitlichen Europäischen Akte.* In: Europarecht, No. 2, 152-192.

Schmitt von Sydow, H. 1980. Organe der erweiterten Europäischen Gemeinschaften. Die Kommission. Baden-Baden.

Schneider, J.-P. 1991. Nachvollziehende Amtsermittlung bei der Umweltverträglichkeitsprüfung: zum Verhältnis zwischen dem privaten Träger des Vorhabens und der zuständigen Behörde bei der Sachverhaltsermittlung nach dem UVPG. Berlin.

Schneider, V./Werle, R. 1992. *Vom Regime zum korporativen Akteur. Zur institutionellen Dynamik der Europäischen Gemeinschaft.* In: B. Kohler-Koch, Ed., Regime in internationalen Beziehungen. Opladen.

Schreiber, H. 1991. Umweltpolitik in Großbritannien und Frankreich. Berlin.

Schumann, W. 1991. *EG-Forschung und Policy-Analyse. Zur Notwendigkeit, den ganzen Elefanten zu erfassen.* In: Politische Vierteljahresschrift, 2, 232-257.

Schumann, W. 1993. *Die EG als neuer Anwendungsbereich für die Policy-Analyse: Möglichkeiten und Perspektiven der konzeptionellen Weiterentwicklung.* In: A. Héritier, Ed., Policy-Analyse. Kritik und Neuorientierung, PVS Sonderheft 24, 394-431.

Seele, G. 1987. *Standort und Politik der Kreise im Bereich des Umweltschutzes.* In: Der Landkreis, No. 57, 375-383.

Shapiro, M. 1992. *The European Court of Justice.* In: A. M. Sbraghia, Ed., Europolitics: Institutions and Policymaking in the New European Community. Washington D.C.

Sheail, J. 1991. Power in Trust. The Environmental History of the Central Electricity Generating Board. Oxford.

Siedentopf, H./Hauschild, C. 1990. *Europäische Integration und die öffentlichen Verwaltungen der Mitgliedstaaten.* In: Die öffentliche Verwaltung, 43, 11, 445-455.

Siedentopf, H./Ziller, J., Eds., 1988. Making European Policies Work. L'Europe des Administrations? Bd 2, National Reports, Rapports Nationaux. London.

Simonis, U.E., Ed., 1988. Präventive Umweltpolitik. Frankfurt/M.

Simonnet, D. 1991. L'écologisme. Paris.

Skowronek, S. 1982. Building a New American State. The Expansion of National Administrative Capacities 1877-1920. Cambridge.

Spanou, C. 1988. *Les associations face à l'information administrative: Le cas de l'environnement.* In: F. Rangeon. Ed., Information et transparence administratives, Centre universitaire de recherches administratives et politiques de Picardie. Paris.

Spelthahn, S. 1990. *Energiesteuern in der Praxis. Am Beispiel Dänemarks.* In: G. Nutzinger and A. Zahrnt, Eds., Für eine ökologische Steuerreform. Energiesteuern als Instrumente der Umweltpolitik. Frankfurt/M.

Spindler, E.A. 1991a. Entstehungsgeschichte und Regelungsgehalt des UVP-Gesetzes. Unpublished paper. Hamm.

Spindler, E.A. 1991b. Anwendung der Umweltverträglichkeitsprüfung (UVP) in der Bundesrepublik.Unpublished paper. Hamm.

Spindler, E.A. 1992. *Neues Denken beim Umweltmanagement. Die Umweltverträglichkeitsprüfung für Industriebetriebe.* In: Entsorgungs-Technik. Zeitschrift für Abfallwirtschaft, Umweltschutz, Recycling, No. 5, 50-56.

Spindler, E.A. 1993a. Umweltschutz als Norm. Unpublished paper. Hamm.

Spindler, E.A. 1993b. *Verbandsschelte. Beim Öko-Audit fühlen sich die Unternehmen vom BDI im Stich gelassen.* In: UVP-Report, 2, 87ff.

Spindler, E.A. 1993c. *Vom UVP-Gedanken zum Umweltmanagement.* In: M. Sietz, A. v. Saldern, Eds., Umweltschutz-Management und Öko-Auditing. Heidelberg. 9-18.

Sprenger, R.-U. 1989. Beschäftigungswirkungen der Umweltpolitik: eine nachfrageorientierte Untersuchung. Berlin.

Steel, D.R. 1979. *Britain.* In: F.F. Ridley, Ed., Government and Administration in Western Europe. Oxford.

Steinberg, R. et al. 1990. Dauer von Genehmigungsverfahren. Projektstudie zum Forschungsvorhaben 23/88, Infratest Industria im Auftrag des Bundesministeriums für Wirtschaft. Munich.

Streeck, W./Schmitter, P. 1991. *From National Corporatism to Transnational Pluralism: Organized Interests in the Single European Market.* In: Politics and Society, Vol. 19, No. 2, 133-164.

Strübel, M. 1992. Internationale Umweltpolitik: Entwicklungen — Defizite — Aufgaben. Opladen.

The Economist Intelligence Unit and Business International, Eds., 1990. French Policy on the Environment. Unsung Achievements, Emerging Anxieties and Possible New Directions, Special Report No. 2063, Okt. 1990.

Tichy, N./Fombrun, C. 1979. *Network Analysis in Organizational Settings.* In: Human Relations, 32, 923-965.

Tinker, J. 1972. *Britain`s Environment — Nanny knows best.* In: New Scientist, Vol. 9, 530-534.

Tsebelis, G. 1990. Nested Games. Berkeley.

Umweltbundesamt 1989. Was Sie schon immer über Luftreinhaltung wissen wollten. Berlin.

Umweltbundesamt 1993. Daten zur Umwelt. Berlin.

Uppendahl, H./Klee, G./Popp, R. 1988. Responsive Demokratie und/oder kommunaler Korporatismus in England. Oldenburg.

Valroff, J. 1985. Pollution atmosphérique et pluies acides: rapport au Premier ministre. Paris.

Vernier, J. 1992. L'environnement. Paris.

Voelzkow, H./Eichener, V. 1990. *Die Harmonisierung technischer Normen in der Europäischen Gemeinschaft: Probleme und Perspektiven.* In: U. von Alemann, R.G. Heinze and B. Hombach, Eds. Die Kraft der Region: Nordrhein-Westfalen in Europa. Bonn.

Vogel, D. 1983. *Cooperative Regulation. Environmental Protection in Great Britain.* In: The Public Interest, 72, 88-106.

Vogel, D. 1986. National Styles of Regulation. Environmental Policy in Great Britain and the United States. Ithaca.

Waarden, F. van 1992. On the Persistence of National Policy Styles and Policy Networks. A Study of the Genesis of their Institutional Bases, Paper presented at the conference of the Society for the Advancement of Socio-Economics in Irvine, California, March 27-29.

Wallace, H./Wallace, W./Webb, C., Eds., 1983. Policy-Making in the European Community. Chichester.

Wallace, W. 1982. *Europe as a Confederation.* In: Journal of Common Market Studies, 21, 57-68.

Wallace, W. 1983. *Less than a Federation, More than a Regime: The Community as a Political System.* In: H. Wallace, W. Wallace and C. Webb., Eds., Policy-Making in the European Community. London.

Weber, A. 1989. Die Umweltverträglichkeitsrichtlinie im deutschen Recht: eine Studie zur Umsetzung der Richtlinie des Rates vom 27. Juni 1985 über die Umweltverträglichkeitsprüfung bei bestimmten öffentlichen und privaten Projekten (85/337/EWG), Cologne.

Wegener, B. 1992. *Umsetzung der EG-Richtlinie über den freien Zugang zu Umweltinformationen.* In: JUR, 211ff.

Wegener, B. 1993. *Die unmittelbare Geltung der EG-Richtlinie über den freien Zugang zu Umweltinformationen.* In: Zeitschrift für Umweltrecht, No. 1, 17-21.

Weidenfeld, W./Wessels, W., Eds., 1992. Europa von A-Z. Taschenbuch der europäischen Integration. Bonn.

Weidner, H. 1987. Clean Air Policy in Great Britain. Problem Shifting as Best Practicable Means. Berlin.

Weidner, H. 1989. Die Umweltpolitik der konservativ-liberalen Regierung im Zeitraum 1983 bis 1988: Versuch einer politikwissenschaftlichen Bewertung. Berlin.

Weidner, H. 1989a. *Die Umweltpolitik der konservativ-liberalen Regierung. Eine vorläufige Bilanz.* In: Aus Politik und Zeitgeschichte, Vol. 39, No. 2, 16-28.

Weiler, J.H.H. 1988. *The White Paper and the Application of Community Law.* In: R. Bieber et al., Eds., One European Market? Baden-Baden. 337-358.

Wessels, W. 1990. *Administrative Interaction.* In: W. Wallace, Ed., The Dynamics of European Integration. London. 229-241.

Wessels, W. 1992. *Staat und (westeuropäische) Integration. Die Fusionsthese.* In: M. Kreile, Ed., Die Integration Europas, PVS Sonderheft 23, 36-61.

Wetstone, G.S./Rosencranz, A. 1983. Acid Rain in Europe and North America: National Responses to an International Problem. Washington D.C.

White-Grove, R. 1992. *Länderbericht Großbritannien.* In: C. Hey und U. Brendle, Eds., Umweltverbände und EG. Handlungsmöglichkeiten der Umweltverbände für die Verbesserung des Umweltbewußtseins und der Umweltpolitik in der Europäischen Gemeinschaft. Freiburg. 101-129.

Wilke, M./Wallace, H. 1992. Subsidiarity: Approaches to Power-Sharing in the European Community, Institute of International Affairs, Discussion Paper No. 27. London.

Wilks, S./Wright, M. 1989. Comparative Government-Industry Relations. Western Europe, the United States and Japan. Oxford.

Wilson, J.Q. 1989. Bureaucracy. What Government Agencies Do and Why They Do It. New York.

Windhoff-Héritier, A. 1991. *Gegen den Strom nationaler Politik: Zum Verhältnis zwischen Metropolen und Zentralstaat — das Beispiel London und New York.* In: B. Blanke, Ed., Staat und Stadt. Systematische, vergleichende und problemorientierte Analyse "dezentraler" Politik, PVS Sonderheft 22, 259-282.

Windhoff-Héritier, A. 1993. Problem-Definition, Agenda-Setting and Policy-Making in the European Community: The Interlinking of National and Supranational "Rationalities" in Clean Air Policy — a Comparison of Great Britain and FR Germany. Paper prepared for the joint sessions of the ECPR, Leiden, 2.-8. April 1993.

Winter, G. 1990. *Akteneinsicht in Frankreich.* In: G. Winter, Ed., Öffentlichkeit von Umweltinformationen. Europäische und nordamerikanische Rechte und Erfahrungen. Baden-Baden.

Wirtschafts- und Sozialausschuß der Europäischen Gemeinschaften 1992. *Stellungnahme zu der Entschließung des Rates der Europäischen Gemeinschaften über ein Programm der Europäischen Gemeinschaften für Umweltpolitik und Maßnahmen im Hinblick auf eine dauerhafte und umweltgerechte Entwicklung* In: Amtsblatt der Europäischen Gemeinschaften Nr. C 287, 27-35.

Woyke, W. 1984. *Kommission der Europäischen Gemeinschaften.* In: W. Woyke, Ed., Europäische Gemeinschaft. Problemfelder-Institutionen-Politik. Munich. 363-369.

Wright, V. 1983. The Government and Politics in France. London.

Wright, V. 1984. Continuity and Change in France. London.

Walter de Gruyter
Berlin • New York

Christian Lahusen

The Rhetoric of Moral Protest

Public Campaigns, Celebrity Endorsement and Political Mobilization

1996. 23 x 15.5 cm. XVI, 425 pages.
With 24 figures, 1 table, and numerous music scores.
Cloth.   ISBN 3-11-015093-X
(de Gruyter Studies in Organization, 76)

Rock-for-a-cause campaigns like the Mandela show, Human Rights Now!, Amnesty and Greenpeace concerts are modern international mobilization strategies. Presenting celebrities and stars, they form a medium and an arena for political campaigns.

The author presents the first analysis of the rhetoric of moral protest, analyzing the narrative, the graphic design, and the music of specific events.

The book puts forward a culturalist approach which includes the role of cultural industries, multi-media events, infotainment and an analysis of the societal context: audiences, markets, and institutions.

This highly topical study provides new insights for our understanding of organizational strategies and mobilization processes.

Dr. *Christian Lahusen*, Assistant Professor, Department of Sociology II, University of Bamberg, Bamberg, Germany

From the Contents
Introduction
Part I: Towards a theory of political mobilization
Part II: Investing in popular music: the opportunities for campaigning
Part III: Designing and composing protest simulacra: the campaign events and artifacts
Part IV: Understanding and explaining mobilization: campaign strategies and organized collective action
Part V: The globalization of collective action: international campaigns in context
Epilogue / Music scores / References

Walter de Gruyter & Co., P.O. Box 30 34 21, D - 10728 Berlin, Tel.: +49-30-260-05-0, Fax +49-30-260-05-222
Walter de Gruyter Inc., 200 Saw Mill River Road, Hawthorne, N.Y. 10532, Phone: (914) 747-0110, Fax (914) 747-1326
Please visit us in the World Wide Web at http://www.deGruyter.de